# Biomass Chars: Elaboration, Characterization and Applications

Special Issue Editors

**Mejdi Jeguirim**
**Lionel Limousy**

MDPI • Basel • Beijing • Wuhan • Barcelona • Belgrade

MDPI

*Special Issue Editors*
Mejdi Jeguirim
University of Haute Alsace
France

Lionel Limousy
University of Haute Alsace
France

*Editorial Office*
MDPI AG
St. Alban-Anlage 66
Basel, Switzerland

This edition is a reprint of the Special Issue published online in the open access journal *Energies* (ISSN 1996-1073) in 2016–2017 (available at: http://www.mdpi.com/journal/energies/ special_issues/biomass_chars_2016).

For citation purposes, cite each article independently as indicated on the article page online and as indicated below:

Lastname, F.M.; Lastname, F.M. Article title. *Journal Name*. **Year**. *Article number*, page range.

**First Edition 2018**

**ISBN 978-3-03842-690-5 (Pbk)**
**ISBN 978-3-03842-691-2 (PDF)**

# Table of Contents

# About the Special Issue Editors

**Mejdi Jeguirim** is associate professor at the University of Haute Alsace (France) since 2005 in the field of energy and process engineering. He dedicates most of his career to the biomass valorization through thermochemical conversion and the chars elaboration for energy and syngas production, pollutants removal, soil amendment. These research topics were performed in the frame of international collaborations and industrial contracts. He acted as PhD advisor for 10 students and he co-authored 85 referred international publications. He is editorial boards member of international scientific journals (Energy, Energy for Sustainable development, Energies) and has organized several special issues as guest editors. He belongs to the scientific committee of international congresses and involved as a scientific expert for for several national and international research programs. He has received the French National Research Excellence Award for researcher with high level scientific activity for the 2009–2012, 2013–2016 and 2017–2020 periods.

**Lionel Limousy** is associate professor at the University of Haute Alsace since 2010. He teaches in chemical and environmental engineering, risk management and assessment. He was recruited in 2000 at the University of South Brittany and was in charge on the Chemical Engineering and Process department from 2003 and 2005. He has created the Prodiabio platform in 2005 and was the director until 2010. In 2013, he joined the Institute of Material Science of Mulhouse where he became responsible of the industrial research partnership. He has experience in biological and chemical treatment of wastewater, membrane filtration, biomass valorisation and material characterisation. He is involved in several national and international academic and industrial projects. He has 65 scientific publications and collaborated in the organization of special issues. He has received the French National Research Excellence Award for researcher with high level scientific activity for the 2012–2015 and 2016–2019 periods.

*energies*

MDPI

*Editorial*

# Biomass Chars: Elaboration, Characterization and Applications

**Mejdi Jeguirim *** and **Lionel Limousy**

Institut de Sciences des Matériaux de Mulhouse, 15 rue Jean Starcky, 68057 Mulhouse CEDEX, France; lionel.limousy@uha.fr
* Correspondence: mejdi.jeguirim@uha.fr

Received: 28 November 2017; Accepted: 1 December 2017; Published: 3 December 2017

## 1. Introduction

This book contains the successful invited submissions [1–15] to a Special Issue of *Energies* on the subject area of "Biomass Chars: Elaboration, Characterization, and Applications". The invited editors have decided to focus the Special Issue on the specific topic of biomass transformation and use. In fact, biomass can be converted to energy, biofuels, and bioproducts, via thermochemical conversion processes such as combustion, pyrolysis, and gasification. Combustion technology is most widely applied on an industrial scale. However, biomass gasification and pyrolysis processes are still in the research and development stage. The major products from these processes are syngas, bio-oil, and char (called also biochar for agronomic applications). Among these products, biomass chars have been receiving increasing attention for different applications such as gasification, co-combustion, catalyst or adsorbent precursors, soil amendment, carbon fuel cells, and supercapacitors.

This Special Issue provides an overview for biomass chars production methods (pyrolysis, hydrothermal carbonization, etc.), the characterization techniques (scanning electronic microscopy, X-ray fluorescence, nitrogen adsorption, Raman spectroscopy, nuclear magnetic resonance spectroscopy, X-ray photoelectron spectroscopy, temperature programmed desorption, mass spectrometry, etc.), their properties and their suitable recovery processes.

Topics of interest for the call included, but were not limited to the production of biochar for:

- Biofuel production
- Soil amendment
- Carbon sequestration
- Heterogeneous catalysis
- Syngas production
- Pollutant adsorption

Responses to our call had the following statistics:

- Submissions (25);
- Publications (15);
- Rejections (10);
- Article types: research article (15).

The authors' geographical distribution (published papers) is:

- China (4)
- USA (2)
- Canada (2)

- France (2)
- Italy (1)
- Spain (1)
- Korea (1)
- Lithuania (1)
- Tunisia (1)

Published submissions are related to biochar elaboration, characterization and application. The different papers show the diversity of research conducted recently on biochars. Agriculture, biofuel, adsorbent, catalyst, these versatile materials can be used for various applications. Recent developments have shown that the return to soil is a promising issue since the carbon balance is negative due to the carbon sequestration. Other benefits are described below, new developments are still under progress, encouraging the organization of a second edition of this special issue.

## 2. Short Review of the Contributions in This Issue

Concerns about climate change and food productivity have spurred interest in biochar, a form of charred organic material typically used in agriculture to improve soil productivity and as a means of carbon sequestration. Zambon et al. [1] have shown an innovative approach in agriculture by the use of agro-forestry waste for the production of soil fertilizers for agricultural purposes and as a source of energy. A common agricultural practice is to burn crop residues in the field to produce ashes that can be used as soil fertilizers. This approach is able to supply plants with certain nutrients, such as Ca, K, Mg, Na, B, S, and Mo. However, the low concentration of N and P in the ashes, together with the occasional presence of heavy metals (Ni, Pb, Cd, Se, Al, etc.), has a negative effect on soil and, therefore, crop productivity. Their work describes the opportunity to create an innovative supply chain from agricultural waste biomass. Olive (*Olea europaea*) and hazelnut (*Corylus avellana*) pruning residues represent a major component of biomass waste in the area of Viterbo (Italy). In this study, the authors have evaluated the production of biochar from these residues. Furthermore, a physicochemical characterization of the produced biochar was performed to assess the quality of the two biochars according to the standards of the European Biochar Certificate (EBC). The results of this study indicate possible cost-effective production of high-quality biochar from olive and hazelnut biomass residues.

Characteristics of biochar vary with pyrolysis temperature. Chloropicrin (CP) is an effective fumigant for controlling soil-borne pests. Liu et al. [2] have investigated the characteristics of biochars prepared at 300°C, 500°C, and 700 °C by *Michelia alba* (*Magnolia denudata*) wood and evaluated their capacity to adsorb CP. The study also determined the potential influence of biochar, which was added to sterilized and unsterilized soils at rates of 0%, 1%, 5%, and 100%, on CP degradation. The specific surface area, pore volume, and micropores increased considerably with an increase in the pyrolytic temperature. The adsorption rate of biochar for CP increased with increasing pyrolytic temperature. The maximum adsorption amounts of CP were similar for the three biochars. Next, this study examined the degradation ability of the biochar for CP. The degradation rate constant (k) of CP increased when biochar was added to the soil, and k increased with increased amendment rate and pyrolysis temperature. The results indicate that biochar can accelerate CP degradation in soil. The findings of this study will be instructive in using biochar as a new fertilizer in fumigating soil with CP.

Biochar is increasingly applied in agriculture; however, due to its adsorption and degradation properties, biochar may also affect the efficacy of fumigant in amended soil. In their research, Fang et al. [3] have intended to study the effects of two types of biochars (BC-1 and BC-2) on the efficacy and emission of methyl isothiocyanate (MITC) in biochar amendment soil. Both types of biochars can significantly reduce MITC emission losses, but, at the same time, decrease the concentration of MITC in the soil. The efficacy of MITC for controlling soil-borne pests (*Meloidogyne* spp., *Fusarium* spp. *Phytophthora* spp., *Abutilon theophrasti*, and *Digitaria sanguinalis*) was reduced when the biochar

(BC-1 and BC-2) was applied at a rate higher than 1% and 0.5% (on a weight basis) (on a weight basis), respectively. However, increased doses of dazomet (DZ) were able to offset decreases in the efficacy of MITC in soils amended with biochars. Biochars with strong adsorption capacity (such as BC-1) substantially reduced the MITC degradation rate by 6.2 times, and increased the MITC degradation rate by 4.1 times following amendment with biochar with high degradability (e.g., BC-2), compared to soil without biochar amendment. This was attributed to the adsorption and degradation of biochar that reduces MITC emission losses and pest control.

Long and Boyette [4] have chosen Yellow poplar (*Liriodendron tulipifera*) as the woody biomass for the production of charcoal for use in a liquid fuel slurry. Charcoal produced from this biomass resulted in a highly porous structure similar to the parent material. Micronized particles were produced from this charcoal using a multi-step milling process and verified using a scanning electron microscope and laser diffraction system. Charcoal particles greater than 50 μm exhibited long needle shapes much like the parent biomass while particles less than 50 μm were produced with aspect ratios closer to unity. Laser diffraction measurements indicated D10, D50, and D90 values of 4.446, 15.83, and 39.69 μm, respectively. Moisture content, ash content, absolute density, and energy content values were also measured for the charcoal particles produced. Calculated volumetric energy density values for the charcoal particles exceeded the no. 2 diesel fuel that would be displaced in a liquid fuel slurry.

Biomass pyrolysis and the valorization of co-products (biochar, bio-oil, syngas) could be a sustainable management solution for agricultural and forest residues. Depending on its properties, biochar amended to soil could improve fertility. Moreover, biochar is expected to mitigate climate change by reducing soil greenhouse gas emissions, if its C/N ratio is lower than 30, and sequestrating carbon if its O/C and H/C ratios are lower than 0.2 and 0.7, respectively. However, the yield and properties of biochar are influenced by biomass feedstock and pyrolysis operating parameters. The objective of the research study carried by Brassard et al. [5] was to validate an approach based on the response surface methodology, to identify the optimal pyrolysis operating parameters (temperature, solid residence time, and carrier gas flowrate), in order to produce engineered biochars for carbon sequestration. The pyrolysis of forest residues, switchgrass, and the solid fraction of pig manure, was carried out in a vertical auger reactor following a Box–Behnken design, in order to develop response surface models. The optimal pyrolysis operating parameters were estimated to obtain biochar with the lowest H/C and O/C ratios. Validation pyrolysis experiments confirmed that the selected approach can be used to accurately predict the optimal operating parameters for producing biochar with the desired properties to sequester carbon.

In the work of Fuente-Hernández et al. [6], the liquid phase hydrogenation of furfural has been studied using a biochar-supported platinum catalyst in a batch reactor. Reactions were performed between 170 °C and 320 °C, using 3 wt % and 5 wt % of Pt supported on a maple-based biochar under hydrogen pressure varying from 500 psi to 1500 psi for reaction times between 1 h and 6 h in various solvents. Under all reactive conditions, furfural conversion was significant, whilst under specific conditions furfuryl alcohol (FA) was obtained in most cases as the main product, showing a selectivity around 80%. Other products as methylfuran (MF), furan, and trace of tetrahydrofuran (THF) were detected. Results showed that the most efficient reaction conditions involved a 3% Pt load on biochar and operations for 2 h at 210 °C and 1500 psi using toluene as solvent. When used repetitively, the catalyst showed deactivation, although only a slight variation in selectivity toward FA at the optimal experimental conditions was observed.

Investigation into clean energies has been focused on finding an alternative to fossil fuels in order to reduce global warming while at the same time satisfying the world's energy needs. Biomass gasification is seen as a promising thermochemical conversion technology, as it allows useful gaseous products to be obtained from low-energy-density solid fuels. Air–steam mixtures are the most commonly used gasification agents. The gasification performances of several biomass samples and their mixtures were compared by González-Vázquez et al. [7] in the present work. One softwood (pine) and one hardwood (chestnut), their torrefied counterparts, and other Spanish-based biomass wastes such as almond shells, olive stones, grape and olive pomaces, or cocoa shells were tested, and their behaviors at several different stoichiometric ratios (SR) and steam/air ratios (S/A) were compared. The optimum SR was found to be in the 0.2–0.3 range for S/A = 75/25. At these conditions a syngas stream with 35% of $H_2$ + CO and a gas yield of 2 L gas/g fuel were obtained, which represents a cold-gas efficiency of almost 50%. The torrefaction process does not significantly affect the quality of the product syngas. Some of the obtained chars were analyzed to assess their use as precursors for catalysts, combustion fuel, or for agricultural purposes such as soil amendment.

Waste residues produced by agricultural and forestry industries can generate energy and are regarded as a promising source of sustainable fuels. Pyrolysis, where waste biomass is heated under low-oxygen conditions, has recently attracted attention as a means to add value to these residues. The material is carbonized and yields a solid product known as biochar. In the study presented by Yang et al. [8], eight types of biomass were evaluated for their suitability as raw material to produce biochar. Material was pyrolyzed at either 350 °C or 500 °C and changes in ash content, volatile solids, fixed carbon, higher heating value (HHV), and yield were assessed. For pyrolysis at 350 °C, significant correlations ($p < 0.01$) between the biochars' ash and fixed carbon content and their HHVs were observed. Masson pine wood and Chinese fir wood biochars pyrolyzed at 350 °C and the bamboo sawdust biochar pyrolyzed at 500 °C were suitable for direct use in fuel applications, as reflected by their higher HHVs, higher energy density, greater fixed carbon, and lower ash contents. Rice straw was a poor substrate as the resultant biochar contained less than 60% fixed carbon and a relatively low HHV. Of the suitable residues, carbonization via pyrolysis is a promising technology to add value to pecan shells and Miscanthus.

Material dielectric properties are important for understanding their response to microwaves. Carbonaceous materials are considered good microwave absorbers and can be mixed with dry biomasses, which are otherwise low-loss materials, to improve the heating efficiency of biomass feedstocks. In the work of Ellison et al. [9], dielectric properties of pulverized biomass and biochar mixtures were presented from 0.5 GHz to 20 GHz at room temperature. An open-ended coaxial-line dielectric probe and vector network analyzer were used to measure dielectric constant and dielectric loss factor. Results have shown a quadratic increase of dielectric constant and dielectric loss with increasing biochar content. In measurements on biochar, a strong dielectric relaxation was observed at 8 GHz as indicated by a peak in dielectric loss factor at that frequency. Biochar was found to be a good microwave absorber and mixtures of biomass and biochar can be utilized to increase microwave heating rates for high temperature microwave processing of biomass feedstocks. These data can be utilized for design, scale-up, and simulation of microwave heating processes of biomass, biochar, and their mixtures.

Wood pellets are a form of solid biomass energy and a renewable energy source. In 2015, the new and renewable energy (NRE) portion of wood pellets was 4.6% of the total primary energy in Korea. Wood pellets account for 6.2% of renewable energy consumption in Korea, the equivalent of 824,000 TOE (ton of oil equivalent, 10 million kcal). The burning phases of a wood pellet can be classified into three modes: (1) gasification; (2) flame burning; and (3) charcoal burning. At each wood pellet burning mode, the volume and weight of the burning wood pellet can drastically change, these parameters are important to understand the wood pellet burning mechanism. Kang et al. [10] have developed a new method for measuring the volume of a burning wood pellet that involves no contact. To measure the volume of a wood pellet, they have taken pictures of the wood pellet in each burning mode. The volume of a burning wood pellet was calculated by image processing. The difference between the calculation method using image processing and the direct measurement of a burning wood pellet in gasification mode was less than 8.8%. In gasification mode, mass reduction of the wood pellet is 37% and volume reduction of the wood pellet is 7%. Whereas in charcoal burning mode, mass reduction of the wood pellet was 10% and volume reduction of the wood pellet was 41%. By measuring volume using image processing, continuous and non-interruptive volume measurements for various solid fuels were possible and could provide more detailed information for CFD (computational fluid dynamics) analysis.

Annually, the olive oil industry generates a significant amount of by-products, such as olive pomace, olive husks, tree prunings, leaves, pits, and branches. Therefore, the recovery of these residues has become a major challenge in Mediterranean countries. The utilization of olive industry residues has received much attention in recent years, especially for energy purposes. Accordingly, the primary experimental study carried out by Tamošiūnas et al. [11] aimed at investigating the potential of olive biomass waste for energy recovery in terms of synthesis gas (or syngas) production using the thermal arc plasma gasification method. The olive charcoal made from the exhausted olive solid waste (olive pomace) was chosen as a reference material for primary experiments with known composition from the performed proximate and ultimate analysis. The experiments were carried out at various operational parameters: raw biomass and water vapor flow rates and the plasma generator power. The producer gas principally involved CO, $H_2$, and $CO_2$ with the highest concentrations of 41.17, 13.06, and 13.48%, respectively. The produced synthesis gas has a lower heating value of 6.09 MJ/nm$^3$ at the $H_2O$/C ratio of 3.15 and the plasma torch had a power of 52.2 kW.

Solid char is a product of biomass pyrolysis. It contains a high proportion of carbon, and lower contents of H, O, and minerals. This char can have different valorization pathways such as combustion for heat and power, gasification for syngas production, activation for adsorption applications, or use as a soil amendment. The optimal recovery pathway of the char depends highly on its physical and chemical characteristics. In the study presented by Guizani et al. [12], different chars were prepared from beech wood particles under various pyrolysis operating conditions in an entrained flow reactor (500–1400 °C). Their structural, morphological, surface chemistry properties, as well as their chemical compositions, were determined using different analytical techniques, including elementary analysis, scanning electronic microscopy (SEM) coupled with an energy dispersive X-ray spectrometer (EDX), Fourier transform infrared spectroscopy (FTIR), and Raman spectroscopy. The biomass char reactivity was evaluated in air using thermogravimetric analysis (TGA). The yield, chemical composition, surface chemistry, structure, morphology, and reactivity of the chars were highly affected by the pyrolysis temperature. In addition, some of these properties related to the char structure and chemical composition were found to be correlated to the char reactivity.

Zhao et al. [13] have studied the structure and physicochemical properties of biochar derived from apple tree branches (ATBs), whose valorization is crucial for the sustainable development of the apple industry. ATBs were collected from apple orchards located on the Weibei upland of the Loess Plateau and pyrolyzed at 300, 400, 500, and 600 °C (BC300, BC400, BC500, and BC600), respectively. Different analytical techniques were used for the characterization of the different biochars. In particular, proximate and element analyses were performed. Furthermore, the morphological, and textural properties were investigated using scanning electron microscopy (SEM), Fourier-transform infrared (FTIR) spectroscopy, Boehm titration, and nitrogen manometry. In addition, the thermal stability of biochars was also studied by thermogravimetric analysis. The results indicated that the increasing temperature increased the content of fixed carbon (C), the C content and inorganic minerals (K, P, Fe, Zn, Ca, Mg), while the yield, the content of volatile matter (VM), O and H, cation exchange capacity, and the ratios of O/C and H/C decreased. Comparison between the different samples has shown that highest pH and ash content were observed in BC500. The number of acidic functional groups decreased as a function of pyrolysis temperature, especially for the carboxylic functional groups. In contrast, a reverse trend was found for the basic functional groups. At a higher temperature, the Brunauer–Emmett–Teller (BET) surface area and pore volume are higher mostly due to the increase of the micropore surface area and micropore volume. In addition, the thermal stability of biochars also increased with the increasing temperature. Hence, pyrolysis temperature had a strong effect on biochar properties, and therefore biochars could be produced by changing the pyrolysis temperature in order to better meet their applications.

Wanassi et al. [14] have investigated the use of phenolic resin and waste cotton fiber as green precursors for the successful synthesis using a soft template approach of a composite carbon with carbon nanofibers embedded in a porous carbon network with ordered pore structure. The optimal composite carbon (PhR/NC-1), exhibited a specific surface area of 394 $m^2 \cdot g^{-1}$ with the existence of both microporosity and mesoporosity. PhR/NC-1 carbon was evaluated as an adsorbent of Alizarin Red S (ARS) dye in batch solution. Various operating conditions were examined and the maximum adsorption capacity of 104 $mg \cdot g^{-1}$ was achieved under the following conditions, i.e., T = 25 °C, pH = 3, contact time = 1440 min. The adsorption and desorption heat was assessed by flow micro-calorimetry (FMC), and the presence of both exothermic and endothermic peaks with different intensities were shown, indicating a partially reversible nature of ARS adsorption. A pseudo-second-order model proved to be the most suitable kinetic model to describe the ARS adsorption according to the linear regression factor. In addition, the best isotherm equilibrium has been achieved with a Freundlich model. The results have shown that the eco-friendly composite carbon derived from green phenolic resin mixed with waste cotton fibers improved the removal of ARS dye from textile effluents.

The treatment of $NO_x$ from automotive gas exhaust has been widely studied, however the presence of low concentrations of $NO_x$ in confined areas is still under investigation. As an example, the concentration of $NO_2$ can approximate 0.15 ppmv inside vehicles when people are driving on highways. This interior pollution becomes an environmental problem and a health problem. In the work carried out by Ghouma et al. [14], the abatement of $NO_2$ immission was studied at room temperature. Three activated carbons (ACs) prepared by physical ($CO_2$ or $H_2O$) or chemical activation ($H_3PO_4$) were tested as adsorbents. The novelty of this work consisted in studying the adsorption of NO2 at low concentrations that approach real life emission concentrations and was experimentally realizable. The ACs presented different structural and textural properties, as well as functional surface groups, which induced different affinities with $NO_2$. The AC prepared using water vapor activation presented the best adsorption capacity, which may originate from a more basic surface. The presence of a mesoporosity may also influence the diffusion of $NO_2$ inside the carbon matrix. The high reduction activity of the AC prepared from $H_3PO_4$ activation was explained by the important concentration of acidic groups on its surface.

We found the edition and selections of papers for this book very inspiring and rewarding. We also thank the editorial staff and reviewers for their efforts and help during the process.

**Author Contributions:** Authors contribute equally to the manuscript.

**Conflicts of Interest:** The authors declare no conflict of interest.

## References

1. Zambon, I.; Colosimo, F.; Monarca, D.; Cecchini, M.; Gallucci, F.; Proto, A.R.; Lord, R.; Colantoni, A. An Innovative Agro-Forestry Supply Chain for Residual Biomass: Physicochemical Characterisation of Biochar from Olive and Hazelnut Pellets. *Energies* **2016**, *9*, 526. [CrossRef]
2. Liu, P.; Wang, Q.; Yan, D.; Fang, W.; Mao, L.; Wang, D.; Li, Y.; Ouyang, C.; Guo, M.; Cao, A. Effects of Biochar Amendment on Chloropicrin Adsorption and Degradation in Soil. *Energies* **2016**, *9*, 869. [CrossRef]
3. Fang, W.; Cao, A.; Yan, D.; Han, D.; Huang, B.; Li, J.; Liu, X.; Guo, M.; Wang, Q. The Effect of Two Types of Biochars on the Efficacy, Emission, Degradation, and Adsorption of the Fumigant Methyl Isothiocyanate. *Energies* **2017**, *10*, 16. [CrossRef]
4. Long, J.M.; Boyette, M.D. Analysis of Micronized Charcoal for Use in a Liquid Fuel Slurry. *Energies* **2017**, *10*, 25. [CrossRef]
5. Brassard, P.; Godbout, S.; Raghavan, V.; Palacios, J.H.; Grenier, M.; Zegan, D. The Production of Engineered Biochars in a Vertical Auger Pyrolysis Reactor for Carbon Sequestration. *Energies* **2017**, *10*, 288. [CrossRef]
6. Fuente-Hernández, A.; Lee, R.; Béland, N.; Zamboni, I.; Lavoie, J.-M. Reduction of Furfural to Furfuryl Alcohol in Liquid Phase over a Biochar-Supported Platinum Catalyst. *Energies* **2017**, *10*, 286. [CrossRef]
7. González-Vázquez, M.P.; García, R.; Pevida, C.; Rubiera, F. Optimization of a Bubbling Fluidized Bed Plant for Low-Temperature Gasification of Biomass. *Energies* **2017**, *10*, 306. [CrossRef]

8.  Yang, X.; Wang, H.; Strong, P.J.; Xu, S.; Liu, S.; Lu, K.; Sheng, K.; Guo, J.; Che, L.; He, L.; et al. Thermal Properties of Biochars Derived from Waste Biomass Generated by Agricultural and Forestry Sectors. *Energies* **2017**, *10*, 469. [CrossRef]

9.  Ellison, C.; McKeown, M.S.; Trabelsi, S.; Boldor, D. Dielectric Properties of Biomass/Biochar Mixtures at Microwave Frequencies. *Energies* **2017**, *10*, 502. [CrossRef]

10. Kang, S.B.; Sim, B.S.; Kim, J.J. Volume and Mass Measurement of a Burning Wood Pellet by Image Processing. *Energies* **2017**, *10*, 603. [CrossRef]

11. Tamošiūnas, A.; Chouchène, A.; Valatkevičius, P.; Gimžauskaitė, D.; Aikas, M.; Uscila, R.; Ghorbel, M.; Jeguirim, M. The Potential of Thermal Plasma Gasification of Olive Pomace Charcoal. *Energies* **2017**, *10*, 710. [CrossRef]

12. Guizani, C.; Jeguirim, M.; Valin, S.; Limousy, L.; Salvador, S. Biomass Chars: The Effects of Pyrolysis Conditions on Their Morphology, Structure, Chemical Properties and Reactivity. *Energies* **2017**, *10*, 796. [CrossRef]

13. Zhao, S.-X.; Ta, N.; Wang, X.-D. Effect of Temperature on the Structural and Physicochemical Properties of Biochar with Apple Tree Branches as Feedstock Material. *Energies* **2017**, *10*, 1293. [CrossRef]

14. Wanassi, B.; Hariz, I.B.; Ghimbeu, C.M.; Vaulot, C.; Jeguirim, M. Green Carbon Composite-Derived Polymer Resin and Waste Cotton Fibers for the Removal of Alizarin Red S Dye. *Energies* **2017**, *10*, 1321. [CrossRef]

15. Ghouma, I.; Jeguirim, M.; Sager, U.; Limousy, L.; Bennici, S.; Däuber, E.; Asbach, C.; Ligotski, R.; Schmidt, F.; Ouederni, A. The Potential of Activated Carbon Made of Agro-Industrial Residues in $NO_x$ Immissions Abatement. *Energies* **2017**, *10*, 1508. [CrossRef]

# energies

MDPI

*Article*

# The Production of Engineered Biochars in a Vertical Auger Pyrolysis Reactor for Carbon Sequestration

Patrick Brassard [1,2,*], Stéphane Godbout [1], Vijaya Raghavan [2], Joahnn H. Palacios [1], Michèle Grenier [1] and Dan Zegan [1]

[1] Research and Development Institute for the Agri-Environment (IRDA), 2700 Einstein Street, Quebec City, QC G1P 3W8, Canada; stephane.godbout@irda.qc.ca (S.G.); joahnn.palacios@irda.qc.ca (J.H.P.); michele.grenier@irda.qc.ca (M.G.); dan.zegan@irda.qc.ca (D.Z.)

[2] Department of Bioresource Engineering, MacDonald Campus, McGill University, 2111 Lakeshore, Sainte-Anne-de-Bellevue, QC H9X 3V9, Canada; vijaya.raghavan@mcgill.ca

* Correspondence: patrick.brassard@mail.mcgill.ca; Tel.: +1-581-997-8670

Academic Editor: Mejdi Jeguirim
Received: 12 January 2017; Accepted: 21 February 2017; Published: 28 February 2017

**Abstract:** Biomass pyrolysis and the valorization of co-products (biochar, bio-oil, syngas) could be a sustainable management solution for agricultural and forest residues. Depending on its properties, biochar amended to soil could improve fertility. Moreover, biochar is expected to mitigate climate change by reducing soil greenhouse gas emissions, if its C/N ratio is lower than 30, and sequestrating carbon if its $O/C_{org}$ and $H/C_{org}$ ratios are lower than 0.2 and 0.7, respectively. However, the yield and properties of biochar are influenced by biomass feedstock and pyrolysis operating parameters. The objective of this research study was to validate an approach based on the response surface methodology, to identify the optimal pyrolysis operating parameters (temperature, solid residence time, and carrier gas flowrate), in order to produce engineered biochars for carbon sequestration. The pyrolysis of forest residues, switchgrass, and the solid fraction of pig manure, was carried out in a vertical auger reactor following a Box-Behnken design, in order to develop response surface models. The optimal pyrolysis operating parameters were estimated to obtain biochar with the lowest $H/C_{org}$ and $O/C_{org}$ ratios. Validation pyrolysis experiments confirmed that the selected approach can be used to accurately predict the optimal operating parameters for producing biochar with the desired properties to sequester carbon.

**Keywords:** pyrolysis; auger reactor; engineered biochar; forest residues; agricultural biomass; response surface methodology

## 1. Introduction

In 2014, the Intergovernmental Panel on Climate Change reported that "global emissions of greenhouse gas (GHG) have risen to unprecedented levels despite a growing number of policies to reduce climate change" [1]. GHG emissions need to be lowered by 40% to 70% compared to the 2010 values by mid-century, and to near-zero by the end of the century, if we are to limit the increase in global mean temperature to two degrees Celcius [1].

Pyrolysis, the thermochemical decomposition of biomass under oxygen-limiting conditions at temperatures between 300 and 700 °C, can be a sustainable management solution for agricultural and forest biomasses, and is proposed as a strategy to mitigate climate change. The resulting co-products of pyrolysis are: a liquid bio-oil, non-condensable gases, and a solid biochar. The yields and characteristics of the products depend on pyrolysis operating parameters and biomass feedstock properties. Non-condensable gases are generally used to heat the pyrolysis unit. Bio-oils have heating values of 40%–50% of that of hydrocarbon fuels [2], and could be used to replace fossil heating oil.

Biochar can be used as a soil amendment to improve soil fertility and has been proposed as a tool for mitigating climate change [3], because of its potential for carbon (C) sequestration. When biomass is converted into biochar and is applied to soil, C can be stored for more than 1000 years [4,5]. In other words, biochar production is a way for C to be drawn from the atmosphere, and is a solution for reducing the global impact of farming [6]. Woolf et al. [7] reported that biochar and its storage in soil can contribute to a reduction of up to 12% of current anthropogenic $CO_2$ emissions. Moreover, there is evidence that biochar amendment to soil can help reduce GHG emissions, and particularly $N_2O$ [8], a powerful GHG, with a global warming potential of 298 [9]. Specifically, agriculture is a major source of $N_2O$, contributing approximately 70% of Canadian anthropogenic $N_2O$ emissions. Agricultural soils contribute to about 82% of these emissions [10]. Despite the many potential benefits of soil amendment with biochar, special attention must be paid to the negative side effects. For example, heavy metals (e.g., Cu, Zn, and Mo) could be found in biochar and accumulate in soil, leading to phytotoxicity problems.

The yield and characteristics of pyrolysis products are influenced by different factors, including biomass feedstock and pyrolysis operating parameters (solid residence time, vapor residence time, temperature, heating rate, and carrier gas flowrate). Thus, not all biochars are created equal and biochars should be designed with special characteristics for their use in environmental or agronomic settings [11,12]. Biochars with a low N content, and consequently a high C/N ratio (>30), could be more suitable for the mitigation of $N_2O$ emissions from soils [8,13]. Moreover, biochars produced at a higher pyrolysis temperature and with an $O/C_{org}$ ratio < 0.2, $H/C_{org}$ ratio < 0.4, and volatile matter below 80%, may have a high C sequestration potential [13]. In fact, a $H/C_{org}$ ratio < 0.4 would indicate a $BC_{+100}$ of 70% (i.e., at least 70% of the C in biochar is predicted to remain in soil for more than 100 years), as an $H/C_{org}$ ratio in the range 0.4–0.7 would indicate a $BC_{+100}$ of 50% [14].

It is also important to select the proper pyrolysis technology to obtain the desired yield and properties of the product. Among all the existing pyrolysis technologies, the auger reactor is one of the most attractive designs that has been developed [15]. It enjoys some popularity because of its simplicity of construction and operation [16]. In an auger reactor, biomass is continuously fed to a screw, where it is heated in oxygen-free conditions, and then the auger rotation moves the product along the auger axis to the end of the reactor. The gases and organic volatiles leave the reactor at the end of the reactor, and the biochar is collected at the bottom. Gas exits may also be performed along the auger reactor wall, in order to decrease the vapor residence time. The yield of bio-oil (condensed gases) in auger reactors is variable, depending on the operating parameters, but is typically in the range of 40 to 60 wt % of the feedstock, which is lower than what is normally achieved with fluidized-bed reactors. This is because the heat transfer in an auger reactor is lower. Therefore, small-diameter reactor tubes which have a limited distance between the inner reactor tube surface and the internal auger shaft, are needed. In order to increase the heat transfer, some auger reactors combine a small inert solid particulate heat carrier (usually sand or steel shot) with relatively small particles of biomass (1 to 5 mm). The residence time of the vapors is much longer in auger reactors than in fluidized beds, which increases the likelihood of secondary reactions and consequently increases the yield of char [16].

The hypothesis of this research project is that it is possible to produce a biochar with beneficial characteristics from an environmental perspective, when pyrolysis operating parameters are suitably selected in a vertical auger reactor. Thus, the main objective was to validate a response surface methodology approach used to identify the optimal pyrolysis operating parameters (temperature, solid residence time, and nitrogen flowrate), in order to produce engineered biochars with the ideal characteristics for C sequestration.

## 2. Materials and Methods

### 2.1. Description of the Response Surface Methodology Approach

#### 2.1.1. Development of the Statistical Models

Response surface methodology (RSM) was selected as an approach to determine the optimal pyrolysis operating parameters, in order to produce engineered biochars that can be used to sequester carbon. RSM is a collection of statistical and mathematical techniques for developing, improving, and optimizing processes [17], and is used to illustrate the relationship between the response variables (dependent variables) and the process variables (independent variables). In this study, the selected independent variables were the pyrolysis temperature, solid residence time in the heater block, and $N_2$ flowrate, which are three parameters known to influence the yields and characteristics of products in an auger pyrolysis reactor [18]. The biochar yield and three indicators of biochar potential for climate change mitigation (C/N, H/C$_{org}$, and O/C$_{org}$ ratios), were the response variables studied. Biochars with the highest C/N ratio are expected to reduce soil GHG emissions, and those with the lowest H/C$_{org}$ and O/C$_{org}$ molar ratios are expected to have a high C sequestration potential [13].

The Box-Behnken design was selected for collecting data. For an experiment of three factors, this incomplete factorial design requires three evenly spaced levels for each factor, coded $-1$, 0, and +1 (Table 1). Two variables ($-1$ and +1 levels) are paired together in a $2^2$ factorial, while the third factor remains fixed at the center (level 0). A total of 15 experiments run in a random order are necessary, including three repetitions of an experiment, with the three independent variables fixed at their central point.

The method of least squares from the RSREG procedure of SAS [19] was used to estimate the parameters of the quadratic response surface regression models (Equation (1)), fitted to the experimental data obtained from the Box-Behnken design:

$$Y = \beta_0 + \beta_1 T + \beta_2 R + \beta_3 N + \beta_4 T^2 + \beta_5 (R \times T) + \beta_6 R^2 + \beta_7 (N \times T) + \beta_8 (N \times R) + \beta_9 N^2 \quad (1)$$

where $Y$ is the studied response variable (biochar yield, C/N, H/C$_{org}$, and O/C$_{org}$ ratios); $\beta_0, \dots \beta_9$ are the regression coefficients to be estimated; and $T$, $R$, and $N$ are the values of the independent variables (temperature, solid residence time, and $N_2$ flowrate, respectively). The significance of each independent variable was determined by the analysis of variance (ANOVA). A lack of fit test was performed to check the adequacy of the model.

#### 2.1.2. Determination of the Stationary Points

A canonical analysis [19] was used to determine the nature of the stationary point (or the point on the surface where the partial derivatives are equal to zero), which can be a point of maximum response, a point of minimum response, or a saddle point. In the case of a saddle point, a RIDGE statement [19] was used to indicate the direction in which further experimentation should be performed, to produce the fastest decrease or increase in the estimated response, starting at the stationary point.

#### 2.1.3. Validation of the Statistical Models

In order to validate the quadratic response surface regression models, a biochar was produced with the pyrolysis operating parameters determined from the response surface analysis, for producing a biochar with the optimal properties to maximize C sequestration (i.e., the lowest O/C$_{org}$ and H/C$_{org}$ ratios). A second biochar with the opposite characteristics (highest O/C$_{org}$ and H/C$_{org}$ ratios) was produced from each biomass. Predicted values from the response surface models vs. the actual values of the O/C$_{org}$, H/C$_{org}$, C/N ratios and yield, were compared using linear regression.

## 2.2. Pyrolysis Experimental Setup and Procedure

### 2.2.1. Description of the Vertical Auger Pyrolysis Reactor

In order to validate the selected approach, pyrolysis tests were carried out in a vertical auger pyrolysis reactor (Patent CA 2830968), developed by the *Institut de recherche et de développement en agroenvironnement* (IRDA) in collaboration with the *Centre de recherche industriel du Québec* (CRIQ). The pyrolysis unit (Figure 1) was installed at the IRDA research facility (Deschambault, QC, Canada). It included a hopper, a horizontal feed screw, a vertical screw passing through a 25.4 cm long heater block, a canister for the biochar recovery, and a condensation system. The feedstock in the hopper was fed to the heater block by a horizontal and vertical feed screw in a 2.54 cm diameter steel tube. The rotation speed of the two screws was controlled separately by gear motors, thus controlling the biomass flow rate. An agitator was installed and fixed at the hopper lid in order to facilitate and ensure the supply to the horizontal screw when using materials with a low density. Then, the feedstock was transported through the 25.4 cm long reactor within the vertical screw. The feedstock residence time in the reactor was set by controlling the rotation speed of the vertical screw, and was calculated in relation to the pitch of the screw (3.8 cm). Thermal power was supplied by two heating elements of 1500 Watts, inserted in a copper block surrounding the tube in the reaction zone. A thermocouple inserted in the middle of the cooper block registered the outside tube temperature and was used as the set point to control the heating elements. Temperatures were acquired every 10 s by a data logger (CR10X, Campbell Scientific, Edmonton, AB, Canada). At the exit of the vertical screw, the solid product of the pyrolysis (char) dropped into the canister (31.4 cm high and 16.8 cm diameter). A pot (15.2 cm high) was placed into the canister in order to recover the accumulated char. A flange at the bottom of the canister gave access to the pot. Moreover, the fine particles were separated from the gas by an inner baffle (10.2 cm diameter and 10.5 cm long) placed at the exit of the vertical screw. The gas was evacuated by an opening in the upper part of the canister and was directed to the condensation system. Every flange was tightened with a high temperature graphite gasket (1034 kPa) in order to prevent the entry of oxygen into the system. The air flowing into the system was purged with nitrogen, which was injected from the hopper's lid at volumetric flowrates ranging from 1 to 5 L·min$^{-1}$, controlled by a flowmeter (Aalborg Instruments, New York, NY, USA; accuracy ±2%). While the nitrogen flow ensured that the pyrolysis reaction occurred in a non-oxygen environment, it also helped to evacuate the pyrolysis gas.

### 2.2.2. Biomass Selection and Analysis

The type of feedstock utilized for pyrolysis (e.g., woody biomass, crop residues, grasses, and manures) influences the yield and characteristics of the biochar, including the concentrations of elemental constituents, density, porosity, and hardness [20]. Moreover, the yield of the biochar from biomass can be influenced by its lignin, holocellulose, and extractives contents [21]. Three biomasses with different physico-chemical properties were selected for the pyrolysis experiments: wood pellets made from a mixture of Black Spruce (*Picea mariana*) and Jack Pine (*Pinus banksiana*), the solid fraction of pig manure (SFPM), and switchgrass (*Panicum Virgatum* L.). In Canada, forest biomass residues such as logging residues are present in large quantities. Moreover, forest biomass is the most common feedstock used for pyrolysis. Woody biomass has a high C content and low N content, which can lead to a biochar with a high C/N ratio. Switchgrass, a perennial grass, shows great characteristics for bioenergy production, because of its medium to high productivity (8 to 12 t DM·ha$^{-1}$·yr$^{-1}$), its sustainability, its great ability to use water and nutrients, its adaptation to the climate of Eastern Canada, and its relatively high gross calorific value (GCV), of between 18.2 to 19.1 MJ·kg$^{-1}$ [22]. SFPM was selected because pyrolysis could be a sustainable management solution for the surplus of pig manure in some regions, where phosphorus (P) spreading in fields is restricted by regulations. Pyrolysis of the solid fraction of pig manure concentrates P in biochar [23], which facilitates its exportation outside of the region in surplus.

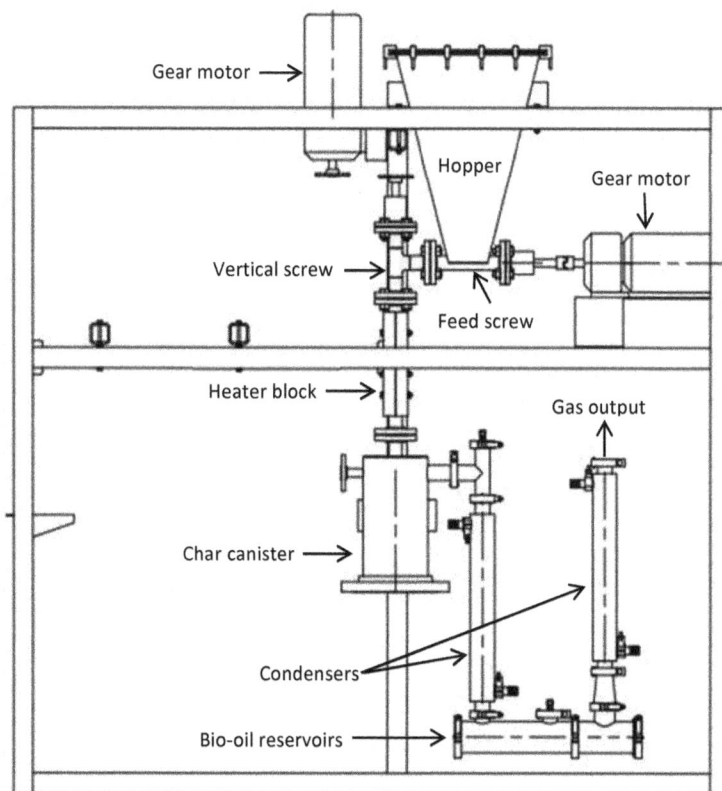

**Figure 1.** Schematic view of the vertical auger pyrolysis reactor.

All biomasses were ground and sieved to a particle size between 1.0 and 3.8 mm, prior to pyrolysis. The chemical properties of biomasses (proximate and ultimate analysis) were analysed at the IRDA laboratory (Quebec City, QC, Canada). The C, H, N, and ash content of the biomass were evaluated by dry combustion (Leco TruSpec, St. Joseph, MI, USA). The O content was calculated by subtracting the C, H, N, and ash contents from 100 wt %. Chlorine (Cl) extraction with water and dosage by titration with silver nitrate ($AgNO_3$) was used to determine the Cl content. Cellulose, hemicellulose, and lignin contents were analysed according to the AFNOR method [24].

## 2.2.3. Pyrolysis Experiments

Preliminary pyrolysis tests and a review of the literature were carried out in order to identify the range of pyrolysis operating parameters needed to obtain typical biochar yields in the pyrolysis auger reactor, ranging from 15% to 45%. For the three selected biomasses, the range of the $N_2$ flowrate selected was between 1 and 5 L·min$^{-1}$, and the range for the solid residence time was between 60 and 120 s. The range of the pyrolysis temperature found for wood and SFPM was between 500 and 650 °C, and between 450 and 600 °C for switchgrass. For the selected solid residence times, the biomass flowrate in the pyrolysis reactor depended on the biomass properties, and varied from 0.61 to 1.08 kg·h$^{-1}$ for wood, from 0.42 to 0.8 kg·h$^{-1}$ for SFPM, and was fixed at 0.57 kg·h$^{-1}$ for switchgrass.

The Box-Behnken design was carried out for each biomass with the defined range of pyrolysis operating conditions (Table 1), for a total of 45 experiments (Tables A1–A3).

**Table 1.** Box-Behnken design: list of independent variables and levels.

| Independent Variable | Biomass | Values of the Coded Levels | | |
|---|---|---|---|---|
| | | −1 | 0 | +1 |
| Temperature (°C) | Wood | 500 | 575 | 650 |
| | SFPM | 500 | 575 | 650 |
| | Switchgrass | 450 | 525 | 600 |
| Solid residence time (s) | Each biomass | 60 | 90 | 120 |
| $N_2$ flowrate (L·min$^{-1}$) | Each biomass | 1 | 3 | 5 |

### 2.2.4. Products Yield and Biochar Characteristics

Bio-oil (Equation (2)) and biochar (Equation (3)) yields were calculated on a wet biomass basis, the non-condensable gas (Equation (4)) yield was calculated by the difference, and the liquid organic yield (Equation (5)) was calculated by subtracting the water content from the bio-oil yield:

$$Yield_{bio-oil}(wt\%) = \frac{m_{B1} + m_{B2}}{m_f} \times 100 \tag{2}$$

$$Yield_{biochar}(wt\%) = \frac{m_{Biochar}}{m_f} \times 100 \tag{3}$$

$$Yield_{gas}(wt\%) = \frac{m_f - m_{Biochar} - m_{B1} - m_{B2}}{m_f} \times 100 \tag{4}$$

$$Yield_{liquid\ organics}(wt\%) = \frac{100 - water\ content}{100 \times yield\ bio - oil} \tag{5}$$

where $m_{B1}$ is the mass of bio-oil produced in the first condenser (g), $m_{B2}$ is the mass of bio-oil produced in the second condenser (g), $m_{biochar}$ is the mass of biochar collected in the canister (g), $m_f$ is the mass of feedstock pyrolysed (g), and the water content is the water content of bio-oil (wt %) measured following the Karl-Fischer titration method [25].

Biochar samples were analysed for moisture, volatile matter, and ash contents, based on the ASTM D1762-84 standard [26]. The organic carbon ($C_{org}$), total carbon ($C_{tot}$), hydrogen (H), nitrogen (N), and oxygen (O) were also analysed, using the same method as that employed for the analysis of biomasses.

## 3. Results and Discussion

### 3.1. Analysis of Biomass

The physicochemical properties of wood, SFPM, and switchgrass, are presented in Table 2. An ultimate analysis (C, H, N, O) shows large variations between the biomasses. The C content of wood is the highest, at 47.7%, and is the lowest for SFPM (40.0%). This is inversely proportional to the ash content, which is highest for the SFPM (11.5%), and lowest for wood (0.57%). SFPM has high N and Cl contents (2.96% and 3609 mg·kg$^{-1}$, respectively) when compared to wood and switchgrass. The O content is low for SFPM (28.2%), when compared to wood (40.0%) and switchgrass (42.5%). The H content ranges from 3.23% (switchgrass) to 6.39% (wood). The water content of SFPM (13.0%) is higher than switchgrass (7.2%) and wood (6.5%).

Based on an analysis of the lignocellulosic components, wood could necessitate a higher temperature to decompose because its lignin content (24%) is higher than that of SFPM and switchgrass (12.9% and 11.2%, respectively). In fact, the proportion of cellulose, hemicellulose, and lignin in biomass, will influence the degree to which the physical structure is modified during processing [27]. Hemicellulose and cellulose, which are more volatile during thermal degradation [28], are degraded at 200–300 and 300–400 °C, respectively, and lignin is degraded between 200 and 700 °C, representing a wide range in temperatures [29].

**Table 2.** Biomasses physicochemical properties.

|  | Unit | Wood | SFPM | Switchgrass |
|---|---|---|---|---|
| $C_{tot}$ | wt % | 47.7 | 40.0 | 45.8 |
| N | wt % | 0.128 | 2.96 | 0.425 |
| O | wt % | 40.0 | 28.2 | 42.5 |
| H | wt % | 6.39 | 5.85 | 3.23 |
| Water content | wt % | 6.5 | 13.0 | 7.2 |
| Ash | d.b.% | 0.57 | 11.5 | 1.6 |
| Cl | $mg \cdot kg^{-1}$ | 10 | 3 609 | 28 |
| Lignin | wt % | 24.0 | 12.9 | 11.2 |
| Cellulose | wt % | 30.4 | 11.9 | 42.9 |
| Hemicellulose | wt % | 29.9 | 22.0 | 30.1 |

*3.2. Response Surface Models*

3.2.1. Biochar Yield

The yields of products from the 45 pyrolysis tests carried out following the Box-Behnken design, are presented in Appendix A for wood (Table A1), switchgrass (Table A2), and SFPM (Table A3). The highest bio-oil yields were obtained from wood (48.6% to 63.6%) and switchgrass (44.8% to 61.4%), and pyrolysis of these materials was associated with low biochar yields (17.5% to 31.2% and 16.8% to 26.4%, respectively). Conversely, the pyrolysis of SFPM produced lower yields of bio-oil (38.3% to 46.7%) and higher yields of biochar (32.1% to 40.4%). The canonical analysis indicated that the stationary points of the three response surface models are saddle points. Thus, results from the RIDGE analysis, indicating the direction toward which further pyrolysis experiments should be performed, in order to obtain the minimal and maximal estimated values of biochar yield, are presented in Table 3. It is known that biochar yield decreases as pyrolysis temperature increases [30]. Based on the results of the analysis of variance for the models, the biochar yield is significantly dependent on the pyrolysis temperature for the three biomass feedstocks ($Pr < 0.05$; Appendix B), as the solid residence time is only significant for the switchgrass biochar. In contrast to what is reported in some studies [18,31], the biochar yield was not significantly influenced by the $N_2$ flowrate, which influences the vapor residence time. The predicted biochar yield is the highest for the pyrolysis of SFPM (maximum of 40%), due to the high ash content of the feedstock, which is found in biochar after pyrolysis. The biochar yield from switchgrass and wood pyrolysis are similar. However, the predicted highest value for wood (27.8%) is higher than for switchgrass (25.2%), despite the highest pyrolysis temperature being demonstrated for wood. It reflects the higher lignin content of wood, which preferentially forms char during pyrolysis [21].

**Table 3.** Estimated values of biochar properties and estimation of optimal pyrolysis operating parameters from the response surface models.

|  | Biochar Yield (wt %) | | $H/C_{org}$ | | $O/C_{org}$ | | C/N | |
|---|---|---|---|---|---|---|---|---|
| Wood | Min | Max | Min | Max | Min | Max | Min | Max |
| Estimated value | 17.2 | 27.8 | 0.54 | 0.81 | 0.14 | 0.25 | 477 | 539 |
| Temperature (°C) | 646 | 507 | 646 | 515 | 642 | 517 | 639 | 522 |
| Residence time (s) | 89 | 79 | 99 | 79 | 103 | 80 | 75 | 90 |
| $N_2$ Flowrate ($L \cdot min^{-1}$) | 3.6 | 3.4 | 2.9 | 3.9 | 2.8 | 4.1 | 2.8 | 4.4 |
| Switchgrass | | | | | | | | |
| Estimated value | 17.4 | 25.2 | 0.47 | 0.77 | 0.10 | 0.23 | 100 | 108 |
| Temperature (°C) | 593 | 451 | 588 | 456 | 594 | 462 | 588 | 466 |
| Residence time (s) | 78 | 88 | 106 | 80 | 102 | 75 | 74 | 72 |
| $N_2$ Flowrate ($L \cdot min^{-1}$) | 3.3 | 2.8 | 3.1 | 3.4 | 2 | 3.4 | 3.3 | 3.1 |

Table 3. *Cont.*

|  | Biochar Yield (wt %) | | H/C$_{org}$ | | O/C$_{org}$ | | C/N | |
|---|---|---|---|---|---|---|---|---|
| **SFPM** | | | | | | | | |
| Estimated value | 32.2 | 40 | 0.66 | 0.90 | 0.14 | 0.21 | 11.5 | 12.8 |
| Temperature (°C) | 649 | 507 | 628 | 508 | 631 | 543 | 594 | 614 |
| Residence time (s) | 95 | 79 | 94 | 79 | 94 | 73 | 84 | 103 |
| N$_2$ Flowrate (L·min$^{-1}$) | 3 | 3.4 | 1.6 | 3.6 | 1.7 | 4.4 | 4.9 | 1.5 |

### 3.2.2. H/C$_{org}$ and O/C$_{org}$ Ratios

The minimum values of H/C$_{org}$ and O/C$_{org}$ indicate a high biochar C stability [32–35], and thus, a maximum potential for C sequestration. H/C$_{org}$ and O/C$_{org}$ ratios of biochars produced from the 45 pyrolysis tests significantly varied for a single biomass, depending on the pyrolysis operating parameters (Tables A1–A3). The response surface models demonstrated that the biochar produced from the three biomasses only demonstrates a good potential for C sequestration if the operating parameters are properly selected. A minimum stationary point was only found for the O/C$_{org}$ molar ratio of biochar made from switchgrass; otherwise, saddle points were found. Minimum and maximum values of H/C$_{org}$ and O/C$_{org}$, predicted from the RIDGE analysis, are presented in Table 3. The minimum predicted H/C$_{org}$ ratios are 0.47, 0.54, and 0.66 for biochars produced from switchgrass, wood, and SFPM, respectively. This means that, for the optimal pyrolysis operational parameters, at least 50% of the C in biochar is expected to remain in the soil for more than 100 years [14]. The predicted minimum O/C$_{org}$ ratio below 0.2 (0.10, 0.14, and 0.14 for switchgrass, wood, and SFPM, respectively) confirms the C sequestration potential of biochars produced with similar pyrolysis operating parameters. In fact, the pyrolysis operating parameters needed to obtain the minimum H/C$_{org}$ and O/C$_{org}$ ratios for each biomass, are similar. Conversely, the maximum predicted H/C$_{org}$ and O/C$_{org}$ values for the three biomasses are always above 0.7 and 0.2, respectively. Harvey et al. [36] found that pyrolysis conditions are the primary factors controlling the thermal stability of the resulting biochar. More specifically, Zhao et al. [37] demonstrated that biochar recalcitrance (i.e., its ability to resist decomposition) is mainly determined by pyrolysis temperature. The ANOVA analysis confirmed this fact: the pyrolysis temperature always significantly influenced (Pr < 0.05) the H/C$_{org}$ and O/C$_{org}$ ratios (Tables A4–A6). Moreover, the solid residence time significantly impacted the indicators of C stability for the pyrolysis of switchgrass: as the residence time increased, the H/C$_{org}$ and O/C$_{org}$ ratios decreased. Di blasi [38] also reported that the solid residence time has an influence on the physical and chemical characteristics of biochar. The addition of a heat carrier material in an auger reactor could decrease the solid residence time required to provide sufficient reaction heat and time [18]. Finally, Antal and GrØnli [21] reported that biochar characteristics can also be modified with a change in the sweeping gas flow rate, which has an impact on the vapor residence time. Statistical analysis revealed that the N$_2$ flowrate has a significant impact on the O/C$_{org}$ ratio of SFPM and wood biochars. A lower O/C$_{org}$ ratio is obtained with lower N$_2$ flowrates.

### 3.2.3. C/N Ratio

Biochars with a C/N ratio higher than 30 could help in decreasing the N$_2$O emissions from soil [13]. Results of the experimental Box-Behnken design showed that the C/N ratio markedly varies among biomasses, from 430 to 541 for wood, 95 to 115 for switchgrass, and 11.0 to 13.0 for SFPM. The Canonical analysis of the response surface models shows that a maximum stationary point was found for the C/N ratio of wood biochar, and that saddle points were identified for switchgrass and SFPM biochars. The minimum and maximum values estimated from the RIDGE analysis are presented in Table 3. The ANOVA (Tables A4–A6) showed that none of the pyrolysis operating conditions significantly influenced the C/N ratio of biochar. In fact, because the N content of biomasses is low, particularly for wood and switchgrass (0.128% and 0.454%), the impact of pyrolysis operating parameters on the N content of biochar, and consequently on its C/N ratio, is low. Even if the C/N ratio for a single biomass does not significantly vary, depending on the pyrolysis operating parameters, there are large variations among the biomasses. In the literature, it was found that the C/N ratio is highly dependent on the type of biomass feedstock used for pyrolysis [8,39]. In the present study, the biomass C/N ratio (13.5, 108, and 372 for SFPM, switchgrass, and wood, respectively) is similar to the C/N ratio of biochar produced from the corresponding biomass, and the C/N ratios of biochars produced from wood pyrolysis are the highest (430 to 565), and ranged from 95 to 115 for switchgrass pyrolysis. Thus, based on their chemical composition, biochars made from these two biomasses have the potential to mitigate $N_2O$ emissions from soil. Biochars produced from the pyrolysis of SFPM have a C/N ratio lower than 30 (11.0–13.0) and could potentially increase $N_2O$ emissions from soil, due to their high N content [39] and low C content.

## 3.3. Experimental Validation of the Models

In order to validate the quadratic response surface regression models, two biochars were produced from wood (B1 and B2), switchgrass (B3 and B4), and SFPM (B5 and B6) (Table 4). B2, B4 (two replicates), and B6 were produced with the pyrolysis operating parameters (temperature, residence time, and $N_2$ flowrate) determined from the response surface analysis for producing a biochar with the optimal properties in order to maximize the C sequestration potential (i.e., the lowest $O/C_{org}$ and $H/C_{org}$ ratios). B1, B3, and B5 were produced using the optimal parameters for producing a biochar with the opposite characteristics (highest $O/C_{org}$ and $H/C_{org}$ ratio). In fact, because the predicted optimal pyrolysis parameters needed to obtain the optimal $O/C_{org}$ and $H/C_{org}$ ratios are similar, the selected temperature, residence time, and $N_2$ flowrate, were average values. For example, the lowest $H/C_{org}$ and $O/C_{org}$ ratios predicted for wood biochar would be obtained at 646 °C and 642 °C, respectively (Table 3). Thus, the selected temperature for the production of biochar with the best C sequestration potential was 644 °C (Table 4). The pyrolysis operating parameters for biochar production that were used to validate the models, and the corresponding yields and properties of the resulting biochars are presented in Table 4. B2, B4, and B6 were produced at a higher temperature, during a longer residence time, and with a lower $N_2$ flowrate than B1, B3, and B5, respectively. Their ash contents are higher, whereas their H and O contents are lower. Moreover, the C and N contents of B2 and B4 are higher than B1 and B3, respectively. The water content is always low (about 1%), whereas the biochars produced at higher temperatures are more alkaline.

**Table 4.** Products yields and physicochemical properties of biochars produced with optimal pyrolysis operating conditions.

| | Unit | B1 | B2 | B3 | B4 [1] | B4 [2] | B5 | B6 |
|---|---|---|---|---|---|---|---|---|
| *Pyrolysis parameters* | | | | | | | | |
| Biomass | | Wood | Wood | SG [3] | SG | SG | SFPM [4] | SFPM |
| Temperature | °C | 516 | 644 | 459 | 591 | 591 | 526 | 630 |
| Res. time | s | 80 | 101 | 78 | 104 | 104 | 76 | 94 |
| $N_2$ flowrate | L·min$^{-1}$ | 4.0 | 2.9 | 3.4 | 2.6 | 2.6 | 4.0 | 1.7 |
| *Products yields* | | | | | | | | |
| Biochar | % (w.b.) | 26.4 | 18.5 | 26.9 | 18.9 | 18.6 | 46.4 | 34.9 |
| Bio-oil | % (w.b.) | 58.2 | 51.5 | 60.2 | 49.4 | 49.0 | 37.9 | 41.5 |
| *Biochar properties* | | | | | | | | |
| $C_{total}$ | % (w.b.) | 71.6 | 80.0 | 67.1 | 79.5 | 80.2 | 51.5 | 49.2 |
| $C_{org}$ | % (w.b.) | 70.7 | 76.0 | 64.9 | 79.1 | 79.9 | 47.4 | 45.2 |
| H | % (w.b.) | 4.8 | 3.73 | 4.85 | 3.36 | 3.35 | 3.73 | 3.36 |
| O | % (w.b.) | 21.6 | 13.4 | 22.9 | 10.0 | 9.59 | 15.6 | 13.7 |
| N | % (w.b.) | 0.141 | 0.166 | 0.641 | 0.828 | 0.780 | 4.40 | 4.05 |
| $P_{soluble}$ | mg·kg$^{-1}$ | 13.7 | 7.16 | 109 | 26.7 | 32.1 | 165 | 55.7 |
| Water content | % (w.b.) | 0.9 | 1.2 | 1.5 | 1.0 | 1.8 | 0.9 | 0.9 |
| Ash (750 °C) | % (d.b.) | 1.4 | 2.1 | 4.1 | 5.6 | 5.4 | 23.6 | 28.1 |
| pH | | 6.8 | 7.6 | 6.4 | 8.7 | 8.9 | 8.6 | 9.3 |

[1] First pyrolysis test for B4 production; [2] Second pyrolysis test for B4 production; [3] Switchgrass; [4] Solid fraction of pig manure.

The observed vs. predicted values for the biochar yield, C/N, $H/C_{org}$, and $O/C_{org}$ ratios, are illustrated in Figure 2. A comparison of the linear regressions with the 1:1 line indicates that the models fit the experimental data for the yield ($R^2 = 0.97$), C/N ($R^2 = 1.0$), $H/C_{org}$ ($R^2 = 0.88$), and $O/C_{org}$ ($R^2 = 0.73$). B2 and B4 are expected to have a better potential for mitigating climate change, have a high C sequestration potential ($H/C_{org} < 0.7$; $O/C_{org} < 0.2$), and have the potential to reduce soil GHG emissions (C/N ratio > 30).

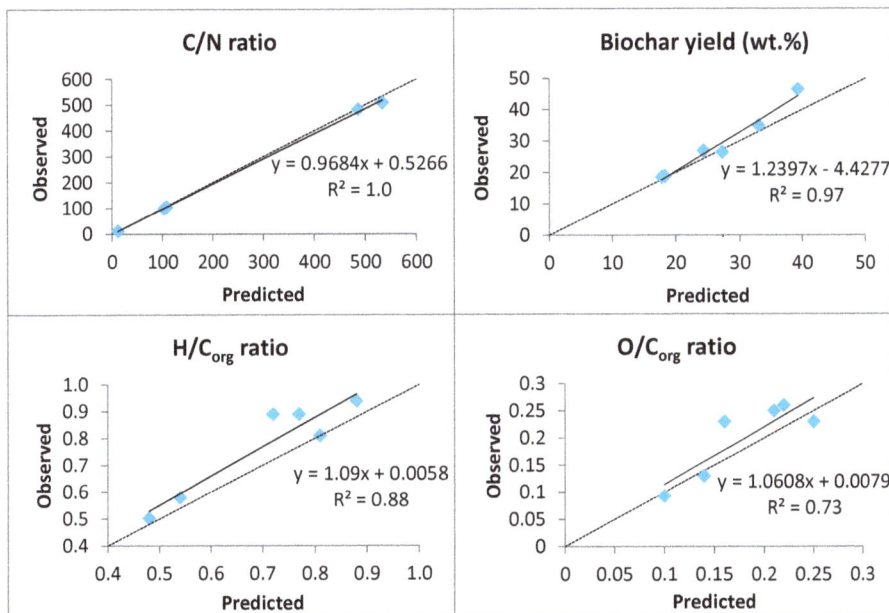

**Figure 2.** Biochar yield, C/N, $H/C_{org}$, and $O/C_{org}$ ratios: observed vs. predicted values.

## 4. Conclusions

Results from this study demonstrated that the response surface methodology approach can be used to accurately predict the optimal operating parameters of a vertical auger reactor (temperature, solid residence time, and nitrogen flowrate), required to produce engineered biochars with specific characteristics for C sequestration. It was highlighted that the pyrolysis products' yields and biochar characteristics highly depend on the pyrolysis operating conditions and biomass feedstock. Biochar produced from wood and switchgrass can only present a high potential for C sequestration if the pyrolysis operating parameters are properly selected. In fact, the minimum $H/C_{org}$ and $O/C_{org}$ ratios predicted from the response surface models reached values lower or equal to 0.54 and 0.14, respectively, for a pyrolysis temperature ranging from 588 to 646 °C, a solid residence time from 99 to 106 s, and a $N_2$ flowrate from 2.0 to 3.1 L·min$^{-1}$. Moreover, regardless of the pyrolysis operating conditions, the biochars produced from the pyrolysis of wood and switchgrass could help to decrease soil $N_2O$ emissions, because their C/N ratios are higher than 30. Further experiments have to be carried out with the produced biochars, in order to evaluate their effect on soil GHG emissions and C sequestration, and to validate the hypothesis made in this study.

**Acknowledgments:** The authors thank the *"Fonds de recherche du Québec—Nature et technologie"* (FQRNT), the *"Programme de soutien à l'innovation en agroalimentaire"* (grant number IA113109), the IRDA and McGill University for their financial support. Special thanks are also addressed to Jean-Pierre Larouche, Cédric Morin, Étienne Le Roux, Salha Elcadhi, and Martin Brouillard for their help during the implementation and the realization of the experiments.

**Author Contributions:** All the authors contributed to the conception and design of the experiments; Patrick Brassard performed the experiments; Patrick Brassard, Michèle Grenier, and Joahnn H. Palacios analyzed the data; Patrick Brassard wrote the paper and all of the co-authors revised it.

**Conflicts of Interest:** The authors declare no conflict of interest.

## Appendix A. Experimental Data: Box-Behnken Design

**Table A1.** Pyrolysis of wood—Experimental data.

| Operational Parameters | | | Products Yields | | | | Biochar Properties | | |
|---|---|---|---|---|---|---|---|---|---|
| T | Res. Time | $N_2$ | Bio-Oil | Liquid Organics | Biochar | Syngas | C/N | $H/C_{org}$ | $O/C_{org}$ |
| °C | s | $L \cdot min^{-1}$ | % | | % | % | | | |
| 500 | 60 | 3 | 57.6 | 39.0 | 31.2 | 10.9 | 517 | 0.84 | 0.25 |
| 500 | 90 | 1 | 61.9 | 39.9 | 24.6 | 13.2 | 491 | 0.68 | 0.19 |
| 500 | 90 | 5 | 55.2 | 36.3 | 30.2 | 14.2 | 531 | 0.92 | 0.29 |
| 500 | 120 | 3 | 63.6 | 41.9 | 23.4 | 12.4 | 541 | 0.68 | 0.19 |
| 575 | 60 | 1 | 49.1 | 31.8 | 22.6 | 28.0 | 483 | 0.68 | 0.19 |
| 575 | 60 | 5 | 56.8 | 37.8 | 22.2 | 20.5 | 512 | 0.74 | 0.22 |
| 575 | 90 | 3 | 60.0 | 38.1 | 20.7 | 18.8 | 565 | 0.65 | 0.19 |
| 575 | 90 | 3 | 60.6 | 40.6 | 20.6 | 18.2 | 487 | 0.65 | 0.18 |
| 575 | 90 | 3 | 61.5 | 39.4 | 20.2 | 17.8 | 504 | 0.62 | 0.17 |
| 575 | 120 | 1 | 58.8 | 34.4 | 21.2 | 19.6 | 503 | 0.60 | 0.15 |
| 575 | 120 | 5 | 54.4 | 35.2 | 19.9 | 25.2 | 500 | 0.63 | 0.18 |
| 650 | 60 | 3 | 56.0 | 36.8 | 18.3 | 25.2 | 430 | 0.59 | 0.16 |
| 650 | 90 | 1 | 52.4 | 31.3 | 18.0 | 29.0 | 491 | 0.51 | 0.13 |
| 650 | 90 | 5 | 48.8 | 27.8 | 17.5 | 33.1 | 497 | 0.57 | 0.15 |
| 650 | 120 | 3 | 48.6 | 27.4 | 17.6 | 33.3 | 466 | 0.53 | 0.13 |

T: temperature; Res. Time: solid residence time; $N_2$: Nitrogen flowrate.

**Table A2.** Pyrolysis of Switchgrass—Experimental data.

| Operational Parameters | | | Products Yields | | | | Biochar Properties | | |
|---|---|---|---|---|---|---|---|---|---|
| T | Res. Time | $N_2$ | Bio-Oil | Liquid Organics | Biochar | Syngas | C/N | $H/C_{org}$ | $O/C_{org}$ |
| °C | s | $L \cdot min^{-1}$ | % | % | % | % | | | |
| 450 | 60 | 3 | 57.8 | 35.4 | 25.6 | 16.4 | 114 | 0.81 | 0.25 |
| 450 | 90 | 1 | 59.2 | 34.3 | 26.4 | 14.0 | 106 | 0.77 | 0.21 |
| 450 | 90 | 5 | 60.1 | 37.1 | 24.9 | 14.4 | 102 | 0.82 | 0.24 |
| 450 | 120 | 3 | 59.4 | 34.1 | 24.4 | 15.9 | 101 | 0.69 | 0.19 |
| 525 | 60 | 1 | 61.4 | 34.7 | 20.5 | 17.9 | 100 | 0.64 | 0.18 |
| 525 | 60 | 5 | 55 | 33.4 | 19.9 | 24.5 | 105 | 0.72 | 0.21 |
| 525 | 90 | 3 | 58.3 | 37.2 | 20.2 | 21.2 | 115 | 0.60 | 0.16 |
| 525 | 90 | 3 | 58.5 | 31.0 | 21.3 | 19.9 | 95 | 0.61 | 0.16 |
| 525 | 90 | 3 | 59 | 30.8 | 20.0 | 20.6 | 99 | 0.58 | 0.14 |
| 525 | 120 | 1 | 56.8 | 42.3 | 21.9 | 21.1 | 102 | 0.57 | 0.14 |
| 525 | 120 | 5 | 54.5 | 27.9 | 20.9 | 24.1 | 103 | 0.54 | 0.14 |
| 600 | 60 | 3 | 51.5 | 30.8 | 16.8 | 30.5 | 98 | 0.58 | 0.15 |
| 600 | 90 | 1 | 48.9 | 21.3 | 18.7 | 31.9 | 105 | 0.48 | 0.10 |
| 600 | 90 | 5 | 44.8 | 20.4 | 17.3 | 37.2 | 99 | 0.49 | 0.11 |
| 600 | 120 | 3 | 48.1 | 21.8 | 18.5 | 32.9 | 102 | 0.46 | 0.10 |

T: temperature; Res. Time: solid residence time; $N_2$: Nitrogen flowrate.

**Table A3.** Pyrolysis of SFPM—Experimental data.

| Operational Parameters | | | Products Yields | | | | Biochar Properties | | |
|---|---|---|---|---|---|---|---|---|---|
| T | Res. Time | N$_2$ | Bio-Oil | Liquid Organics | Biochar | Syngas | C/N | H/C$_{org}$ | O/C$_{org}$ |
| °C | s | L·min$^{-1}$ | % | % | % | % | | | |
| 500 | 60 | 3 | 42.8 | 12.5 | 41.6 | 14.9 | 11.6 | 0.92 | 0.21 |
| 500 | 90 | 1 | 45.7 | 12.4 | 38.8 | 15.1 | 12.4 | 0.80 | 0.16 |
| 500 | 90 | 5 | 39.3 | 10.6 | 40.4 | 19.5 | 12.0 | 0.91 | 0.21 |
| 500 | 120 | 3 | 41.7 | 10.8 | 39.6 | 17.0 | 12.5 | 0.85 | 0.18 |
| 575 | 60 | 1 | 46.7 | 10.8 | 36.7 | 15.0 | 12.3 | 0.75 | 0.16 |
| 575 | 60 | 5 | 40.1 | 11.7 | 38.5 | 20.6 | 11.5 | 0.85 | 0.23 |
| 575 | 90 | 3 | 42.3 | 11.7 | 35.8 | 21.0 | 12.7 | 0.78 | 0.18 |
| 575 | 90 | 3 | 43.7 | 12.1 | 36.0 | 19.4 | 12.4 | 0.76 | 0.16 |
| 575 | 90 | 3 | 43.6 | 11.9 | 34.8 | 19.8 | 11.4 | 0.74 | 0.17 |
| 575 | 120 | 1 | 45.7 | 12.0 | 34.7 | 17.7 | 12.9 | 0.65 | 0.14 |
| 575 | 120 | 5 | 38.6 | 9.2 | 35.9 | 24.5 | 12.1 | 0.72 | 0.16 |
| 650 | 60 | 3 | 42.7 | 10.5 | 33.8 | 21.8 | 12.6 | 0.66 | 0.14 |
| 650 | 90 | 1 | 44.0 | 7.7 | 32.4 | 22.8 | 13.0 | 0.61 | 0.13 |
| 650 | 90 | 5 | 38.3 | 9.3 | 32.1 | 28.8 | 11.0 | 0.74 | 0.18 |
| 650 | 120 | 3 | 39.1 | 8.5 | 32.6 | 27.2 | 12.8 | 0.68 | 0.14 |

T: temperature; Res. Time: solid residence time; N$_2$: Nitrogen flowrate.

## Appendix B. ANOVA Tables

**Table A4.** ANOVA for the model of wood biochar.

| Wood | Factor | DF | Mean Squares | F Value | Pr > F |
|---|---|---|---|---|---|
| | Temperature | 4 | 53.001 | 29.96 | 0.0011 * |
| Yield | Res. time | 4 | 8.0950 | 4.580 | 0.0632 |
| | N$_2$ flowrate | 4 | 2.9350 | 1.660 | 0.2936 |
| | Temperature | 4 | 0.0287 | 18.78 | 0.0033 * |
| H/C$_{org}$ | Res. time | 4 | 0.0063 | 4.120 | 0.0763 |
| | N$_2$ flowrate | 4 | 0.0070 | 4.580 | 0.0631 |
| | Temperature | 4 | 0.0043 | 22.04 | 0.0022 * |
| O/C$_{org}$ | Res. time | 4 | 0.0010 | 4.930 | 0.0552 |
| | N$_2$ flowrate | 4 | 0.0014 | 7.430 | 0.0247 * |
| | Temperature | 4 | 1452.1 | 1.250 | 0.3972 |
| C/N | Res. time | 4 | 471.35 | 0.410 | 0.7982 |
| | N$_2$ flowrate | 4 | 304.41 | 0.260 | 0.8904 |

DF: Degrees of freedom; Res. Time: solid residence time; * Significant at Pr < 0.05.

**Table A5.** ANOVA for the model of switchgrass biochar.

| Switchgrass | Parameter | DF | Mean Squares | F Value | Pr > F |
|---|---|---|---|---|---|
| | Temperature | 4 | 29.441 | 87.23 | <0.0001 * |
| Yield | Res. time | 4 | 0.8077 | 2.390 | 0.1822 |
| | N$_2$ flowrate | 4 | 0.7911 | 2.340 | 0.1876 |
| | Temperature | 4 | 0.0368 | 45.51 | 0.0004 * |
| H/C$_{org}$ | Res. time | 4 | 0.0083 | 10.30 | 0.0124 * |
| | N$_2$ flowrate | 4 | 0.0014 | 1.700 | 0.2847 |
| | Temperature | 4 | 0.0061 | 72.32 | 0.0001 * |
| O/C$_{org}$ | Res. time | 4 | 0.0017 | 20.26 | 0.0027 * |
| | N$_2$ flowrate | 4 | 0.0003 | 3.000 | 0.1298 |
| | Temperature | 4 | 29.954 | 0.530 | 0.7194 |
| C/N | Res. time | 4 | 21.608 | 0.380 | 0.8125 |
| | N$_2$ flowrate | 4 | 2.1106 | 0.040 | 0.9964 |

DF: Degrees of freedom; Res. Time: solid residence time; * Significant at Pr < 0.05.

**Table A6.** ANOVA for the model of SFPM biochar.

| SFPM | Parameter | DF | Mean Squares | F Value | Pr > F |
|---|---|---|---|---|---|
| | Temperature | 4 | 27.624 | 96.31 | <0.0001 * |
| Yield | Res. time | 4 | 2.7895 | 9.730 | 0.0141 * |
| | $N_2$ flowrate | 4 | 0.8267 | 2.880 | 0.1381 |
| | Temperature | 4 | 0.0207 | 18.07 | 0.0036 * |
| $H/C_{org}$ | Res. time | 4 | 0.0030 | 2.630 | 0.1592 |
| | $N_2$ flowrate | 4 | 0.0054 | 4.680 | 0.0606 |
| | Temperature | 4 | 0.0009 | 5.020 | 0.0533 * |
| $O/C_{org}$ | Res. time | 4 | 0.0008 | 4.470 | 0.0661 |
| | $N_2$ flowrate | 4 | 0.0014 | 8.040 | 0.021 * |
| | Temperature | 4 | 0.2138 | 0.850 | 0.5509 |
| C/N | Res. time | 4 | 0.1987 | 0.790 | 0.5793 |
| | $N_2$ flowrate | 4 | 0.6988 | 2.770 | 0.1466 |

DF: Degrees of freedom; Res. Time: solid residence time; * Significant at Pr < 0.05.

## References

1. Intergovernmental Panel on Climate Change (IPCC). *Climate Change 2014: Synthesis Report. Contribution of working Groups I, II and III to the Fifth Assessment Report of the Intergovernmental Panel on Climate Change*; Pachauri, R.K., Meyer, L.A., Eds.; IPCC: Geneva, Switzerland, 2014; pp. 1–151.
2. Jahirul, M.I.; Rasul, M.G.; Chowdhury, A.A.; Ashwath, N. Biofuels Production through Biomass Pyrolysis—A Technological Review. *Energies* **2012**, *5*, 4952–5001. [CrossRef]
3. Wang, J.; Pan, X.; Liu, Y.; Zhang, X.; Xiong, Z. Effects of biochar amendment in two soils on greenhouse gas emissions and crop production. *Plant Soil* **2012**, *360*, 287–298. [CrossRef]
4. Haefele, S.M.; Konboon, Y.; Wongboon, W.; Amarante, S.; Maarifat, A.A.; Pfeiffer, E.M.; Knoblauch, C. Effects and fate of biochar from rice residues in rice-based systems. *Field Crop. Res.* **2011**, *121*, 430–440. [CrossRef]
5. Kuzyakov, Y.; Bogomolova, I.; Glaser, B. Biochar stability in soil: Decomposition during eight years and transformation as assessed by compound-specific 14C analysis. *Soil Biol. Biochem.* **2014**, *70*, 229–236. [CrossRef]
6. Brar, S.K.; Dhillon, G.S.; Soccol, C.R. *Biotransformation of Waste Biomass into High Value Biochemicals*; Springer: New York, NY, USA, 2014; pp. 1–504.
7. Woolf, D.; Amonette, J.E.; Street-Perrott, F.A.; Lehmann, J.; Joseph, S. Sustainable biochar to mitigate global climate change. *Nat. Commun.* **2010**, *1*, 56. [CrossRef] [PubMed]
8. Cayuela, M.L.; Van Zwieten, L.; Singh, B.P.; Jeffery, S.; Roig, A.; Sánchez-Monedero, M.A. Biochar's role in mitigating soil nitrous oxide emissions: A review and meta-analysis. *Agric. Ecosyst. Environ.* **2014**, *191*, 5–16. [CrossRef]
9. Foster, P.; Ramaswamy, V.; Artaxo, P.; Berntsen, T.; Betts, R.; Fahey, D.W.; Haywood, J.; Lean, J.; Lowe, D.C.; Myhre, G.; et al. Changes in Atmospheric Constituents and in Radiative Forcing. In *Climate Change 2007: The Physical Science Basis, Contribution of Working Group I to the Fourth Assessment Report of the Intergovernmental Panel on Climate Change*; Solomon, S., Qin, D., Manning, M., Chen, Z., Marquis, M., Averyt, K.B., Tignor, M., Miller, H.L., Eds.; Cambridge University Press: Cambridge, UK, 2007; pp. 129–234.
10. Liang, C.; MacDonald, D. Chapter 5: Agriculture (CRF Sector 3). In *National Inventory Report 1990–2013: Greenhouse Gas Sources and Sinks in Canada Part 1*; Environment and Climate Change Canada: Gatineau, QC, Canada, 2015; pp. 123–146.
11. Novak, J.M.; Busscher, W.J. Selection and Use of Designer Biochars to Improve Characteristics of Southeastern USA Coastal Plain Degraded Soils. In *Advanced Biofuels and Bioproducts*; Lee, J.W., Ed.; Springer: New York, NY, USA, 2013; pp. 69–96.
12. Sun, Y.; Gao, B.; Yao, Y.; Fang, J.; Zhang, M.; Zhou, Y.; Chen, H.; Yang, L. Effects of feedstock type, production method, and pyrolysis temperature on biochar and hydrochar properties. *Chem. Eng. J.* **2014**, *240*, 574–578. [CrossRef]

13. Brassard, P.; Godbout, S.; Raghavan, V. Soil biochar amendment as a climate change mitigation tool: Key parameters and mechanisms involved. *J. Environ. Manag.* **2016**, *181*, 484–497. [CrossRef] [PubMed]
14. Budai, A.; Zimmerman, A.R.; Cowie, A.L.; Webber, J.B.W.; Singh, B.P.; Glaser, B.; Masiello, C.A.; Andersson, D.; Shields, F.; Lehmann, J.; et al. Biochar Carbon Stability Test Method: An assessment of methods to determine biochar carbon stability. *Int. Biochar Initiat.* **2013**, 1–10. Available online: http://www.biochar-international. org/sites/default/files/IBI_Report_Biochar_Stability_Test_Method_Final.pdf (accessed on 11 January 2017).
15. Garcia-Perez, M.; Lewis, T.; Kruger, C.E. *Methods for Producing Biochar and Advanced Biofuels in Washington State. Part 1: Literature Review of Pyrolysis Reactors*; First Project Report; Washington State Department of Ecology: Pullman, WA, USA, 2010; pp. 1–137.
16. Resende, F.L.P. Chapter 1: Reactor Configurations and Design Parameters for Thermochemical Conversion of Biomass into Fuels, Energy, and Chemicals. In *Reactor and Process Design in Sustainable Energy Technology*; Elsevier: Amsterdam, The Netherlands, 2014; pp. 1–25.
17. Myers, R.H.; Anderson-Cook, C.M.; Montgomery, D.C. *Wiley Series in Probability and Statistics: Response Surface Methodology: Process and Product Optimization Using Designed Experiments*, 3rd ed.; Wiley: Somerset, NJ, USA, 2016; p. 20.
18. Brown, J.N.; Brown, R.C. Process optimization of an auger pyrolyzer with heat carrier using response surface methodology. *Bioresour. Technol.* **2012**, *103*, 405–414. [CrossRef] [PubMed]
19. SAS Institute Inc. *SAS/STAT® 12.1 User's Guide*; SAS Institute Inc.: Cary, NC, USA, 2007.
20. Spokas, K.A.; Cantrell, K.B.; Novak, J.M.; Archer, D.W.; Ippolito, J.A.; Collins, H.P.; Boateng, A.A.; Lima, I.M.; Lamb, M.C.; McAloon, A.J.; et al. Biochar: A synthesis of its agronomic impact beyond carbon sequestration. *J. Environ. Qual.* **2012**, *41*, 973–989. [CrossRef] [PubMed]
21. Antal, M.J., Jr.; Grønli, M. The art, science, and technology of charcoal production. *Ind. Eng. Chem. Res.* **2003**, *42*, 1619–1640. [CrossRef]
22. Brassard, P.; Palacios, J.H.; Godbout, S.; Bussières, D.; Lagacé, R.; Larouche, J.-P.; Pelletier, F. Comparison of the gaseous and particulate matter emissions from the combustion of agricultural and forest biomasses. *Bioresour. Technol.* **2014**, *155*, 300–306. [CrossRef] [PubMed]
23. Cantrell, K.B.; Martin, J.H. Stochastic state-space temperature regulation of biochar production. Part II: Application to manure processing via pyrolysis. *J. Sci. Food Agric.* **2012**, *92*, 490–495. [CrossRef] [PubMed]
24. Association Française de Normalisation (AFNOR). *Organic Soil improvers and Growing Media: Biochemical Fractionning and Estimation of Biological Stability: Method of Organic Matter Characterisation by Successive Solubilisations*; Groupe AFNOR: La Plaine Saint-Denis, France, 2005; pp. 1–16.
25. American Society for Testing and Materials (ASTM). *E 203–16: Standard Test Method for Water Using Using Volumetric Karl-Fischer Titration*; ASTM International: West Conshohocken, PA, USA, 2016; pp. 1–9.
26. American Society for Testing and Materials (ASTM). *D 1762–84: Standard Test Method for Chemical Analysis of Wood Charcoal*; ASTM International: West Conshohocken, PA, USA, 2011; pp. 1–2.
27. Lehmann, J.; Joseph, S. *Biochar for Environmental Management: An Introduction*; Earthscan: London, UK, 2009; pp. 1–12.
28. Yang, H.; Yan, R.; Chen, H.; Lee, D.H.; Zheng, C. Characteristics of hemicellulose, cellulose and lignin pyrolysis. *Fuel* **2007**, *86*, 1781–1788. [CrossRef]
29. Kim, K.H.; Kim, J.Y.; Cho, T.S.; Choi, J.W. Influence of pyrolysis temperature on physicochemical properties of biochar obtained from the fast pyrolysis of pitch pine (Pinus rigida). *Bioresour. Technol.* **2012**, *118*, 158–162. [CrossRef] [PubMed]
30. Scott, D.S.; Piskorz, J.; Bergougnou, M.A.; Overend, R.P.; Graham, R. The role of temperature in the fast pyrolysis of cellulose and wood. *Ind. Eng. Chem. Res.* **1988**, *27*, 8–15. [CrossRef]
31. Liaw, S.-S.; Wang, Z.; Ndegwa, P.; Frear, C.; Ha, S.; Li, C.-Z.; Garcia-Perez, M. Effect of pyrolysis temperature on the yield and properties of bio-oils obtained from the auger pyrolysis of Douglas Fir wood. *J. Anal. Appl. Pyrolysis* **2012**, *93*, 52–62. [CrossRef]
32. Spokas, K.A.; Baker, J.M.; Reicosky, D.C. Ethylene: Potential key for biochar amendment impacts. *Plant Soil* **2010**, *333*, 443–452. [CrossRef]
33. Enders, A.; Hanley, K.; Whitman, T.; Joseph, S.; Lehmann, J. Characterization of biochars to evaluate recalcitrance and agronomic performance. *Bioresour. Technol.* **2012**, *114*, 644–653. [CrossRef] [PubMed]
34. Schimmelpfennig, S.; Glaser, B. One step forward toward characterization: Some important material properties to distinguish biochars. *J. Environ. Qual.* **2012**, *41*, 1001–1013. [CrossRef] [PubMed]

35. Manyà, J.J.; Ortigosa, M.A.; Laguarta, S.; Manso, J.A. Experimental study on the effect of pyrolysis pressure, peak temperature, and particle size on the potential stability of vine shoots-derived biochar. *Fuel* **2014**, *133*, 163–172. [CrossRef]

36. Harvey, O.R.; Kuo, L.-J.; Zimmerman, A.R.; Louchouarn, P.; Amonette, J.E.; Herbert, B.E. An index-based approach to assessing recalcitrance and soil carbon sequestration potential of engineered black carbons (biochars). *Environ. Sci. Technol.* **2012**, *46*, 1415–1421. [CrossRef] [PubMed]

37. Zhao, L.; Cao, X.; Mašek, O.; Zimmerman, A. Heterogeneity of biochar properties as a function of feedstock sources and production temperatures. *J. Hazard. Mater.* **2013**, *256–257*, 1–9. [CrossRef] [PubMed]

38. Di Blasi, C. Modeling intra- and extra-particle processes of wood fast pyrolysis. *AIChE J.* **2002**, *48*, 2386–2397. [CrossRef]

39. Zheng, J.; Stewart, C.E.; Cotrufo, M.F. Biochar and nitrogen fertilizer alters soil nitrogen dynamics and greenhouse gas fluxes from two temperate soils. *J. Environ. Qual.* **2012**, *41*, 1361–1370. [CrossRef] [PubMed]

energies

MDPI

Article

# An Innovative Agro-Forestry Supply Chain for Residual Biomass: Physicochemical Characterisation of Biochar from Olive and Hazelnut Pellets

Ilaria Zambon [1], Fabrizio Colosimo [2], Danilo Monarca [1], Massimo Cecchini [1], Francesco Gallucci [3], Andrea Rosario Proto [4], Richard Lord [2] and Andrea Colantoni [1,*]

[1] Department of Agricultural and Forestry Sciences (DAFNE), Tuscia University, Via San Camillo de Lellis snc, Viterbo 01100, Italy; ilaria.zambon@unitus.it (I.Z.); monarca@unitus.it (D.M.); cecchini@unitus.it (M.C.)

[2] Department of Civil and Environmental Engineering, University of Strathclyde, 75 Montrose Street, Glasgow G1 1XJ, UK; fabrizio.colosimo@strath.ac.uk (F.C.); richard.lord@strath.ac.uk (R.L.)

[3] Council for Agricultural Research and Agricultural Economy Analysis (CREA) Research Unit for Agricultural Engineering, Via della Pascolare 16, Monterotondo, Rome 00015, Italy; francesco.gallucci@crea.gov.it

[4] Department of Agriculture, Mediterranean University of Reggio Calabria, Feo di Vito, Reggio Calabria 89122, Italy; andrea.proto@unirc.it

* Correspondence: colantoni@unitus.it; Tel.: +39-07-6135-7356

Academic Editor: Mejdi Jeguirim
Received: 9 April 2016; Accepted: 1 July 2016; Published: 9 July 2016

**Abstract:** Concerns about climate change and food productivity have spurred interest in biochar, a form of charred organic material typically used in agriculture to improve soil productivity and as a means of carbon sequestration. An innovative approach in agriculture is the use of agro-forestry waste for the production of soil fertilisers for agricultural purposes and as a source of energy. A common agricultural practice is to burn crop residues in the field to produce ashes that can be used as soil fertilisers. This approach is able to supply plants with certain nutrients, such as Ca, K, Mg, Na, B, S, and Mo. However, the low concentration of N and P in the ashes, together with the occasional presence of heavy metals (Ni, Pb, Cd, Se, Al, etc.), has a negative effect on soil and, therefore, crop productivity. This work describes the opportunity to create an innovative supply chain from agricultural waste biomass. Olive (*Olea europaea*) and hazelnut (*Corylus avellana*) pruning residues represent a major component of biomass waste in the area of Viterbo (Italy). In this study, we evaluated the production of biochar from these residues. Furthermore, a physicochemical characterisation of the produced biochar was performed to assess the quality of the two biochars according to the standards of the European Biochar Certificate (EBC). The results of this study indicate the cost-effective production of high-quality biochar from olive and hazelnut biomass residues.

**Keywords:** biochar; biomass; soil fertiliser; olive; hazelnut

## 1. Introduction

Biochar is a carbon-rich material produced by thermal decomposition of biomass under oxygen-limited conditions [1]. According to the International Biochar Initiative (IBI), biochar is primarily used for soil applications for both agricultural and environmental gains [2]. The IBI definition differentiates biochar from charcoal, whose use is as a fuel for heat, as an absorbent material, or as a reducing agent in metallurgical processes [1,3]. Thermo-chemical processes include (i) slow pyrolysis (conventional carbonization); (ii) fast pyrolysis; (iii) flash carbonization; and (iv) gasification [4–7]. During the last two decades, pyrolysis process received more attention from the scientific community, since it is an efficient method for converting biomass into bio-fuel [5,8]. The pyrolysis process and its parameters, such as final temperature, pressure, heating rate, and residence time, greatly influence biochar quality [5]. The advantage of slow pyrolysis is to retain up to 50% of the carbon (C) feedstock in stable biochar [9], which makes it suitable as soil fertiliser. High-temperature pyrolysis (>550 °C) produces biochar with high aromatic content and, therefore, recalcitrant to decomposition [10]. Biochars produced through low-temperature processes (<550 °C) typically have a less-condensed C structure and are expected to give a better contribution to soil fertility [11]. The nature of the biomass feedstock also influences the properties of the produced biochar [3,12]. The relation between biochar properties and its potential to improve agricultural soils is a nascent focus area and the appropriate pyrolysis conditions are still unclear [13]. Numerous recent studies focused on methodologies for the chemical characterisation of biochars [13–15], other studies investigated the intrinsic potential of biochar as a soil amendment [16,17], although further efforts are required to obtain biochar with suitable properties [3]. One of the attractive characteristics of biochar as a soil amendment is its porous structure, which improves water retention and increases soil surface area [2]. Moreover, the concentration of biochar into soil has been related to an improved nutrient use efficiency, either through nutrients contained in biochar or through physicochemical processes that allow a better uptake of soil-inherent or fertiliser-derived nutrients [2]. The application of biochar increases physical and chemical qualities of soils, resulting in greater productivity of the agro-ecosystem [18]. Biochar, due to its biological and chemical stability, can also act as a C sink. The recalcitrance of biochar to microbial degradation enables the long-term sequestration of C in soil [2,19].

Biochar application in agriculture, positively affects the water holding capacity; this property derives from the distribution and the degree of cohesion of the pores in biochar, which depends on the particles size and aggregation, as well as the organic matter (OM) content. The effect of biochar on water holding capacity is dependent on both the high internal surface area of biochar and the capability to aggregate soil particles with OM, minerals, and microorganisms. The increase in soil porosity also allows a better percolation of excess water towards the deeper layers of the soil, therefore increasing ventilation.

This work aims to determine the opportunity to create an innovative supply chain from agricultural waste biomass, especially regarding olive (*Olea europaea*) and hazelnut (*Corylus avellana*) in order to evaluate the production of biochar from their pruning residues. Biomass residues in Mediterranean areas come mainly from agricultural and agro-industrial activities, as well as forest by-products. Only a few woody residues are used to produce fertilisers and as renewable energy resources [20]. In contrast, typical management strategies in the agricultural industry do not provide any valorisation of these biomasses, which are burnt in the field to prevent proliferation of plant diseases [20]. However, this landfill choice affects the soil structure since OM in woody biomass residues must be completely decomposed before used as fertiliser.

## 2. Materials and Methods

### 2.1. Biomass from Olive and Hazelnut Prunings

In the area of Viterbo, pruning residues from olive and hazelnut are rarely utilised as a source of energy in burning stoves or boilers; they are, instead, burnt in situ, therefore reducing the formation of soil organic carbon. During summer, besides pruning residues, suckers are removed before the harvest, representing another significant loss of biomass. Approximately 15 m$^3$ of biomass samples from both olive and hazelnut have been collected in farms of the Viterbo province. Recent studies [21,22] have investigated the possibility of enhancing olive and hazelnut residue waste management as a means to produce soil fertilisers and energy, therefore reducing the environmental impact of such residual organic wastes. Biomass from pruning crop operations (Figure 1a,b) represents an attractive resource that could be exploited for (i) fuel production (combustion and/or gasification) and (ii) biochar production (pyrolysis) that can be used as soil fertiliser.

**Figure 1.** Pruning residues from (**a**) olive and (**b**) hazelnut after crop operations; and (**c**) bio-shredding.

A pelletization procedure was developed and applied on bio-shredding obtained from olive and hazelnut residues (Figure 1c). Pruning residues were collected on site and immediately transferred to the laboratory for sifting and exsiccation (Figure 2) until a water content of 15% was achieved. Final water content as low as 15% is necessary for further refining of the product and pellet production. The humidity concentration in the prunings is very notable, because we can improve the technical process for pellet production by biomass. In Italy there are not many companies and total supply chains that work the prunings for pellet production and for use of these residual agriculture sub-products (Figure 3).

**Figure 2.** Schematic of the pelletization process showing (**a**) pellet mill; (**b**) olive and hazelnut pellet; and (**c**) packaging.

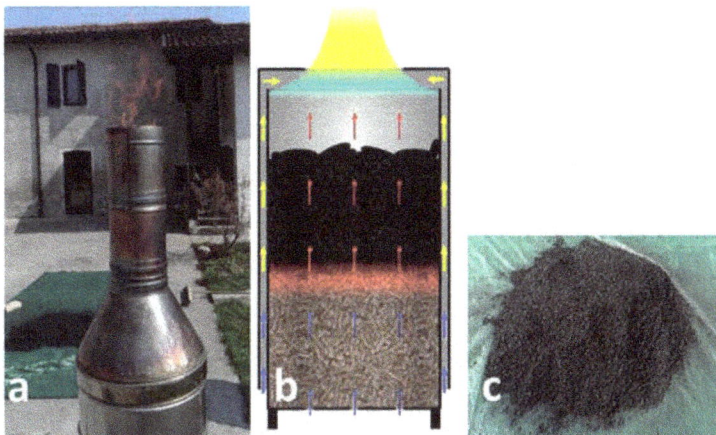

**Figure 3.** Biochar production from pellets showing (**a**) the Elsa Research carbonization system; (**b**) a schematic representation of the conversion process; and (**c**) the final product (biochar).

## 2.2. Pyrolysis Process

Pyrolysis of biomass is commonly considered as a thermo-chemical conversion process [7,23]. Pyrolysis is carried out under partial (or complete) absence of oxygen and relies on capturing the off-gases from thermal decomposition of the organic materials [19]. The physicochemical characteristics of biochar are determined by the type of feedstock and by the temperature of pyrolysis. For example, higher salt and ash contents are expected in wheat straw than in wood-derived biochar [24], and C content and N content are greater in pine chips than in poultry litter-derived biochar [25]. A higher pyrolysis temperature results in greater surface area, lower biochar recovery, higher ash content, elevated pH, minimal total surface charge [26], and lower cation exchange capacity [24]. Removal of volatile compounds at higher pyrolysis temperatures also cause biochars to have higher C content and lower hydrogen (H) and O content [26]. Pyrolysis of agro-forestry residues is typically carried out with temperatures between 400 and 800 °C. With these conditions, the feedstock is converted to liquid products (so-called tar or pyrolysis oil) and/or gas (syngas), which can be used as fuels or raw materials for subsequent chemical transformation. The residual solid carbonaceous material obtained (biochar) could be further refined to products, such as activated carbon.

## 2.3. Biochar Production form Olive and Hazelnut Pellets

The carbonisation system Elsa Research (Blucomb Ltd., based in Udine, Italy) was used to produce the biochar from olive and hazelnut pellets; biomass conversion was achieved by pyrolytic micro-gasification (Figure 3). The Elsa Research carbonisation system works with natural ventilation and does not require being powered by batteries or electricity. A chimney is typically used to increase the air draft for fuels that have difficulties igniting.

Physicochemical characterisation of the biochar obtained from the Elsa Research carbonisation system was performed at the European Biochar Institute, which released the EBC based on the quality of the biochar.

## 3. Results

### 3.1. From Biomass to Biochar: Conversion Rates Analyses

Auto-thermal conversion of biomass was carried out under natural ventilation. Quantitative analyses of pyrolysed biomass and produced biochar, as well as the conversion rates, are reported for 10 and four sessions of pyrolysis, respectively, for olive and hazelnut pellets (Table 1). A statistical comparison between olive and hazelnut performances during pyrolysis is reported in Table 2, showing the total conversion rates, mean, and standard deviation (SD) of the results obtained in the experiments.

**Table 1.** Conversion rates of biomass obtained from each pyrolysis session.

| Olive | | | | Hazelnut | | | |
|---|---|---|---|---|---|---|---|
| Session | Biomass (kg) | Biochar (kg) | Conversion Rate | Session | Biomass (kg) | Biochar (kg) | Conversion Rate |
| 1 | 38.35 | 8.11 | 0.209 | 1 | 37.69 | 8.11 | 0.215 |
| 2 | 39.07 | 8.21 | 0.210 | 2 | 36.25 | 7.96 | 0.220 |
| 3 | 38.88 | 8.19 | 0.211 | 3 | 37.03 | 8.09 | 0.218 |
| 4 | 38.96 | 8.16 | 0.209 | 4 | 37.11 | 8.09 | 0.218 |
| 5 | 34.09 | 7.10 | 0.208 | | | | |
| 6 | 39.02 | 8.23 | 0.211 | | | | |
| 7 | 38.89 | 8.19 | 0.211 | | | | |
| 8 | 38.93 | 8.19 | 0.210 | | | | |
| 9 | 38.97 | 8.20 | 0.210 | | | | |
| 10 | 38.81 | 8.13 | 0.209 | | | | |

**Table 2.** Comparisons and statistical values of conversion rates of olive and hazelnut.

| | Olive | | Hazelnut | |
|---|---|---|---|---|
| | Biochar (kg) | Conversion Rate | Biochar (kg) | Conversion Rate |
| Total | 384.47 | 80.71 | 148.08 | 32.25 |
| Mean | | 0.210 | | 0.218 |
| SD | | 0.00088 | | 0.00188 |

Further analyses were carried out to investigate the calorific power of the two biochars produced. Composition, structure, heat value of the gas, tar liquid, and semi-char solid products depend on the pyrolysis temperature [7]. Quantity and quality of resulting outputs from biomass pyrolysis are related to the chemical composition of the operating temperature and the feedstock [7,27]. The calorific values calculated were compared with those provided by the producers in order to make energy considerations on the process. The results obtained are consistent with other pyrolysis processes. The latter led to the volatilisation of a fraction of biomass with a calorific value ranging between 75% and 85% of the starting biomass. The calorific value is measured in terms of the high calorific value [28]. Table 3 distinguishes two types of calorific value (usually expressed in MJ/kg): (i) the higher calorific value that it is the amount of heat produced by a complete combustion of a mass unit of a sample, at constant volume, in an atmosphere rich of oxygen at standard conditions (25 °C, 101.3 kPa); and (ii) the lower calorific value (PCI) that does not include the heat of the condensation of water [28].

**Table 3.** Analysis of the calorific power of pyrolysis reaction for the two biochars produced in this study.

| Olive Wood | Units | Pellet | Biochar |
|---|---|---|---|
| Higher calorific value | MJ/kg | 19.47 | 31.71 |
| Lower calorific value | MJ/kg | 16.17 | 30.48 |
| Calorific value from pyrolysis | MJ/kg | 12.37 | |
| Percentage of calorific value from pyrolysis | % | 0.76 | |
| Hazelnut Wood | Units | Pellet | Biochar |
| Higher calorific value | MJ/kg | 19.02 | 26.62 |
| Lower calorific value | MJ/kg | 16.71 | 25.66 |
| Calorific value from pyrolysis | MJ/kg | 14.21 | |
| Percentage of calorific value from pyrolysis | % | 0.85 | |

Pyrolysis does not produce energy from heat; rather, it leads to the production of gas from biomass. In general, pyrolysis involves the heating of biomass to temperatures greater than 400 °C in the absence of oxygen [29]. At these temperatures, biomass thermally decomposes releasing a vapour phase and biochar (solid phase). On cooling the pyrolysis vapour, polar and high-molecular-weight compounds condense out as bio-oil (liquid phase) while low-molecular-weight volatile compounds remain in the gas phase (syngas) [6]. The physics and chemistry of pyrolysis process results are extremely complex, and are dependant depending on both the rector conditions and the nature of the biomass [29]. The combustion of gas in the Elsa Research system occurs in "close-coupled combustion" (micro-gasification). Biochars produced by Blucomb Ltd. (Udine, Italy) (spin-off) for the European project were analysed by Eurofins laboratories, accredited for the certification of the EBC. International biochar experts developed the EBC in order to consider it in the European context as a voluntary industrial standard [30]. The EBC guarantees a sustainable biochar production, with a low-risk use in agronomic systems. Biochar produced in accordance with the standards of the EBC fulfils all of the requirements of sustainable production and environmental impact by certifying (i) sustainable production and provision of biomass feedstock; (ii) energy efficient, low emission pyrolysis technique; (iii) low contaminant level in the biochar; and (iv) low hazard use and application of the biochar. These standards are in compliance with current environmental European regulations [31].

## 3.2. Elemental Analysis

The chemical composition of biochar is determined also by the source of biomass employed. Biochar produced from wood, for example, is denser and has higher C content (~80%) [32]. These properties reflect the chemical complexity of lignin, which makes it more resistant to thermal degradation. The elemental composition, plotted as H/C vs. O/C ratios (Figure 4), is often used to describe maturity, decomposition rate, and combustion behaviour of fossil chars and coal [33,34]. When applied to biochar, the H/C and O/C ratios can be suitable indicators of the degree of carbonisation. High ratios typically point to primary plant macromolecules, such as simple carbohydrates and cellulose [35]. An H/C ratio of ⩽0.2 indicates C of plant origin with elevated carbonisation [36].

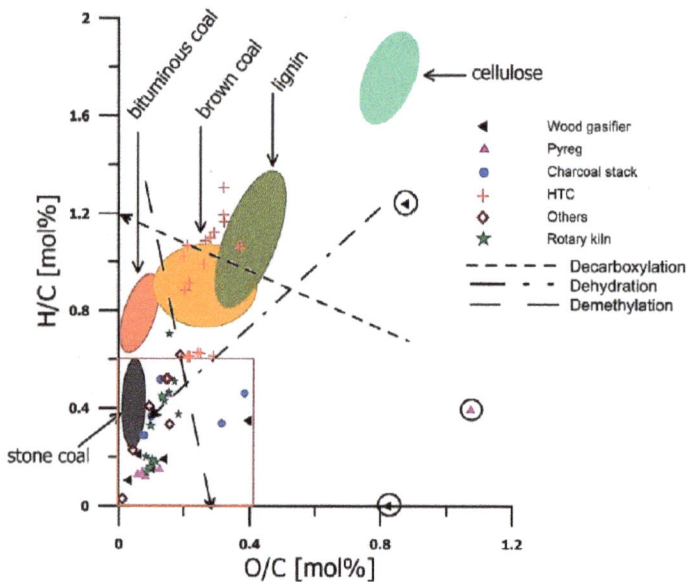

**Figure 4.** Example of Van Krevelen diagram of biochars obtained through different pyrolisis processes. The red square shows the optimum elemental ratio values of H/C and O/C for biochar production.

The O/C ratio is an indicator of the presence of polar functional groups, which influence the stability of biochar by preventing a dense, graphite-like structure of the material [37]. Therefore, the O/C ratio is useful to assess hydrophilicity and hydrophobicity of the charred material. The ratios of H, O, and C can also be used to differentiate between materials obtained by different processes. In the view of C sequestration and for material with complex aromatic structure and low presence of functional groups, optimum ratios of H/C and O/C are approximately ⩽0.6 and ⩽0.4, respectively [38]. Nitrogen in biochar is an important nutrient, its concentration is related to the concentration in the starting material, with values between 1.8 and 56 g· kg$^{-1}$, although N in biochar is in a form often not readily bioavailable [39]. The C/N ratio, an indicator of the bioavailability of an organic compound, is highly variable and ranges between 7 and 500 [38].

The results of elemental analyses of the two biochars investigated in this study are reported in Table 4. Both biochars are characterised by values well below the limits established by the EBC, in particular the olive and hazelnut biochars have high values of C and low H/C and O/C ratios. A low H/C ratio indicates that the produced biochars are also recalcitrant to microbial degradation. These results indicate that our production process yield high-quality biochars with a level of carbonisation that makes it suitable for C sequestration, as confirmed by the H/C ratios.

Table 4. Elemental analyses from EBC (Method DIN 51732).

| Elements | Units | Hazelnut Biochar | Olive Biochar | EBC Biochar Base | EBC Biochar Premium |
|---|---|---|---|---|---|
| H (Hydrogen) | % w/w | 1.21 | 1.58 | - | - |
| C (Carbon, total) | % w/w | 78.1 | 90.1 | >50 | >50 |
| N (Nitrogen, total) | % w/w | 0.64 | 0.42 | - | - |
| O (Oxygen) | % w/w | 1.2 | 1.7 | - | - |
| Carbonate as $CO_2$ | % w/w | 2.62 | 1.17 | - | - |
| Carbonate (organic) | | 75.5 | 89.8 | | |
| H/C ratio (molar) | | 0.18 | 0.21 | <0.6 | <0.6 |
| O/C rate (molar) | | 0.012 | 0.014 | <0.4 | <0.4 |
| Sulphur (total) | % w/w | 0.07 | <0.03 | | |

## 3.3. Nutrients and Trace Elements

Biomass residues containing high concentrations of minerals, such as those obtained from herbaceous plants produce biochars with high ash content [32], maintain in the biochar matrix most of the nutrients present in the starting material (Table 5). These types of biochar have a lower total carbon (TC) content and cohesion than those obtained from wood-pruning biomass. The low C content, together with elevated concentrations of nutrients, makes biochars from herbaceous material more readily available for microorganisms [2]. The concentration of phosphorus (P) and potassium (K) in the biochar is related to the initial content in the feedstock. The content of P and K are typically between 2.7 and 480 g· kg$^{-1}$ and 10 to 58 g· kg$^{-1}$, respectively [39].

Table 5. Determination from microwave digestion (method: DIN 22022-1).

| Elements | Units | Methods | Hazelnut Biochar | Olive Biochar | EBC Biochar Base | EBC Biochar Premium |
|---|---|---|---|---|---|---|
| P (Phosphorus) | mg/kg | ISO 11885 | 590 | 330 | - | - |
| Mg (Magnesium) | mg/kg | ISO 11885 | 2900 | 1400 | - | - |
| Ca (Calcium) | mg/kg | ISO 11885 | 38,000 | 11,000 | - | - |
| K (Potassium) | mg/kg | ISO 11885 | 5500 | 3500 | - | - |
| Na (Sodium) | mg/kg | ISO 11885 | 2100 | 260 | - | - |
| Fe (Iron) | mg/kg | ISO 11885 | 6500 | 1500 | - | - |
| Si (Silicon) | mg/kg | ISO 11885 | 25,000 | 9700 | - | - |
| S (Sulphur) | mg/kg | ISO 11885 | 910 | 200 | - | - |
| Pb (Lead) | mg/kg | ISO 17294-2 | 66 | 20 | <150 | <120 |
| Cd (Cadmium) | mg/kg | ISO 17294-2 | <0.2 | <0.2 | <1.5 | <1 |
| Cu (Copper) | mg/kg | ISO 17294-2 | 100 | 6 | <100 | <100 |
| Ni (Nickel) | mg/kg | ISO 17294-2 | 9 | 8 | <50 | <30 |
| Hg (Mercury) | mg/kg | DIN EN 1483 | <0.07 | <0.07 | <1 | <1 |
| Zn (Zinc) | mg/kg | ISO 17294-2 | 340 | 84 | <400 | <400 |
| Cr (Chromium total) | mg/kg | ISO 17294-2 | 22 | 15 | <90 | <80 |
| B (Boron) | mg/kg | ISO 17294-2 | 32 | 10 | - | - |
| Mn (Manganese) | mg/kg | ISO 17294-2 | 350 | 380 | - | - |

EBC biochar base and premium report the limits required by the EBC protocol of certification. The total ash content ranged between 6.2% and 18.8% (w/w) for biochar from pellets of olive and hazelnut wood. The nutrient content is much greater in hazelnut biochar than olive, which was evident especially for Mg, Ca, Fe, S, Cu, and Zn. Biochar from hazelnut pellets could bring a greater contribution of nutrients to the soil and, therefore, be less resistant to microbial decomposition. Heavy metal content in both biochars was well below the EBC limits. Only Cu in the hazelnut biochar was close to the maximum value established by the EBC.

## 3.4. PAHs (Polycyclic Aromatic Hydrocarbons) Composition

PAHs are ubiquitous in the environment, being by-products of the incomplete combustion of organic material [40]. The chemical structure of PAHs makes them highly resistant to biodegradation and oxidation [41]. The presence of PAHs in pyrolytic reactions above 700 °C is well established [42], although they can be produced in pyrolysis reactions of less than 700 °C at low concentration [43]. It is, therefore, critical to ensure PAH concentrations remain below the limits established by the EBC. The 16 priority US EPA PAHs are typically used to assess the total PAH content; the limits established by the EBC are of <12 and <4 mg/kg for biochar standard and premium, respectively. The PAH composition of the two biochars analysed in this study (Table 6), shows that both biochars are well below the EBC limits, with values ranging from <0.1 to 1.1 mg/Kg.

**Table 6.** PAHs determination from toluene extract.

| Elements | Units | Methods | Limits | | Hazelnut Biochar | Olive Biochar |
|---|---|---|---|---|---|---|
| | | | GW 1 * | GW 2 * | | |
| Naphthalene | mg/kg | DIN EN 15527 | - | - | 0.9 | 1.1 |
| Acenaphthylene | mg/kg | DIN EN 15527 | - | - | <0.1 | <0.1 |
| Acenaphthene | mg/kg | DIN EN 15527 | - | - | <0.1 | <0.1 |
| Fluorene | mg/kg | DIN EN 15527 | - | - | <0.1 | <0.1 |
| Phenanthrene | mg/kg | DIN EN 15527 | - | - | 0.3 | 0.3 |
| Anthracene | mg/kg | DIN EN 15527 | - | - | <0.1 | <0.1 |
| Fluoranthene | mg/kg | DIN EN 15527 | - | - | 0.1 | 0.1 |
| Pyrene | mg/kg | DIN EN 15527 | - | - | 0.1 | 0.1 |
| Benz(a)anthracene | mg/kg | DIN EN 15527 | - | - | <0.1 | <0.1 |
| Chrysene | mg/kg | DIN EN 15527 | - | - | <0.1 | <0.1 |
| Benzo(b)fluoranthene | mg/kg | DIN EN 15527 | - | - | <0.1 | <0.1 |
| Benzo(k)fluoranthene | mg/kg | DIN EN 15527 | - | - | <0.1 | <0.1 |
| Benzo(a)pyrene | mg/kg | DIN EN 15527 | - | - | <0.1 | <0.1 |
| Indeno(1,2,3-cd)pyrene | mg/kg | DIN EN 15527 | - | - | <0.1 | <0.1 |
| Dibenz(a,h)anthracene | mg/kg | DIN EN 15527 | - | - | <0.1 | <0.1 |
| Benzo(g,h,i)perylene | mg/kg | DIN EN 15527 | - | - | <0.1 | <0.1 |
| SUM PAHs (EPA) | mg/kg | calculated | <12 | <4 | 1.20 | 1.60 |

\* (GW 1 = quality level basic related dry bases; GW 2 = quality level premium related dry bases).

Total PAH content of the two biochars are 1.2 and 1.6 mg/kg for olive and hazelnut, respectively. Therefore, both biochars can be considered suitable for soil applications, since both are well below the EBC threshold limit of 4 mg/kg for biochar premium.

## 3.5. pH, Electrical Conductivity (EC), and Density

In general, the pH of biochar is relatively homogeneous and varies from neutral to basic pH. Feedstock of various origins produce biochar with an average pH between 6.2 and 9.6 [39]. Lower pH is typically found for biochars obtained from green pruning feedstock and organic waste, while the highest values are to be attributed to poultry litter biochar. The Table 7 reports the elements values, according to their pH, electrical conductivity, salt content and density.

The two biochar have a pH of 8.4 and 9.9 for olive and hazelnut, respectively. The EBC indicates a maximum limit of 10; therefore, biochar produced from these types of wood residues is slightly below the limit established by the certification. The EC is of particular importance when adding biochar to soils with high EC and salinity. The two biochars had an EC of 217 and 332 mS/cm, respectively, for olive and hazelnut (as shown in Table 7). Both values are very low and do not represent a real risk for the addition to soil even under conditions of high EC. In general, biochar has a lower density than soil, with an average of 0.4 g· cm$^{-3}$ compared to a soil of medium texture, with average of 1.3 g· cm$^{-3}$. When adding biochar to soils with little ventilation, this property can help to reduce the density by mitigating issues related to the compaction of soil. The olive and hazelnut biochars produced in this study have a density of 0.45 and 0.44 g· cm$^{-3}$, respectively.

**Table 7.** Elements value (pH, electrical conductivity, and density).

| Elements | Units | Hazelnut Biochar | Olive Biochar |
|---|---|---|---|
| pH values (CaCl$_2$) | - | 9.9 | 8.4 |
| Electrical conductivity | µS/cm | 332 | 217 |
| Salt content | g/kg | 0.655 | 1.18 |
| Salt content cal. with bulk density | g/L | 0.287 | 0.527 |

## 4. Conclusions

The two biochars analysed in this study show excellent physicochemical properties, which makes them suitable for agricultural applications. Both biochars can be certified as Biochar Premium according to the regulations of the EBC; this allows a potential commercialisation of the biochars, with higher prices than Biochar Base, typically less expensive, but with a higher content of PAHs. The benefits of using Biochar Premium as soil fertiliser includes improved productivity, increased water holding capacity of the soil (e.g., [44–46]), and a better retention of nutrients and agrochemicals in soils, all of which should offset initial investment and provide added profits per application. Biochar fuel commands a high-value application, offering numerous benefits, and an authentic alternative to develop the biomass utilization efficiency [4,47]. The added value of biochar is also linked to other issues, such as those involving agricultural and environmental sustainability. As claimed by many studies [1,16], biochar as a soil amendment can improve soil health and increase agricultural productivity with further environmental benefits related to global warming mitigation [16,48–52]. Based on our results, we intend to define an agro-forestry chain to use the residual waste biomass for the production of high quality biochar for agronomic and commercial purposes. We are proceeding to evaluate the properties of biochar for soil improvement.

**Acknowledgments:** This work was supported during my visiting professor at Strathclyde University.

**Author Contributions:** All authors contributed equally to the work done.

**Conflicts of Interest:** The authors declare no conflict of interest.

## References

1. Lehmann, J.; Stephen, J. *Biochar for Environmental Management—Science and Technology*; Routledge: Abingdon, UK, 2009.
2. Sohi, S.; Lopez-Capel, E.; Krull, E.; Bol, R. Biochar, climate change and soil: A review to guide future research. *Civ. Eng.* **2009**, 6618–6664.
3. Manyà, J.J. Pyrolysis for biochar purposes: A review to establish current knowledge gaps and research needs. *Environ. Sci. Technol.* **2012**, *46*, 7939–7954. [CrossRef] [PubMed]
4. Anupam, K.; Sharma, A.K.; Lal, P.S.; Dutta, S.; Maity, S. Preparation, characterization and optimization for upgrading Leucaena leucocephala bark to biochar fuel with high energy yielding. *Energy* **2016**, *106*, 743–756.
5. Jahirul, M.I.; Rasul, M.G.; Chowdhury, A.A.; Ashwat, N. Biofuels production through biomass pyrolysis—A technological review. *Energies* **2012**, *5*, 4952–5001. [CrossRef]
6. Hoogwijk, M.; Faaij, A.P.C.; van den Broek, R.; Berndes, G.; Gielen, D.; Turkenburg, W. Exploration of the ranges of the global potential of biomass for energy. *Biomass Bioenergy* **2003**, *25*, 119–133. [CrossRef]
7. Panwar, N.L.; Kothari, R.; Tyagi, V.V. Thermo chemical conversion of biomass – Eco friendly energy routes. *Renew. Sustain. Energy Rev.* **2012**, *16*, 1801–1816. [CrossRef]
8. Demirbas, A. Partly chemical analysis of liquid fraction of flash pyrolysis products from biomass in the presence of sodium carbonate. *Energy Convers. Manag.* **2002**, *43*, 1801–1809. [CrossRef]
9. Gaunt, J.L.; Lehmann, J. Energy balance and emissions associated with biochar sequestration and pyrolysis bioenergy production. *Environ. Sci. Technol.* **2008**, *42*, 4152–4158. [CrossRef] [PubMed]

10. Singh, B.P.; Cowie, A.L. A novel approach, using [13]C natural abundance, for measuring decomposition of biochars in soil. In Proceedings of the Carbon and Nutrients Management in Agriculture, Palmerston North, New Zealand, 13–14 February 2008; Currie, L.D., Yates, L., Eds.

11. Steinbeiss, S.; Gleixner, G.; Antonietti, M. Effect of biochar amendment on soil carbon balance and soil microbial activity. *Soil Biol. Biochem.* **2009**, *41*, 1301–1310. [CrossRef]

12. Di Blasi, C.; Branca, C.; Lombardi, V.; Ciappa, P.; di Giacomo, C.; Chimica, I. Effects of particle size and density on the packed-bed pyrolysis of wood. *Energy Fuels* **2013**, *27*, 6781–6791. [CrossRef]

13. Brewer, C.; Schmidt-Roht, K.; Satrio, J.; Brown, R. Characterization of Biochar from fast pyrolysis and gasification systems. *Environ. Prog.* **2009**, *28*, 386–396. [CrossRef]

14. Sánchez, M.E.; Lindao, E.; Margaleff, D.; Martínez, O.; Morán, A. Pyrolysis of agricultural residues from rape and sunflowers: Production and characterization of bio-fuels and biochar soil management. *J. Anal. Appl. Pyrolysis* **2009**, *85*, 142–144. [CrossRef]

15. Keiluweit, M.; Nico, P.S.; Johnson, M.G.; Kleber, M. Dynamic molecular structure of plant biomass-derived black carbon (biochar). *Environ. Sci. Technol.* **2010**, *44*, 1247–1253. [CrossRef] [PubMed]

16. Agegnehu, G.; Bass, A.M.; Nelson, P.N.; Muirhead, B.; Wright, G.; Bird, M.I. Biochar and biochar-compost as soil amendments: Effects on peanut yield, soil properties and greenhouse gas emissions in tropical North Queensland, Australia. *Agric. Ecosyst. Environ.* **2015**, *21*, 372–385. [CrossRef]

17. Beesleya, L.; Moreno-Jiménez, E.; Gomez-Eyles, J.L.; Harris, E.; Robinsond, B.; Sizmure, T. A review of biochars' potential role in the remediation, revegetation and restoration of contaminated soils. *Environ. Pollut.* **2011**, *159*, 3269–3282. [CrossRef] [PubMed]

18. Laird, D.A.; Fleming, P.; Davis, D.D.; Horton, R.; Wang, B.; Karlen, D.L. Impact of biochar amendments on the quality of a typical Midwestern agricultural soil. *Geoderma* **2010**, *158*, 443–449. [CrossRef]

19. Lehmann, J. A handful of carbon. *Nature* **2007**, *447*, 143–144. [CrossRef] [PubMed]

20. Di Giacomo, G.; Taglieri, L. Renewable energy benefits with conversion of woody residues to pellets. *Energy* **2009**, *34*, 724–731. [CrossRef]

21. Colantoni, A.; Longo, L.; Evic, N.; Gallucci, F.; Delfanti, L. Use of Hazelnut's pruning to produce biochar by gasifier small scale plant. *Int. J. Renew. Energy Res.* **2015**, *5*, 873–878.

22. Di Giacinto, S.; Longo, L.; Menghini, G.; Delfanti, L.; Egidi, G.; de Benedictis, L.; Riccioni, S.; Salvati, L. Model for estimating pruned biomass obtained from Corylus avellana L. *Appl. Math. Sci.* **2014**, *8*, 6555–6564.

23. White, J.E.; Catall, W.J.; Legendre, B.L. Biomass pyrolysis kinetics: A comparative critical review with relevant agricultural residue case studies. *Anal. Appl. Pyrol.* **2011**. [CrossRef]

24. Kloss, S.; Zehetner, F.; Dellantonio, A.; Hamid, R.; Ottner, F.; Liedtke, V.; Schwanninger, M.; Gerzabek, M.H.; Soja, G. Characterization of slow pyrolysis biochars: Effects of feedstocks and pyrolysis temperature on biochar properties. *J. Environ. Qual.* **2012**, *41*, 990–1000. [CrossRef] [PubMed]

25. Gaskin, J.W.; Steiner, C.; Harris, K.; Das, K.C.; Bibens, B. Effect of low-temperature pyrolysis conditions on biochar for agricultural use. *Trans. Asabe* **2008**, *51*, 2061–2069. [CrossRef]

26. Novak, J.; Lima, I.; Xing, B. Characterization of designer biochar produced at different temperatures and their effects on a loamy sand. *Ann. Environ. Sci.* **2009**, *3*, 195–206.

27. Ceylan, R.; B-son Bredenberg, J. Hydrogenolysis and hydrocracking of the car-bonoxygen bond. 2. Thermal cleavage of the carbon–oxygen bond in guaiacol. *Fuel* **1982**, *61*, 377–382.

28. Jothibasu, K.; Mathivanan, G. Enhancing combustion efficiency using nano nickel oxide catalyst in biomass gasifier. *Int. J. Core Eng. Manag. (IJCEM)* **2015**, *2*, 31–41.

29. Laird, D.A. Review of the pyrolysis platform for coproducing bio-oil and biochar. *Biofuels Bioprod. Bioref.* **2008**, *3*, 547–562. [CrossRef]

30. European Biochar Certificate (EBC). *Guidelines for a Sustainable Production of Biochar*; European Biochar Foundation: Arbaz, Switzerland, 2012; pp. 1–23.

31. European Parliament, Council of the European Union. Regulation (EC) No. 1907/2006 of the European Parliament and of the Council of 18 December 2006 concerning the Registration, Evaluation, Authorisation and Restriction of Chemicals (REACH), establishing a European Chemicals Agency, amending Directive 1999/45/EC and repealing Council Regulation (EEC) No 793/93 and Commission Regulation (EC) No 1488/94 as well as Council Directive 76/769/EEC and Commission Directives 91/155/EEC, 93/67/EEC, 93/105/EC and 2000/21/EC. Available online: http://eur-lex.europa.eu/legal-content/EN/TXT/?uri=CELEX%3A02006R1907-20140410 (accessed on 10 April 2014).

32. Demirbas, A. Effects of temperature and particle size on bio-char yield from pyrolysis of agricultural residues. *J. Anal. Appl. Pyrolysis* **2004**, *72*, 243–248. [CrossRef]
33. Hammes, K.; Smernik, R.J.; Skjemstad, J.O.; Herzog, A.; Vogt, U.F.; Schmidt, M.W.I. Synthesis and characterisation of laboratory-charred grass straw (Oryza sativa) and chestnut wood (Castanea sativa) as reference materials for black carbon quantification. *Org. Geochem.* **2006**, *37*, 1629–1633. [CrossRef]
34. Krevelen, V. *Coal Science*; Elsevier: Amsterdam, The Netherlands, 1957.
35. Chun, Y.; Sheng, G.; Chiou, G.T.; Xing, B. Compositions and sorptive properties of crop residue-derived chars. *Environ. Sci. Technol.* **2004**, *38*, 4649–4655. [CrossRef] [PubMed]
36. Kuhlbusch, T.A.J.; Crutzen, P.J. Toward a global estimate of black carbon in residues of vegetation fires representing a sink of atmospheric $CO_2$ and a source of $O_2$. *Glob. Biogeochem. Cycles* **1995**, *9*, 491–501. [CrossRef]
37. Laine, J.; Yunes, S. Effect of the preparation method on the pore size distribution of activated carbon from coconut shell. *Carbon N. Y.* **1992**, *30*, 601–604. [CrossRef]
38. Schimmelpfennig, S.; Glaser, B. One step forward toward characterization: Some important material properties to distinguish biochars. *J. Environ. Qual.* **2012**, *41*, 1001–1013. [CrossRef] [PubMed]
39. Chan, K.Y.; Xu, Z. Biochar: Nutrient properties and their enhancement. *Biochar Environ. Manag. Sci. Technol.* **2012**, 67–84.
40. Keith, L.H.; Telliard, W.A. Priority pollutants I-a perspective view. *Environ. Sci. Technol.* **1979**, *13*, 416–423. [CrossRef]
41. Preston, C.M.; Schmidt, M.W.I. Black (pyrogenic) carbon: A synthesis of current knowledge and uncertainties with special consideration of boreal regions. *Biogeosciences* **2006**, *3*, 397–420. [CrossRef]
42. Ledesma, E.B.; Marsh, N.D.; Sandrowitz, A.K.; Wornat, M.J. Global kinetic rate parameters for the formation of polycyclic aromatic hydrocarbons from the pyrolysis of catechol, a model compound representative of solid fuel moieties. *Energy Fuels* **2002**, *16*, 1331–1336. [CrossRef]
43. Garcia-Perez, M. *The Formation of Polyaromatic Hydrocarbons and Dioxins during Pyrolysis: A Review of the Literature with Descriptions of Biomass Composition, Fast Pyrolysis Technologies and Thermochemical Reactions*; Washington State University: Pullman, WA, USA, 2008; pp. 1–58.
44. Fellet, G.; Marchiol, L.; delle Vedove, G.; Peressotti, A. Application of biochar on mine tailings: Effects and perspectives for land reclamation. *Chemosphere* **2011**, *83*, 1262–1267. [CrossRef] [PubMed]
45. Karhu, K.; Mattila, T.; Bergstrom, I.; Regina, K. Biochar addition to agricultural soil increased CH4 uptake and water holding capacity—Results from a short-term pilot field study. *Agric. Ecosyst. Environ.* **2011**, *140*, 309–313. [CrossRef]
46. Streubel, J.D.; Collins, H.P.; Garcia-Perez, M.; Tarara, J.; Granatstein, D.; Kruger, C.E. Influence of contrasting biochar types on five soils at increasing rates of application. *Soil Sci. Soc. Am. J.* **2011**, *75*, 1402–1413. [CrossRef]
47. Liu, Z.; Quek, A.; Hoekman, S.K.; Balasubramanian, R. Production of solid biochar fuel from waste biomass by hydrothermal carbonization. *Fuel* **2013**, *103*, 943–949. [CrossRef]
48. Reza, M.T.; Andert, J.; Wirth, B.; Busch, D.; Pielert, J.; Lynam, J.G.; Mumme, J. Hydrothermal carbonization of biomass for energy and crop production. *Appl. Bioenergy* **2014**, *1*, 11–29. [CrossRef]
49. Hill, J.; Nelson, E.; Tilman, D.; Polasky, S.; Tiffany, D. Environmental, economic, and energetic costs and benefits of biodiesel and ethanol biofuels. *Proc. Natl. Acad. Sci. USA* **2006**, *103*, 11206–11210. [CrossRef] [PubMed]
50. Chandra, R.; Takeuchi, H.; Hasegawa, T. Hydrothermal pretreatment of rice straw biomass: A potential and promising method for enhanced methane production. *Appl. Energy* **2012**, *94*, 129–140. [CrossRef]
51. Kim, D.; Yoshikawa, K.; Park, K.Y. Characteristics of biochar obtained by hydrothermal carbonization of cellulose for renewable energy. *Energies* **2015**, *8*, 14040–14048. [CrossRef]
52. Civitarese, V.; Spinelli, R.; Barontini, M.; Gallucci, F.; Santangelo, E.; Acampora, A.; Scarfone, A.; del Giudice, A.; Pari, L. Open-air drying of cut and windrowed short-rotation poplar stems. *Bioenergy Res.* **2015**, *8*, 1614–1620. [CrossRef]

*energies*

MDPI

*Article*

# Thermal Properties of Biochars Derived from Waste Biomass Generated by Agricultural and Forestry Sectors

Xing Yang [1,2], Hailong Wang [2,3,*], Peter James Strong [4], Song Xu [1,*], Shujuan Liu [1], Kouping Lu [2], Kuichuan Sheng [5,*], Jia Guo [6], Lei Che [7], Lizhi He [2], Yong Sik Ok [8], Guodong Yuan [9,10], Ying Shen [2] and Xin Chen [1]

[1] School of Environment and Chemical Engineering, Foshan University, Foshan 528000, China; yx20080907@163.com (X.Y.); liujuan_407@163.com (S.L.); chenxin20170331@163.com (X.C.)
[2] Key Laboratory of Soil Contamination Bioremediation of Zhejiang Province, School of Environmental and Resource Sciences, Zhejiang Agricultural and Forestry University, Lin'an, Hangzhou 311300, China; kkping111@163.com (K.L.); helizhiyongyuan@163.com (L.H.); shenying@zafu.edu.cn (Y.S.)
[3] Department of Environmental Engineering, Foshan University, Foshan 528000, China
[4] Centre for Solid Waste Bioprocessing, School of Civil Engineering, School of Chemical Engineering, The University of Queensland, St Lucia, QLD 4072, Australia; j.strong2@uq.edu.au
[5] College of Biosystems Engineering and Food Science, Zhejiang University, Hangzhou 310058, China
[6] Zhejiang Chengbang Landscape Co. Ltd., Hangzhou 310008, China; guojiafuture@foxmail.com
[7] School of Engineering, Huzhou University, Huzhou 313000, China; three_stone_cn@163.com
[8] School of Natural Resources and Environmental Science & Korea Biochar Research Center, Kangwon National University, Chuncheon 24341, Korea; soilok@kangwon.ac.kr
[9] Yantai Institute of Coastal Zone Research, Chinese Academy of Sciences, Yantai 264003, China; gdyuan@yic.ac.cn
[10] Guangdong Dazhong Agriculture Science Co. Ltd., Dongguan 523169, China
* Correspondence: nzhailongwang@163.com (H.W.); xuson@yeah.net (S.X.); kcsheng@zju.edu.cn (K.S.)

Academic Editor: S. Kent Hoekman
Received: 18 December 2016; Accepted: 24 March 2017; Published: 2 April 2017

**Abstract:** Waste residues produced by agricultural and forestry industries can generate energy and are regarded as a promising source of sustainable fuels. Pyrolysis, where waste biomass is heated under low-oxygen conditions, has recently attracted attention as a means to add value to these residues. The material is carbonized and yields a solid product known as biochar. In this study, eight types of biomass were evaluated for their suitability as raw material to produce biochar. Material was pyrolyzed at either 350 °C or 500 °C and changes in ash content, volatile solids, fixed carbon, higher heating value (HHV) and yield were assessed. For pyrolysis at 350 °C, significant correlations ($p < 0.01$) between the biochars' ash and fixed carbon content and their HHVs were observed. Masson pine wood and Chinese fir wood biochars pyrolyzed at 350 °C and the bamboo sawdust biochar pyrolyzed at 500 °C were suitable for direct use in fuel applications, as reflected by their higher HHVs, higher energy density, greater fixed carbon and lower ash contents. Rice straw was a poor substrate as the resultant biochar contained less than 60% fixed carbon and a relatively low HHV. Of the suitable residues, carbonization via pyrolysis is a promising technology to add value to pecan shells and Miscanthus.

**Keywords:** biochar; biomass; higher heating value (HHV); proximate analysis; renewable energy

## 1. Introduction

By 2020, the use of petroleum and other liquid fuels is estimated to reach nearly 100 million bpd globally, and this is anticipated to increase a further 10% by 2035 [1]. This increase in energy demand, coupled to the depletion of petroleum resources, has intensified renewable energy research [2]. Biofuel derived from lignocellulosic biomass is one potential source of renewable energy. Biomass energy currently provides almost 14% of the world's primary energy. It is regarded as the renewable fuel with the highest potential for sustainable development in the future, and its adoption can significantly lower fossil fuel use and $CO_2$ emissions [3]. In addition, a providing sustainable energy solution is urgently required in developing countries.

In China, the use of biomass residues as a renewable energy source has increased in importance. Approximately 70% of the Chinese population live in rural areas, where a large amount of agricultural biomass residue is generated. It is a challenge to utilize these residues as a fuel in their original form due to their low bulk density, low heating value and the volume of smoke they generate [4]. Lignocellulosic biomass is an abundant organic material that, in addition to its use as a fuel, can be upgraded to generate biochar [5]. Biochar is a general term for a solid product derived from the pyrolysis of agricultural or forestry biomass [6,7]. Pyrolysis is a process where a substrate is heated in the absence of oxygen [3], and this conversion is a new strategy to potentially add value to biomass residues [8].

Depending on its physical and chemical properties, biochar can range from a high-quality fuel [6] to a soil amendment [9]. It may also be used to remediate contaminated soil [10–12] or sequester carbon [13,14]. The processing method and pyrolysis temperature are the decisive elements affecting biochar yield, but chemical and physical properties of the feedstock cannot be ignored [15,16]. There has been considerable research regarding the use of biochar as a soil amendment, including the effect of pyrolysis conditions and substrate type on biochar quality [8]. Moreover, numerous studies focus on biomass as a substrate for the carbonization of biochar and its application as a fuel [17–19]. The resulting biomass-derived biochar can be further processed into fuel briquettes after carbonization and can partially address the challenges of waste biomass management related to air pollution and improving transport efficiency [20]. For this reason, it is important to examine the thermal characteristics of biochars generated from different biomass types and identify those most suited to biochar production.

In this study, biochars were generated using laboratory-scale pyrolytic carbonization of eight types of waste biomass feedstocks commonly available in subtropical China. These feedstocks were pine wood, Chinese fir wood, Chinese fir bark, bamboo leaves, bamboo sawdust, Miscanthus, pecan shells and rice straw. To date, there is little research evaluating the thermal characteristics of biochar generated from these biomass sources. The aims of this research were to:

(1)  evaluate the influence of biomass types and pyrolysis temperatures on the thermal properties of biochars;
(2)  correlate biochar properties such as ash, volatile solid and fixed carbon contents to the higher heating values (HHVs); and
(3)  identify which waste biomass is best suited for fuel biochar production.

In brief, the outcome of this work would determine which substrates were suitable as fuel biochar and provide technical guidance for using agricultural and forestry residues for biochar production.

## 2. Materials and Methods

### 2.1. Biomass Feedstock Preparation

Eight types of biomass from the agricultural and forestry sectors in subtropical China were assessed. These were Masson pine wood, Chinese fir wood, Chinese fir bark, bamboo leaves, bamboo sawdust, Miscanthus, pecan shells and rice straw. The biomass was obtained from Lin'an City, northwest of Hangzhou, Zhejiang Province, China. The biomass feedstocks were air-dried, chopped using a pulverizer, and finally ground to a size able to pass through 40-mesh sieves for laboratory analyses.

## 2.2. Biochar Production

Biochars were produced using a slow pyrolysis procedure, which was performed under an inert nitrogen (purity $\geq$ 99.99%, flow rate: 200 mL min$^{-1}$) atmosphere in a laboratory-scale (5 L) fixed-bed tubular reactor made of stainless steel. The reactor was heated by an electrical furnace that had the maximum temperature of 800 °C. The reactor was filled to 75% of its capacity and heated to either 350 °C or 500 °C at a heating rate of 5 °C min$^{-1}$. When peak temperature was reached, it was held for 2 h. The pyrolyzed material was then cooled to room temperature under an inert atmosphere. Sixteen biochar samples were generated from the eight kinds of feedstock pyrolyzed at two temperatures. Biochar samples were ground to a size able to pass through 80-mesh sieves for laboratory analyses.

## 2.3. Laboratory Analyses

The moisture, ash, fixed carbon content, volatile solids and the HHV were determined for each feedstock and corresponding biochar samples. Moisture, ash and volatile solids content were determined according to the NY/T1881-2010 Standard (China) [21]. For the moisture content, ground samples were placed in an oven at 105 °C and dried until constant weight. Dried samples were cooled in a desiccator for 1 h prior to weighing. For the ash content, samples were placed in an uncovered crucible in a muffle furnace and the temperature was raised to approximately 275 °C in air with a heating rate lower than 10 °C min$^{-1}$ (i.e., ambient to 275 °C in less than 30 min). This was held for 30 min and then raised to 750 °C and held for 3 h. Samples were cooled in a desiccator for 20 min prior to weighing. For the volatile solids content, the muffle furnace was preheated to 920 °C, samples were placed inside and processed under an inert atmosphere (nitrogen) and kept at 900 $\pm$ 10 °C for 7 min. Samples were cooled in a desiccator for 20 min prior to weighing. The HHV was determined according to the GB/T 213-2008 Standard (China) [22] with a Bomb Calorimeter. The fixed carbon content was calculated from the moisture, ash and volatile solids contents. All analyses were performed in duplicate.

## 2.4. The Calculations of Biochar Yield, Energy Yield and Energy Density

The biochar (mass) yield ($G_m$) was calculated as the ratio of the mass of dry biochar ($M_2$) to the mass of dry biomass ($M_1$) (Equation (1)). The energy yield ($G_e$) represented the energy contained in biomass that was retained in the biochar. It was calculated using $G_m$ and the HHVs of biochar ($Q_2$) and raw biomass ($Q_1$) (Equation (2)). The energy density (*ED*) indicated the ratio of energy yield and biochar yield, as presented below (Equation (3)) [23].

$$G_m = \frac{M_2}{M_1} \times 100\% \tag{1}$$

$$G_e = G_m \times \frac{Q_2}{Q_1} \tag{2}$$

$$ED = \frac{G_e}{G_m} \tag{3}$$

## 2.5. Statistical Analyses

Statistical analyses were performed using the SPSS 17.0 statistical package program. The sample means ($n = 2$) for proximate analyses and HHV were subjected to one-way analysis of variance (ANOVA) and Duncan's multiple range tests. Variability in the data was expressed as the standard error, and the level of significance was set at $p$ value <0.05. The correlations between raw biomass and the biochars, as well as that between ash, volatile and fixed carbon content and HHV were based on Pearson's correlation coefficients ($p < 0.01$ and $p < 0.05$).

## 3. Results and Discussion

### 3.1. Yield Analyses

The biochar and energy yields for the various substrates are displayed in Figure 1. In this study, slow pyrolysis was used as this generally favors biochar production. Slow pyrolysis, which incorporates longer residence times at slow heating rates at lower temperatures, produces primarily charcoal, while at high temperatures it produces primarily gaseous products [24]. For example, Nam et al. [25] obtained much higher mass yields of rice straw biochars using slow pyrolysis (45–48%) compared to fast pyrolysis (27%). Fast pyrolysis, which incorporates short residence times, fast heating rates, and moderate temperatures, favors the production of bio-oil [24].

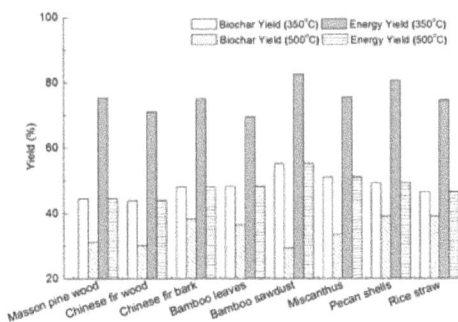

**Figure 1.** Biochar and energy yield of various substrates pyrolyzed at 350 or 500 °C.

The biochar yields from bamboo sawdust (55%) and Miscanthus (51%) were relatively high at 350 °C. All biochar yields decreased at the higher pyrolysis temperature [26], most notably for bamboo sawdust, which decreased from 55% at 350 °C to 29% at 500 °C. However, at 500 °C the biochar yields from pecan shells and rice straw remained relatively high—both at 39%. Yields are known to vary due to the differences in the relative abundance of cellulose, hemicellulose and lignin within different biomass, which have different thermal degradation kinetics [13]. Xiong et al. [27] also observed biochar yields decreased at higher carbonization temperatures. The differences between the resultant biochars were most likely attributable to the thermal resilience of lignin, as opposed to hemicellulose and cellulose that decompose at temperatures lower than 400 °C (220 to 315 °C and 315 to 400 °C, respectively).

The energy yields displayed a similar trend to the biochar yields. For pyrolysis at 350 °C, the biochar energy yields in decreasing order were: bamboo sawdust > pecan shells > Miscanthus > Masson pine wood > Chinese fir bark > rice straw > Chinese fir wood > bamboo leaves. Energy yields were lower after pyrolysis at 500 °C; the biochar energy yields in decreasing order were: bamboo sawdust > Miscanthus > pecan shells > bamboo leaves > Chinese fir bark > rice straw > Masson pine wood > Chinese fir wood. In the current study, the energy density of biochars pyrolyzed at 350 °C ranged from 1.44 to 1.69, with Masson pine wood, Chinese fir wood and pecan shell biochars above 1.60. Although the energy density of bamboo sawdust biochar obtained at 500 °C (1.88) was greatly improved compared that pyrolyzed at 350 °C (1.49), all other biochars generated at the higher temperature had lower energy densities (ranging from 1.19 to 1.52). Masson pine wood, Chinese fir wood and bamboo sawdust biomass are theoretically excellent biofuels because of their low ash content, but environmental factors and transport costs limit their application [28].

Biochars contain less volatile solids and, to a certain point, have an increased energy density. Harsono et al. [29] obtained 0.2 tons of biochar, 0.3 tons of biogas and 0.025 tons of bio-oil from one ton of palm oil residues after slow pyrolysis. The biogas consisted mainly of $CO_2$ (40–75%) and $CH_4$ (15–60%) [30], and trace amounts of water vapor, hydrogen sulfide, siloxanes, hydrocarbons, ammonia, oxygen, carbon monoxide and nitrogen [31]. Biochar is lighter than the original material, making it easier to transport. Biochar is also moisture resistant and resistant to microbial degradation, which can lessen the impact of seasonal variations as they can be stored for much longer periods than raw biomass [5]. Reza et al. [32] also reported that the biochar was about 41–90% of the original mass, but contained 80–95% of the fuel value of the raw feedstock, i.e., biochar had higher energy density than the raw biomass due to the mass lost to volatilization. In China, families used to collect agricultural and forestry biomass such as crop residuals, weeds, branches and leaves for cooking and heating [33], and lots of biomass residues were burnt in the field, which caused serious air pollution [34]. Wang et al. [35] reported that $PM_{2.5}$ concentrations increased because of biomass burning to an average of 134 $\mu g\ m^{-3}$, which was three times worse than clear days in Shanghai (from 2011 to 2013). Therefore, carbonization by pyrolysis represents a better approach to utilize these residues.

### 3.2. Proximate Analyses

Proximate analysis is a measure of total biomass components in terms of moisture content, ash content, volatile solids and fixed carbon of the solid fuel [36]. It is a relatively simple, cheap, robust method that is widely used to describe the properties of biomass fuels [37]. In the current study, ash and volatile solid contents varied significantly for the raw biomass as well as the biochars. In addition to the pyrolysis conditions (peak temperature, heating rate and residence time), structural components of the raw biomass also affected resultant biochars [38].

The ash content varied significantly ($p < 0.05$) among different feedstocks (Figure 2). The ash content of pecan shells, bamboo leaves and rice straw were relatively high, while Chinese fir wood, bamboo sawdust, Chinese fir bark and Masson pine wood had very low ash contents (all below 0.5%). The rice straw had the largest ash content (14%), while the Chinese fir wood had the lowest (0.08%). These results were consistent with Obernberger et al. [39], who found that wood generally had a lower ash content than bark, straw or cereal. Variation in ash content is attributable to the different concentrations of ash-forming elements, such as calcium carbonate, potassium silicates, iron and other metals [40]. The ash content of different feedstocks in this study were consistent with data from Liu et al. [41]. As ash is non-combustible, it negatively impacts a material's calorific value [4]. In addition, processing combustible material with a high ash content requires more frequent residue removal, as well as increased boiler maintenance due to higher dust emissions [42]. This results in unnecessary down time and lowers process efficiency, which is why a biomass with a low ash content is preferred as a fuel source.

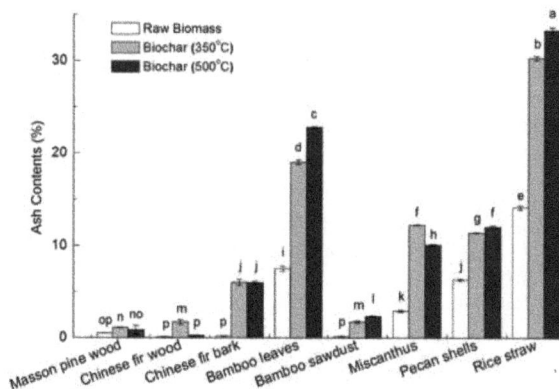

**Figure 2.** Ash contents of biomass and biochars pyrolyzed at 350 or 500 °C. Different letters above the columns indicate a significant difference ($p < 0.05$) between treatments. Error bars represent standard error of the means ($n = 2$).

In general, the pyrolysis (at 350 or 500 °C) increased the relative ash contents of the biochar, but values varied greatly and ranged from 0.3% (Chinese fir wood at 500 °C) to 33.3% (rice straw biochar at 500 °C). There were significant ($p < 0.05$) differences between the raw biomass and their biochars (Figure 2). The majority of mass loss occurred below 350 °C and the ash content increased dramatically at 350 °C (range: 0.6–16.2%), and generally increased (albeit less markedly) at 500 °C (range: 0.3–19.1%). Ash contents were relatively higher for biochars from pecan shells, rice straw, Miscanthus, Chinese fir bark and bamboo leaves (>10%), whereas the biochars made from Masson pine wood, Chinese fir wood and bamboo sawdust were low (<3%). There was a significant ($p < 0.01$) positive correlation between raw biomass and their biochars in terms of the ash content (Table 1). The results indicate that a raw material with a high ash content generates a biochar with a high ash content. Biomass from Masson pine wood, Chinese fir wood and bamboo sawdust had low starting ash contents, which was reflected in their biochars.

**Table 1.** Correlation coefficients between properties of raw biomass and biochars pyrolyzed at 350 or 500 °C ($n = 8$).

| Properties | Samples | Raw Biomass | Biochar (350 °C) | Biochar (500 °C) |
|---|---|---|---|---|
| Ash content | Raw biomass | 1 | 0.966 ** | 0.969 ** |
| | Biochar (350 °C) | 0.966 ** | 1 | 0.991 ** |
| | Biochar (500 °C) | 0.969 ** | 0.991 ** | 1 |
| Volatile solid | Raw biomass | 1 | 0.794 ** | 0.216 |
| | Biochar (350 °C) | 0.794 ** | 1 | −0.059 |
| | Biochar (500 °C) | 0.216 | −0.059 | 1 |
| Fixed carbon | Raw biomass | 1 | 0.518 * | 0.096 |
| | Biochar (350 °C) | 0.518 * | 1 | 0.405 |
| | Biochar (500 °C) | 0.096 | 0.405 | 1 |
| Higher heating value | Raw biomass | 1 | 0.733 ** | 0.743 ** |
| | Biochar (350 °C) | 0.733 ** | 1 | 0.708 ** |
| | Biochar (500 °C) | 0.743 ** | 0.708 ** | 1 |

Notes: * Correlation is significant at 0.05 level; ** Correlation is significant at 0.01 level.

The volatile solids of biomass and biochars are presented in Figure 3. In contrast to ash content, volatile solids (VS) retention was primarily affected by the pyrolysis temperature. The VS contents of the raw biomass were all greater than 50% (Chinese fir wood had the greatest content at 78%) and an increase in pyrolysis temperature from 350 to 500 °C increased VS release. This is in agreement with Sadaka et al. [43], who reported that VS decreased with increasing carbonization temperature for switchgrass (*Panicum virgatum* L.). Bamboo sawdust had the greatest VS loss with increasing temperature: the VS content decreased to 52% at 350 °C and 11% at 500 °C. The Chinese fir bark biochars had the lowest VS content at the tested pyrolysis temperatures, which decreased from 71% in the feedstock to 30% at 350 °C and dropped to 23% at 500 °C. Biomass typically consists of three components: hemicellulose, cellulose and lignin [37]. The differences may be due to thermal stability variances between hemicellulose, cellulose and lignin, and the relative abundance of these three components in each biomass. Generally, hemicellulose is the most volatile, cellulose is less volatile, while lignin is the most difficult to volatilize [37]. Bark is primarily composed of hemicelluloses and cellulose, while the woody biomass may contain more lignin than bark. At 350 °C, almost all hemicellulose and cellulose were pyrolyzed, while the lignin was more stable and continued to be pyrolyzed at 500 °C. The VS for the full set of biochars generated at 500 °C had the smallest variation, ranging from 11% to 23%, with the majority of the values between 14% and 20%. According to Chaney [44], low-grade fuels with low volatile solid may result in smouldering, which can generate a large amount of smoke and even toxic gases. A high VS content indicated easy ignition and that most of the fuel will volatize during combustion [4]. In the present study, although the volatile solids content of biochar mainly depended on pyrolysis temperature, a significant ($p < 0.01$) positive correlation was observed between raw biomass and their biochars from pyrolysis at 350 °C. However, there was no significant correlation between raw biomass and the 500 °C biochar (Table 1).

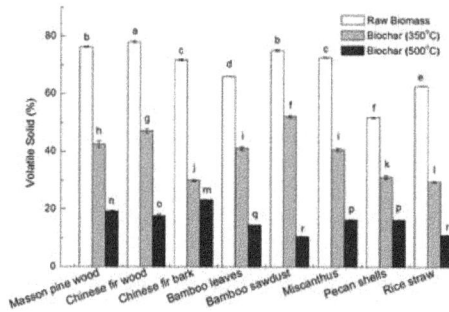

**Figure 3.** Volatile solids contents of biomass and biochars pyrolyzed at 350 or 500 °C. Letters above the columns indicate the significant difference ($p < 0.05$) between treatments. Error bars represent standard error of the means ($n = 2$).

The fixed carbon content of biochar increased significantly ($p < 0.05$) with increasing pyrolysis temperature due to the increased loss of volatile solids. Thus, the fixed carbon negatively correlated to the VS (Figures 3 and 4), which was consistent with a previous study [43]. The fixed carbon content followed the reverse pattern to the VS with all feedstocks. In addition, fixed carbon provided a rough estimate of the heating value of a fuel and was the primary source of heat generated during combustion [4]. In our study, the fixed carbon content ranged from 10% to 26% for the feedstocks, 38% to 62% after pyrolysis at 350 °C, and 52% to 84% for biochars after pyrolysis at 500 °C. The fixed carbon content increased two to four times after pyrolysis at 350 °C. The bamboo sawdust and Chinese fir bark pyrolyzed at 500 °C displayed the largest and smallest changes to fixed carbon content (increasing by 97% and 6%, respectively) in comparison to the pyrolysis products generated at 350 °C. Rice straw had the lowest fixed carbon contents after pyrolysis at 350 °C (38%) and 500 °C (52%), while Chinese fir wood had the highest fixed carbon content at 350 °C (62%) and bamboo sawdust the highest at 500 °C (84%). The Pearson's correlation test confirmed that the fixed carbon content of the 350 °C biochar correlated negatively ($p < 0.05$) with that of the raw material (Table 1). However, there was no significant correlation between 500 °C biochars and the raw biomass. As with volatile solids, pyrolysis temperature was the dominant factor affecting the fixed carbon content of biochar.

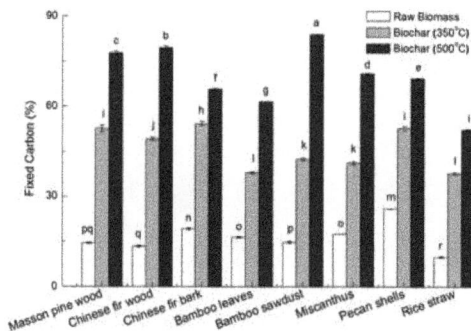

**Figure 4.** Fixed carbon contents of biomass and biochars pyrolyzed at 350 or 500 °C. Letters above the columns indicate the significant difference ($p < 0.05$) between treatments. Error bars represent standard error of the means ($n = 2$).

## 3.3. Higher Heating Value

The HHV is one of the most important parameters related to energy analyses. It is defined as the energy released per unit mass or per unit volume of fuel after complete combustion, including the energy contained in the water vapor in the exhaust gases [45]. The HHV indicates the best use for biomass fuel, as it describes the energy content [36]. As is evident in Figure 5, the HHVs of biomass feedstocks were generally similar (ranging from 18.44 to 20.10 MJ kg$^{-1}$), but the HHVs were lower for pecan shells (16.95 MJ kg$^{-1}$) and rice straw (15.85 MJ kg$^{-1}$). However, the HHVs of the eight feedstocks differed after pyrolysis. The HHVs of bamboo sawdust and Miscanthus biochars pyrolyzed at 500 °C were higher than those at 350 °C, while the opposite trend was observed for bamboo leaves and rice straw biochars. This change was due to variations in structural components, and their ratios, which were volatilized or carbonized to different extents at the two temperatures. During the low-temperature carbonization period, the heating value of the biochar initially increased, and decreased as the carbonized components that contributed to the HHV were most likely volatilized to $CH_4$ and $H_2$ at higher temperatures [46]. This was also observed by Xiong et al. [27], who found peak heating values of cotton stalk and bamboo sawdust biochar peaked after pyrolysis at 550 or 600 °C, respectively. For the 350 °C biochars the HHVs of Masson pine wood, Chinese fir wood and Chinese fir bark were >30 MJ kg$^{-1}$—comparable with the HHVs for coals ranging from brown coal to bituminous coal [47]. According to Qian et al. [36], these HHVs were comparable to high quality coal, which typically has a HHV of 25–35 MJ kg$^{-1}$. Anderson et al. [48] also reported that biochars derived from woody plants had higher HHVs. The Douglas-fir (*Pseudotsuga menziesii*) and Lodgepole pine (*Pinus contorta*) biochars were higher in energy than charcoal, but these were lower in energy compared to medium or high quality coal. In the current study, the bamboo sawdust biochar pyrolysis at 500 °C had the highest HHV at 32.4 MJ kg$^{-1}$, which was comparable with the HHV of bituminous coal [47].

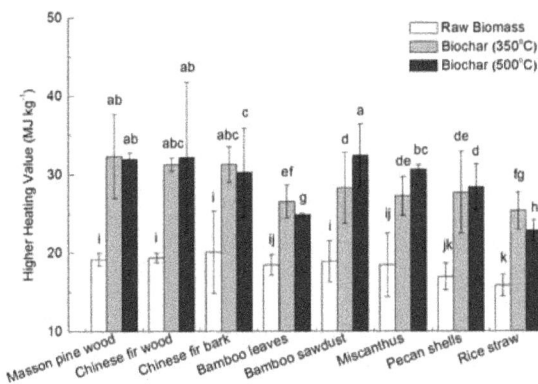

**Figure 5.** Higher heating values of the raw biomass and biochars pyrolyzed at 350 or 500 °C. Letters above the columns indicate the significant difference ($p < 0.05$) between treatments. Error bars represent standard error of the means ($n = 2$).

## 3.4. The Correlation between Proximate Analyses and HHV

There was no significant correlation between volatile solids content and HHV (Figure 6), whereas HHV had significant correlations ($p < 0.01$) with ash and fixed carbon contents. Accordingly, biochar samples with a low ash content or a high fixed carbon content display high HHVs. Therefore, suitable substrates for fuel biochar production need to be selected according to their ash and fixed carbon contents. In addition, there was a strong positive correlation ($p < 0.01$) between the HHVs of raw biomass and their corresponding biochars (Table 1).

The lower ash content (1.1%), higher fixed carbon (52.6%), energy density (1.69) and HHV (32.3 MJ kg$^{-1}$), as well as the higher energy yield (75.3%) indicate that Masson pine wood biochar produced at 350 °C was the most efficient biofuel of the eight substrates tested. In addition, Chinese fir wood biochar pyrolyzed at 350 °C was also a suitable fuel that displayed good thermal properties. After pyrolysis at 500 °C, the thermal properties of bamboo sawdust biochar were enhanced, and the higher HHV (32.4 MJ kg$^{-1}$), energy density (1.88) and fixed carbon content (83.9%) made it a higher quality fuel than that obtained at 350 °C (28.3 MJ kg$^{-1}$ HHV; 1.49 energy density; 42.5% fixed carbon content). The rice straw biomass was a poor energy source after pyrolysis, but it has potential remediation applications as it is an effective adsorbent for soil pollutants [14,49] and a good nutrient source due to its high concentration of inorganic elements [50]. Pecan shells and Miscanthus biochars were well suited as energy sources due to their high energy yield when converted to biochar. In Europe, Miscanthus is widely planted to produce liquid biofuels and chemical precursors, as well as improve soil fertility and land aesthetics [51]. In the current study, carbonization via pyrolysis is an innovative and promising method for the utilization of Miscanthus residues. The conversion efficiency for energy and weight of the pecan shells was considerable, and combined with other data, we considered this the best substrate for biochar production.

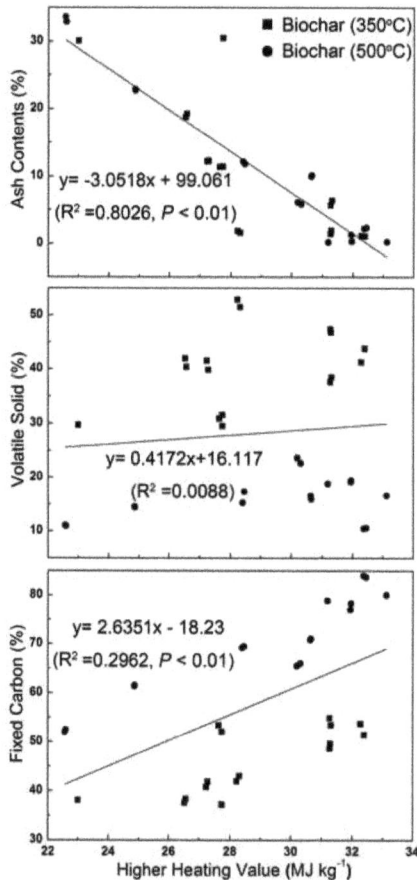

**Figure 6.** Correlation between the higher heating value and the contents of ash, fixed carbon and volatile solids in biochars (*n* = 16).

*Energies* **2017**, *10*, 469

## 4. Conclusions

The results of this study demonstrate that the thermal properties and yields of biochars differ vastly for different biomass feedstocks. The thermal properties of the biochars were dependent on raw material characteristics and the temperature of carbonization during pyrolysis. Similarly, the effects on volatile solid and fixed carbon content were more dependent on pyrolysis temperature, while the ash contents of biochar were dependent on feedstock composition more than pyrolysis temperature. The HHV of the biochars increased after the initial pyrolysis at 350 °C, but the increase tapered or decreased with pyrolysis at 500 °C. Masson pine wood and Chinese fir wood biochars pyrolyzed at 350 °C and bamboo sawdust biochar pyrolyzed at 500 °C were suitable for direct use in fuel applications, as reflected by their higher HHVs, greater fixed carbon contents and lower ash contents. Of the substrates, carbonization via pyrolysis is a promising technology to add value to pecan shells and Miscanthus due to the higher energy and mass retention of the resultant biochars.

**Acknowledgments:** This study was funded by the National Natural Science Foundation of China (21577131, 41401338), the Natural Science Foundation of Zhejiang Province, China (LZ15D010001), Zhejiang Chengbang Landscape Co. Ltd., Hangzhou, China, and the Special Funding for the Introduced Innovative R&D Team of Dongguan (2014607101003), China. We thank Baoqi Chen and Yanhui Lin for their technical assistance.

**Author Contributions:** Xing Yang conceived and wrote this paper, Hailong Wang, Song Xu and Kuichuan Sheng conceived and designed the research, Kouping Lu and Lizhi He analyzed the data, Jia Guo and Ying Shen provided and prepared some of the biomass feedstocks and assisted the experiments, Peter James Strong, Shujuan Liu, Yong Sik Ok, Lei Che, Jia Guo and Xin Chen revised the paper and provided some valuable suggestions.

**Conflicts of Interest:** The authors declare no conflict of interest.

## References

1. *International Energy Outlook 2011*; U.S. Energy Information Administration: Washington, DC, USA, 2011.
2. Kunkes, E.L.; Simonetti, D.A.; West, R.M.; Serrano-Ruiz, J.C.; Gärtner, C.A.; Dumesic, J.A. Catalytic conversion of biomass to monofunctional hydrocarbons and targeted liquid-fuel classes. *Science* **2008**, *322*, 417–421. [CrossRef] [PubMed]
3. Qin, L.; Li, Q.; Meng, A.; Zhang, Y. Pyrolysis properties of potential biomass fuels in southwestern China. In *Cleaner Combustion and Sustainable World*; Qi, H., Zhao, B., Eds.; Tsinghua University Press: Beijing, China, 2013; pp. 457–464.
4. Akowuah, J.O.; Kemausuor, F.; Mitchual, S. Physico-chemical characteristics and market potential of sawdust charcoal briquette. *Int. J. Energy Environ. Eng.* **2012**, *3*, 1–6. [CrossRef]
5. Liu, Z.; Quek, A.; Hoekman, S.K.; Balasubramanian, R. Production of solid biochar fuel from waste biomass by hydrothermal carbonization. *Fuel* **2013**, *103*, 943–949. [CrossRef]
6. Galgani, P.; Voet, E.V.; Korevaar, G. Composting, anaerobic digestion and biochar production in Ghana. Environmental–economic assessment in the context of voluntary carbon markets. *Waste Manag.* **2014**, *34*, 2454–2465. [CrossRef] [PubMed]
7. Kung, C.; Chang, M. Effect of agricultural feedstock to energy conversion rate on bioenergy and GHG emissions. *Sustainability* **2015**, *7*, 5981–5995. [CrossRef]
8. Qian, K.; Kumar, A.; Zhang, H.; Bellmer, D.; Huhnke, R. Recent advances in utilization of biochar. *Renew. Sustain. Energy Rev.* **2015**, *42*, 1055–1064. [CrossRef]
9. Mao, X.; Xu, X.; Lu, K.; Gielen, G.; Luo, J.; He, L.; Donnison, A.; Xu, Z.; Xu, J.; Yang, W.; et al. Effect of 17 years of organic and inorganic fertilizer applications on soil phosphorus dynamics in a rice–wheat rotation cropping system in eastern China. *J. Soils Sediments* **2015**, *15*, 1889–1899. [CrossRef]
10. Zhang, X.; Wang, H.; He, L.; Lu, K.; Sarmah, A.; Li, J.; Bolan, N.S.; Pei, J.; Huang, H. Using biochar for remediation of soils contaminated with heavy metals and organic pollutants. *Environ. Sci. Pollut. Res.* **2013**, *20*, 8472–8483. [CrossRef] [PubMed]
11. He, L.; Gielen, G.; Bolan, N.S.; Zhang, X.; Qin, H.; Huang, H.; Wang, H. Contamination and remediation of phthalic acid esters in agricultural soils in China. *Agron. Sustain. Dev.* **2015**, *35*, 519–534. [CrossRef]
12. Zhang, X.; He, L.; Sarmah, A.; Lin, K.; Liu, Y.; Li, J.; Wang, H. Retention and release of diethyl phthalate in biochar-amended vegetable garden soils. *J. Soils Sediments* **2014**, *14*, 1790–1799. [CrossRef]

13. Novak, J.M.; Lima, I.; Xing, B.; Gaskin, J.W.; Steiner, C.; Das, K.; Ahmedna, M.; Rehrah, D.; Watts, D.W.; Busscher, W.J. Characterization of designer biochar produced at different temperatures and their effects on a loamy sand. *Ann. Environ. Sci.* **2009**, *3*, 195–206.
14. Yang, X.; Liu, J.; McGrouther, K.; Huang, H.; Lu, K.; Guo, X.; He, L.; Lin, X.; Che, L.; Ye, Z.; et al. Effect of biochar on the extractability of heavy metals (Cd, Cu, Pb and Zn) and enzyme activity in soil. *Environ. Sci. Pollut. Res.* **2016**, *23*, 974–984. [CrossRef] [PubMed]
15. Yargicoglu, E.N.; Sadasivam, B.Y.; Reddy, K.R.; Spokas, K. Physical and chemical characterization of waste wood derived biochars. *Waste Manag.* **2015**, *36*, 256–268. [CrossRef] [PubMed]
16. Yang, X.; Lu, K.; McGrouther, K.; Che, L.; Hu, G.; Wang, Q.; Liu, X.; Shen, L.; Huang, H.; Ye, Z.; et al. Bioavailability of Cd and Zn in soils treated with biochars derived from tobacco stalk and dead pigs. *J. Soils Sediments* **2017**, *17*, 751–762. [CrossRef]
17. Zulu, L.C. The forbidden fuel: Charcoal, urban woodfuel demand and supply dynamics, community forest management and woodfuel policy in Malawi. *Energy Policy* **2010**, *38*, 3717–3730. [CrossRef]
18. Mwampamba, T.H.; Owen, M.; Pigaht, M. Opportunities, challenges and way forward for the charcoal briquette industry in Sub-Saharan Africa. *Energy Sustain. Dev.* **2013**, *17*, 158–170. [CrossRef]
19. Liu, L.; Liu, Y.; Xing, Z. Char trace analysis of composite wood floor under different heating conditions. *Procedia Eng.* **2016**, *135*, 217–220. [CrossRef]
20. Lohri, C.R.; Rajabu, H.M.; Sweeney, D.J.; Zurbrügg, C. Char fuel production in developing countries—A review of urban biowaste carbonization. *Renew. Sust. Energy Rev.* **2016**, *59*, 1514–1530. [CrossRef]
21. Ministry of Agriculture of PRC. *Densified Biofuel—Test Methods*; NY/T1881-2010 Standard; China Standards Press: Beijing, China, 2010.
22. *Determination of Calorific Value of Coal, GB/T 213-2008*; General Administration of Quality Supervision, Inspection and Quarantine, Standardization Administration of China: Beijing, China, 2008.
23. Nam, H.; Capareda, S. Experimental investigation of torrefaion of two agricultural wastes of different composition using RSM (response surface methodology). *Energy* **2015**, *91*, 507–516. [CrossRef]
24. Huber, G.W.; Iborra, S.; Corma, A. Synthesis of transportation fuels from biomass: Chemistry, catalysts, and engineering. *Chem. Rev.* **2006**, *106*, 4044–4098. [CrossRef] [PubMed]
25. Nam, H.; Capareda, S.C.; Ashwath, N.; Kongkasawan, J. Experimental investigation of pyrolysis of rice straw using bench-scale auger, batch and fluidized bed reactors. *Energy* **2015**, *93*, 2384–2394. [CrossRef]
26. Mašek, O.; Brownsort, P.; Cross, A.; Sohi, S. Influence of production conditions on the yield and environmental stability of biochar. *Fuel* **2013**, *103*, 151–155. [CrossRef]
27. Xiong, S.; Zhang, S.; Wu, Q.; Guo, X.; Dong, A.; Chen, C. Investigation on cotton stalk and bamboo sawdust carbonization for barbecue charcoal preparation. *Bioresour. Technol.* **2014**, *152*, 86–92. [CrossRef] [PubMed]
28. Roberts, K.G.; Gloy, B.A.; Joseph, S.; Scott, N.R.; Lehmann, J. Life cycle assessment of biochar systems: Estimating the energetic, economic, and climate change potential. *Environ. Sci. Technol.* **2010**, *44*, 827–833. [CrossRef] [PubMed]
29. Harsono, S.S.; Grundmann, P.; Siahaan, D. Role of biogas and biochar palm oil residues for reduction of greenhouse gas emissions in the biodiesel production. *Energy Procedia* **2015**, *65*, 344–351. [CrossRef]
30. Yan, N.; Ren, B.; Wu, B.; Bao, D.; Zhang, X.; Wang, J. Multi-objective optimization of biomass to biomethane system. *Green Energy Environ.* **2016**, *1*, 156–165. [CrossRef]
31. Ryckebosch, E.; Drouillon, M.; Vervaeren, H. Techniques for transformation of biogas to biomethane. *Biomass Bioenergy* **2011**, *35*, 1633–1645. [CrossRef]
32. Reza, M.T.; Lynam, J.G.; Uddin, M.H.; Coronella, C.J. Hydrothemal carbonization: Fate of inorganics. *Biomass Bioenergy* **2013**, *49*, 86–94. [CrossRef]
33. Chen, J.; Li, C.; Ristovski, Z.; Milic, A.; Gu, Y.; Islam, M.S.; Wang, S.; Hao, J.; Zhang, H.; He, C.; et al. A review of biomass burning: Emissions and impacts on air quality, health and climate in China. *Sci. Total Environ.* **2017**, *579*, 1000–1034.
34. Cheng, Z.; Wang, S.; Fu, X.; Watson, J.G.; Jiang, J.; Fu, Q.; Chen, C. Impact of biomass burning on haze pollution in the Yangtze River delta, China: A case study in summer 2011. *Atmos. Chem. Phys.* **2014**, *14*, 4573–4585. [CrossRef]
35. Wang, H.; Qiao, L.; Lou, S.; Zhou, M.; Chen, J.M.; Wang, Q.; Tao, S.K.; Chen, C.H.; Huang, H.Y.; Li, L.; et al. $PM_{2.5}$ pollution episode and its contributors from 2011 to 2013 in urban Shanghai, China. *Atmos. Environ.* **2015**, *123*, 298–305. [CrossRef]

36. Qian, K.; Kumar, A.; Patil, K.; Bellmer, D.; Wang, D.; Yuan, W.; Huhnke, R.L. Effects of biomass feedstocks and gasification conditions on the physiochemical properties of char. *Energies* **2013**, *6*, 3972–3986. [CrossRef]
37. Yang, H.; Yan, R.; Chen, H.; Zheng, C.; Lee, D.H.; Liang, D.T. In-depth investigation of biomass pyrolysis based on three major components: Hemicellulose, cellulose and lignin. *Energy Fuels* **2006**, *20*, 388–393. [CrossRef]
38. Kim, D.; Yoshikawa, K.; Park, K.Y. Characteristics of biochar obtained by hydrothermal carbonization of cellulose for renewable energy. *Energies* **2015**, *8*, 14040–14048. [CrossRef]
39. Obernberger, I.; Biedermann, F.; Widmann, W.; Riedl, R. Concentrations of inorganic elements in biomass fuels and recovery in the different ash fractions. *Biomass Bioenergy* **1997**, *12*, 211–224. [CrossRef]
40. Lewandowski, I.; Kicherer, A. Combustion quality of biomass: Practical relevance and experiments to modify the biomass quality of Miscanthus × giganteus. *Eur. J. Agron.* **1997**, *6*, 163–177. [CrossRef]
41. Liu, L.; Guo, J.; Lu, F. Chemical composition and ash characteristics of several straw stalks and residues. *J. Zhejiang For. Coll.* **2006**, *23*, 388–392. (In Chinese)
42. Hytönen, J.; Nurmi, J. Heating value and ash content of intensively managed stands. *Wood Res-Slovak.* **2015**, *60*, 71–82.
43. Sadaka, S.; Sharara, M.A.; Ashworth, A.; Keyser, P.; Allen, F.; Wright, A. Characterization of biochar from switchgrass carbonization. *Energies* **2014**, *7*, 548–567. [CrossRef]
44. Chaney, J.O. Combustion Characteristics of Biomass Briquettes. Ph.D. Thesis, University of Nottingham, Nottingham, UK, 2010.
45. Ghugare, S.B.; Tiwary, S.; Elangovan, V.; Tambe, S.S. Prediction of higher heating value of solid biomass fuels using artificial intelligence formalisms. *Bioenergy Res.* **2013**, *7*, 681–692. [CrossRef]
46. Imam, T.; Capareda, S. Characterization of bio-oil, syn-gas and bio-char from switchgrass pyrolysis at various temperatures. *J. Anal. Appl. Pyrolysis* **2012**, *93*, 170–177. [CrossRef]
47. Raveendran, K.; Ganesh, A. Heating value of biomass and biomass pyrolysis products. *Fuel* **1996**, *75*, 1715–1720. [CrossRef]
48. Anderson, N.; Jones, J.G.; Page-Dumroese, D.; McCollum, D.; Baker, S.; Loeffler, D.; Chung, W. A comparison of producer gas, biochar, and activated carbon from two distributed scale thermochemical conversion systems used to process forest biomass. *Energies* **2013**, *6*, 164–183. [CrossRef]
49. Wu, W.; Li, J.; Lan, T.; Müller, K.; Niazi, N.K.; Chen, X.; Xu, S.; Zheng, L.; Chu, Y.; Li, J.; et al. Unraveling sorption of lead in aqueous solutions by chemically modified biochar derived from coconut fiber: A microscopic and spectroscopic investigation. *Sci. Total Environ.* **2017**, *576*, 766–774. [CrossRef] [PubMed]
50. Dong, D.; Feng, Q.; McGrouther, K.; Yang, M.; Wang, H.; Wu, W. Effects of biochar amendment on rice growth and nitrogen retention in a waterlogged paddy field. *J. Soils Sediments* **2015**, *15*, 153–162. [CrossRef]
51. Brosse, N.; Dufour, A.; Meng, X.; Sun, Q.; Ragauskas, A. Miscanthus: A fast-growing crop for biofuels and chemicals production. *Biofuels Bioprod. Biorefining* **2012**, *6*, 580–598. [CrossRef]

![energies logo] *energies*

MDPI

*Article*

# Effect of Temperature on the Structural and Physicochemical Properties of Biochar with Apple Tree Branches as Feedstock Material

**Shi-Xiang Zhao, Na Ta and Xu-Dong Wang \***

College of Resources & Environment, Northwest A&F University, 3 Taicheng Road, Yangling 712100, China; zhaoshixiang1989@126.com (S.-X.Z.); tana0214@163.com (N.T.)
\* Correspondence: wangxd@nwsuaf.edu.cn; Tel./Fax: +86-29-8708-0055

Received: 18 April 2017; Accepted: 21 August 2017; Published: 30 August 2017

**Abstract:** The objective of this study was to study the structure and physicochemical properties of biochar derived from apple tree branches (ATBs), whose valorization is crucial for the sustainable development of the apple industry. ATBs were collected from apple orchards located on the Weibei upland of the Loess Plateau and pyrolyzed at 300, 400, 500 and 600 °C (BC300, BC400, BC500 and BC600), respectively. Different analytical techniques were used for the characterization of the different biochars. In particular, proximate and element analyses were performed. Furthermore, the morphological, and textural properties were investigated using scanning electron microscopy (SEM), Fourier-transform infrared (FTIR) spectroscopy, Boehm titration and nitrogen manometry. In addition, the thermal stability of biochars was also studied by thermogravimetric analysis. The results indicated that the increasing temperature increased the content of fixed carbon (C), the C content and inorganic minerals (K, P, Fe, Zn, Ca, Mg), while the yield, the content of volatile matter (VM), O and H, cation exchange capacity, and the ratios of O/C and H/C decreased. Comparison between the different samples show that highest pH and ash content were observed in BC500. The number of acidic functional groups decreased as a function of pyrolysis temperature, especially for the carboxylic functional groups. In contrast, a reverse trend was found for the basic functional groups. At a higher temperature, the brunauer–emmett–teller (BET) surface area and pore volume are higher mostly due to the increase of the micropore surface area and micropore volume. In addition, the thermal stability of biochars also increased with the increasing temperature. Hence, pyrolysis temperature has a strong effect on biochar properties, and therefore biochars can be produced by changing pyrolysis temperature in order to better meet their applications.

**Keywords:** biochar; pyrolysis temperature; apple tree branch; physicochemical properties; structural

---

## 1. Introduction

Biochar is a carbon (C)-rich byproduct produced in an oxygen-limited environment [1], which has been gaining increasing attention over the last decade due to its potential to mitigate global climate change [2,3]. Biochar can be used not only as a soil amendment with the aim of improving soil physical, chemical and biological properties [4,5], but also as an adsorbent to remove organic and inorganic pollutants [6,7]. The functions and applications of biochars mostly depend on their structural and physicochemical properties [8], therefore, it is very important to characterize the structural and physicochemical properties of biochar before its use.

Various types of biomass (wood materials, agricultural residues, dairy manure, sewage sludge, et al.) have been used to produce biochars under different pyrolysis conditions [9,10]. During pyrolysis, biomass undergoes a variety of physical, chemical and molecular changes [11]. Previous studies indicated that pyrolysis condition and feedstock type significantly affect the structural and physicochemical characteristics of the resulting biochar products [12–14]. Generally, woody biomass provides a more C-rich biochar compared to other feedstocks since it contains varying amounts of hemicellulose, cellulose, lignin and small quantities of other organic extractives and inorganic compounds [15]. Xu and Chen [16] suggest that higher lignin and minerals content result in a higher yield of biochar. Therefore, woody biomass is one of the most important sources for biochar production.

The structural and physicochemical properties of biochar, such as surface area, pore structures, surface functional groups and element composition, can also be influenced by varying the pyrolysis condition, such as pyrolysis temperature, heating rate and holding time [17,18]. The pyrolysis temperature is reported to significantly influence the final structural and physicochemical properties of biochar due to the release of volatiles as well as the formation and volatilisation of intermediate melts [18]. Previous studies indicated that higher temperature resulted in a higher C content, while the losses of nitrogen (N), hydrogen (H) and oxygen (O) were also recorded [19]. In addition, increasing the temperature lead to an increase the ash and fixed C contents, and to a decrease the content of volatile materials [20]. Furthermore, the increase in pyrolysis temperature affects H/C and O/C ratios; porosity; surface area; surface functional groups and cation exchange capacity (CEC) and so on [21]. In particular, biochar produced at high temperature has high aromatic content, which is recalcitrant to decomposition [22]. In contrast, biochar produced at low temperature has a less-condensed C structure and, therefore, may improve the fertility of soils [23]. Therefore, the pyrolysis temperature was investigated in this study.

Weibei upland of the Loess Plateau has been recognized as one of the best apple production areas in the world due to its special topography and climate features which are suitable for planting apples [24]. Apple tree branches (ATBs) are a major agricultural residue in this region due to the quick development of the apple-planted area in recent years. During the autumn of each year, the residues might be burned by the fruit grower with the aim of reducing the occurrence of plant diseases and insect pests, which then becomes an important source of atmospheric $CO_2$. Thus, conversion of ATBs into biochar has the potential to be used to mitigate environmental problems. In addition, the biochar products also can be used as a soil amendment, which can improve the soil quality [25,26]. Therefore, the aim of this study was to examine the structure and physicochemical properties of biochar produced at different pyrolysis temperatures using ATB as feedstock. Such understanding is crucial for the sustainable development of apple industry on the Weibei upland.

## 2. Materials and Methods

### 2.1. Feedstock Preparation

ATBs were collected from apple orchards located on the Weibei upland of the Loess Plateau, Northwest China (34°53′ N, 108°52′ E). Prior to the experiments, ATBs were ground to a particle size of less than 2 mm and then washed with deionized water several times to remove impurities. These pre-prepared samples were dried at 80 °C for 24 h to remove moisture.

### 2.2. Biochar Production

Pyrolysis experiments were carried out in a muffle furnace (Yamato Scientific Co., Ltd, FO410C, Tokyo, Japan) under nitrogen gas stream (at a rate 630 $cm^3 \cdot min^{-1}$, standard temperature and pressure 298 K, 101.2 kPa). The feedstock was placed in a stainless steel reactor of 20.5 cm internal length, 12.2 cm internal width and 7.5 cm internal height with a lid and subjected to pyrolysis at different temperatures (300, 400, 500 and 600 °C, respectively) for 2 h 10 min. The pyrolysis heating rate employed was 10 °C min$^{-1}$. After pyrolysis, the reactor was left inside the furnace to cool to room temperature. The biochars obtained were labeled as BC300, BC400, BC500 and BC600, respectively.

All biochars were weighed and then the generated biochars were milled to pass through a 0.25 mm sieve (60 mesh) for further analysis and use. The yield of biochar was calculated as follows:

$$Yield\ (\%) = \frac{mass\ of\ biochar\ (g)}{oven\ dry\ mass\ of\ feedstock\ (g)} \times 100\%$$

*2.3. Characterization of Biochar*

2.3.1. Proximate Analysis, pH and Cation Exchange Capacity

The contents of volatile matter (VM) and ash were determined using the American Society for Testing and Materials (ASTM) D5142 method [27]. VM content was determined as weight loss after heating the char in a covered crucible to 950 °C and holding for 7 min. Ash content was determined as weight loss after combustion at 750 °C for 6 h with no ceramic cap. Fixed C content was calculated by the following equation:

Fixed carbon % = 100% − (Ash % + Volatile matter %)

The pH of biochars was measured using a pH meter at a 1:5 solid/water ratio after shaking for 30 min. The CEC of biochar was estimated using an $NH_4^+$ replacement method [28]. Briefly, 0.20 g was leached five times with 20 mL of deionized water. Then, the biochar was leached with 20 mL of 1 mol·L$^{-1}$ Na–acetate (pH 7) five times. The biochar samples were then washed with 20 mL of ethanol five times to remove the excess Na$^+$. Afterwards, the Na$^+$ on the exchangeable sites of the biochar was displaced by 20 mL of 1 mol·L$^{-1}$ $NH_4$–acetate (pH 7) five times, and the CEC of the biochar was calculated from the Na$^+$ displaced by $NH_4^+$.

2.3.2. Elemental and Nutrients Analysis

A CHN Elemental Analyzer (Vario EL III, Heraeus, Germany) was used to determine the contents of C, N and H. The O content (%) was calculated by the following equation: O (%) = 100 − (C % + H % + N % + Ash %). The H/C and O/C ratios were also calculated. Total nutrients (K, P, Fe, Mn, Cu, Zn, Ca and Mg) in the biochar were extracted using a wet acid digestion method (concentration $HNO_3$ + 30% $H_2O_2$) [29]. The nutrients in the digestion solution were determined using an ICAP Q ICP-MS spectrometer (ThermoFisher, Waltham, MA, USA).

2.3.3. Surface Properties of Biochars

The surface morphology of these biochars was examined using an environmental scanning electron microscopy (SEM) system (JEOL JSM-6360LV, Tokyo, Japan). Biochars were held onto an adhesive carbon tape on an aluminum stub followed by sputter coating with gold prior to viewing. The surface area and porosity of biochars were measured using a NOVA 2200e analyser (Quantachrome Instruments, Boynton Beach, FL, USA) at liquid nitrogen temperature (77 K). The Brunauer–Emmett–Teller (BET) surface area ($S_{BET}$), micropore surface area ($S_{mic}$) and micropore volume ($V_{mic}$), total pore volume ($V_T$) of the biochars produced at different temperatures were determined using the BET equation, t-plot method and single point adsorption total pore volume analysis, respectively [30]. Fourier-transform infrared (FTIR) spectra peaks of biochars were also obtained on pressed pellets of 1:10 biochar/KBr mixtures using a Tensor27 FTIR spectrometer (Bruker, Karlsruhe, Germany). The spectra were obtained at 4 cm$^{-1}$ resolution from 400 to 4000 cm$^{-1}$. The amount of acidic and basic functional groups was measured by the Boehm method [31].

2.3.4. Thermal Stability Evaluation

The thermal stability evaluation of the biochars was performed by thermogravimetric analysis (STA449F3, NETZSCH, Freistaat Bayern, Germany). Approximately 5 mg of biochar was weighed into an alumina crucible, the sample was subjected to a thermogravimetric analysis in a nitrogen flow (gas flow of 50 mL·min$^{-1}$) at a heating rate of 10 °C·min$^{-1}$, from 50 °C to 1000 °C.

### 2.4. Statistical Analysis of Data

All data were reported as means ± standard deviation. The data were subjected to analysis of variance (ANOVA) using SAS version 8.0 (SAS Institute Inc, Cary, NC, USA). The least significant difference (LSD, $p < 0.05$) test was applied to assess the differences among the means.

## 3. Results and Discussion

### 3.1. Effect of Temperature on the Basic Characteristics of Biochars

#### 3.1.1. Proximate Analysis

Table 1 presents the results of proximate analyses as a function of pyrolysis temperature. During the pyrolysis process, the temperature kept rising and was then held at the peak temperature for 2 h and 10 min before cooling down to room temperature. When the pyrolysis temperature increased from 300 to 500 °C, the biochar yield sharply decreased from 47.94% to 31.71%. This was probably due to most of the lignocellulosic material was decomposed at this temperature range [32]. While, when the pyrolysis temperature further increased from 500 to 600 °C, the biochar yield only decreased from 31.71% to 28.48%. This result indicated that most of the volatile fraction had been removed at lower temperatures.

**Table 1.** Proximate, elemental and nutrients analysis of biochars produced at different temperatures.

| Sample | | BC300 | BC400 | BC500 | BC600 |
|---|---|---|---|---|---|
| Proximate analysis, dry basis | Yeld (%) | 47.94 ± 1.27 a | 35.49 ± 1.39 b | 31.73 ± 1.02 c | 28.48 ± 0.72 d |
| | Ash (%) | 6.72 ± 0.02 d | 7.85 ± 0.04 c | 10.06 ± 0.15 a | 9.40 ± 0.21 b |
| | Volatile matter (%) | 60.77 ± 0.86 a | 29.85 ± 0.90 b | 23.19 ± 0.34 c | 14.86 ± 0.63 d |
| | Fixed carbon (%) | 32.50 ± 0.86 d | 62.30 ± 0.93 c | 66.75 ± 0.28 b | 75.73 ± 0.76 a |
| Elemental analysis, dry basis | C (%) | 62.20 ± 0.85 d | 71.13 ± 2.39 c | 74.88 ± 2.11 b | 80.01 ± 4.58 a |
| | H (%) | 5.18 ± 0.19 a | 4.03 ± 0.21 b | 2.88 ± 0.08 c | 2.72 ± 0.14 c |
| | N (%) | 1.69 ± 0.08 c | 1.94 ± 0.06 a | 1.77 ± 0.08 b | 1.28 ± 0.06 d |
| | O (%) | 24.21 ± 0.62 a | 15.05 ± 2.35 b | 10.41 ± 2.05 c | 6.59 ± 1.38 c |
| Nutrients analysis, dry basis | K (%) | 0.57 ± 0.01 c | 0.89 ± 0.03 b | 1.10 ± 0.02 a | 1.14 ± 0.04 a |
| | P (%) | 0.21 ± 0.01 c | 0.28 ± 0.01 b | 0.34 ± 0.01 a | 0.34 ± 0.01 a |
| | Ca (g·kg$^{-1}$) | 12.90 ± 0.46 d | 16.81 ± 0.34 c | 20.19 ± 0.22 b | 20.89 ± 0.48 a |
| | Mg (g·kg$^{-1}$) | 3.01 ± 0.06 d | 4.04 ± 0.13 c | 4.69 ± 0.10 b | 5.64 ± 0.17 a |
| | Fe (mg·kg$^{-1}$) | 268.35 ± 6.53 d | 361.62 ± 8.99 c | 480.52 ± 10.58 b | 583.50 ± 5.38 a |
| | Mn (mg·kg$^{-1}$) | 56.96 ± 2.30 d | 79.26 ± 0.28 c | 102.89 ± 4.95 a | 89.41 ± 2.77 b |
| | Cu (mg·kg$^{-1}$) | 20.29 ± 0.45 d | 50.53 ± 1.96 c | 85.07 ± 2.27 a | 58.90 ± 1.22 b |
| | Zn (mg·kg$^{-1}$) | 33.06 ± 0.48 c | 53.30 ± 1.41 b | 60.50 ± 0.17 a | 61.68 ± 2.41 a |

Note: Values in the same row followed by the same letter are not significantly different at $p < 0.05$ according to least significant difference test. All data were reported as means ± standard deviation ($n = 3$). Fixed carbon was estimated by difference: Fixed carbon % = 100% − (Ash % + Volatile matter %). O content was estimated by difference: O % = 100% − (C% + H% + N% + Ash %).

The content of VM and fixed C for the generated biochars ranged from 14.86% to 60.77% and 33.60% to 73.50%, respectively. An increase in the pyrolysis temperature decreased the content of VM, exhibiting a similar trend with the biochar yield, while an opposite trend was found for the content of fixed C. This might due to the fact that the increasing temperature resulted in the further crack of the volatiles fractions into low molecular weight liquids and gases instead of biochar [33]. Meanwhile, the dehydration of hydroxyl groups and thermal degradation of cellulose and lignin might also occurred with the increasing temperature [12]. These results confirmed that the increase in temperature enhanced the stability of biochar for the loss of volatile fractions [34]. It was interesting that the ash content remarkably increased from 6.72% to 10.06% with an increase in the pyrolysis temperature from

300 to 500 °C. The increase in the content of ash resulted from progressive concentration of inorganic constituents [35], which also was confirmed by our nutrients analysis (Table 1). When pyrolysis temperature increased from 500 to 600 °C, some inorganic materials might volatilize as gas or liquid, thus, the content of ash decreased at higher temperature (600 °C).

3.1.2. Elemental and Nutrients Analysis

The elemental composition for the generated biochars changed with pyrolysis temperature (Table 1). The C content increased from 62.20% to 80.01%, while the H and O contents decreased from 5.18% to 2.72% and 24.21% to 6.59% as the pyrolysis temperature increased from 300 to 600 °C, respectively. These results were consistent with previous results [36]. The decrease in the contents of H and O at higher temperature was likely due to the decomposition of the oxygenated bonds and release of low molecular weight byproducts containing H and O [15]. Interestingly, the highest N content was observed in BC400 (1.94%). This was attributed to the incorporation of nitrogen into complex structures which were resistant to lower temperature and not easily volatilized [37]. Furthermore, the ratios of H/C (the degree of aromaticity) [36] and O/C (the degree of polarity) [38] varied as a function of pyrolysis temperature. In our study, the H/C and O/C ratios of biochars were significantly decreased from 1.00 to 0.41 and 0.29 to 0.06 with the increasing temperature, respectively (Figure 1). The gradually reduced in the H/C and O/C atomic ratios with the increasing pyrolysis temperature was mainly contributed to the dehydration reactions [39], which could be well described by the Van Krevelen diagram (Figure 1). In addition, the H/C and O/C ratios also indicated that the structural transformations [36] and surface hydrophilicity of biochar [40], the higher extent of carbonization and loss of functional groups containing O and H (such as hydroxyl, carboxyl, et al.) at higher temperature resulted in the lower ratios of H/C and O/C, indicating that the surface of biochar was more aromatic and less hydrophilic [41,42].

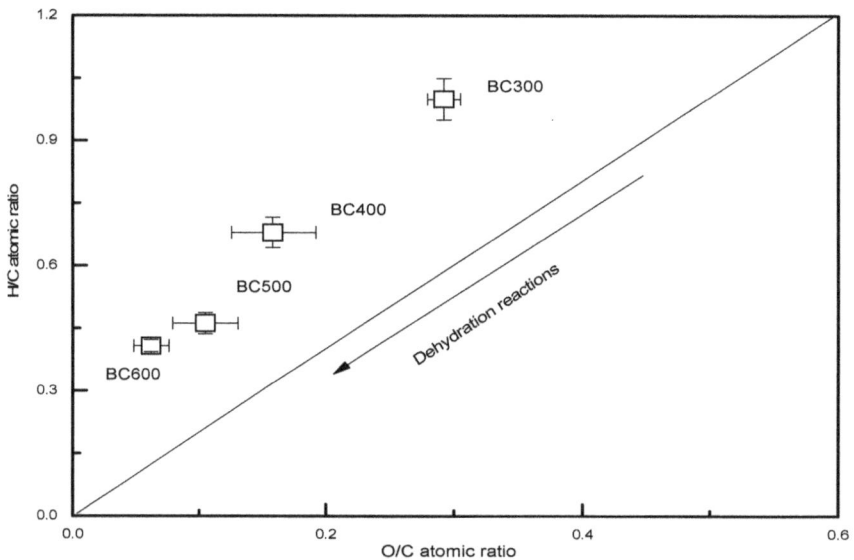

**Figure 1.** The van Krevelen plot of elemental ratios for biochars produced at different temperatures. Thick line represents the direction for dehydration reaction. Individual points are averages (*n* = 3) and error bars are standard deviations.

In addition, the nutrients K, P, Ca, Mg, Fe and Zn were also increased with increasing the pyrolysis temperature (Table 1). While, the highest concentration of Cu (85.07 mg·kg$^{-1}$) and Mn

(102.89 mg·kg$^{-1}$) were found in BC500. Pituello et al. [43] suggested that some metals might volatilize at high temperature. Sun et al. also found a decrease in the concentration of Ca and Mg as the pyrolysis temperature increased from 450 to 600 °C using hickory wood as feedstock [21]. Thus, both the volatility of the nutrients and the influence of temperature on the composition and chemical structure of biochar can significantly affect the concentration of nutrients during the process [13].

### 3.1.3. pH and Cation Exchange Capacity

The generated biochars were generally alkaline (pH > 7) and the pH of biochar significant increased from 7.48 to 11.62 when the pyrolysis temperature increased from 300 to 500 °C, and then decreased to 10.60 in BC600 (Figure 2). This was probably due to the highest ash content in BC500. The pH values and ash content were positively correlated ($R^2$ = 0.97), hence, the minerals, especially for the carbonates formation (such as $CaCO_3$ and $MgCO_3$) and inorganic alkalis (such as K and Na), were probably the main cause of each biochars' inherent alkaline pH [28].

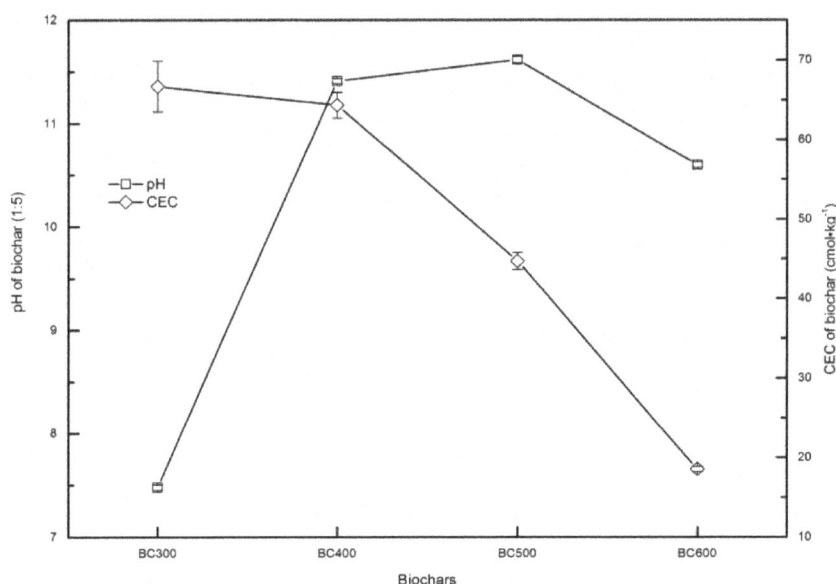

**Figure 2.** The pH and cation exchange capacity (CEC) of biochars produced at different temperatures. Individual points are averages ($n$ = 3) and error bars are standard deviations.

CEC is an important property of biochar indicating the capacity of a biochar to adsorb cation nutrients [20]. In our study, the CEC for the generated biochars significantly decreased from 66.59 to 18.53 cmol·kg$^{-1}$ when the pyrolysis temperature increased from 300 to 600 °C (Figure 2), which was consistent with previous study [44]. The shift in CEC may due to the reduction of functional groups and oxidation of aromatic C with temperature [34], which was well supported by the lower O/C ratio and our FTIR and Boehm titration results at higher temperature.

### 3.2. Effect of Temperature on the Surface Properties of Biochars

### 3.2.1. Surface Morphology (Scanning Electron Microscopy Analysis)

Figure 3 shows SEM micrographs (×4000) of biochars produced at different temperatures. The image of BC300 showed that the biomass had softened, melted and fused into a mass of vesicles (Figure 3a) [45]. The vesicles were the result of volatile gasses released within the biomass. As temperatures increased, more volatile gasses released from the biomass, the vesicles on the surface of BC400 busted after cooling, thus the morphology of BC400 exhibited a number of pore structure (Figure 3b). For the BC500, portions of the skeletal structure appeared brittle because of the decomposition of more components (Figure 3c, ellipse). The fracture phenomenon also appeared within the pore structure for the BC600, the last temperature that samples were collected (Figure 3d, ellipse).

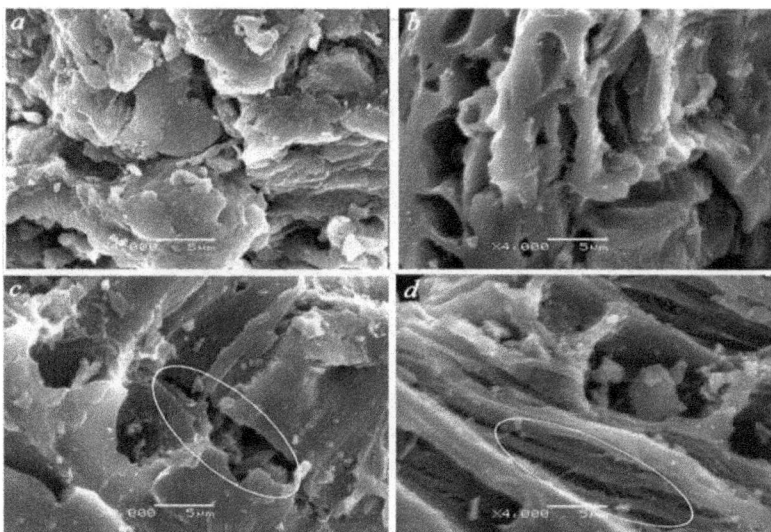

**Figure 3.** Scanning electron microscopy (SEM) micrographs (magnification 4000×) of biochars samples pyrolyzed at: (**a**) 300 °C; (**b**) 400 °C; (**c**) 500 °C; and (**d**) 600 °C, respectively.

### 3.2.2. Surface Area and Pore Volume

The surface area and pore volumes produced at various pyrolysis temperatures were obtained by $N_2$ adsorption and the results shown in Table 2. An increase in the pyrolysis temperature from 300 to 600 °C resulted in a significant increase in the $S_{BET}$ from 2.39 m$^2$·g$^{-1}$ to 108.59 m$^2$·g$^{-1}$ and in the $V_T$ from 2.56 × 10$^{-3}$ cm$^3$·g$^{-1}$ to 58.54 × 10$^{-3}$ cm$^3$·g$^{-1}$. In the same way, $S_{mic}$ and $V_{mic}$ significant increased from 0.10 m$^2$·g$^{-1}$ and 0.13 × 10$^{-3}$ cm$^3$·g$^{-1}$ at 300 °C to 84.44 m$^2$·g$^{-1}$ and 37.87 × 10$^{-3}$ cm$^3$·g$^{-1}$ at 600 °C, respectively. This evolution is somewhat similar to that reported in the literature [15]. The increase in the surface area and pore volumes might be caused by the progressive degradation of the organic materials (hemicelluloses, cellulose and lignin) and the formation of vascular bundles or channel structures during pyrolysis during the process [46,47]. Hemicellulose has a high reactivity during thermal treatment at lower temperature (usually under 300 °C).

**Table 2.** Surface area and pore volumes of biochars produced at different temperatures.

| Sample | BC300 | BC400 | BC500 | BC600 |
|---|---|---|---|---|
| $S_{BET}$ (m$^2$·g$^{-1}$) | 2.39 ± 0.12 d | 7.00 ± 0.25 c | 37.24 ± 0.80 b | 108.59 ± 4.11 a |
| $S_{mic}$ (m$^2$·g$^{-1}$) | 0.10 ± 0.01 d | 1.47 ± 0.01 c | 9.33 ± 0.73 b | 84.44 ± 6.76 a |
| $V_T$ (10$^{-3}$·cm$^3$·g$^{-1}$) | 2.56 ± 0.25 d | 6.52 ± 0.64 c | 12.41 ± 0.32 b | 58.54 ± 3.44 a |
| $V_{mic}$ (10$^{-3}$·cm$^3$·g$^{-1}$) | 0.13 ± 0.01 d | 0.52 ± 0.03 c | 1.58 ± 0.10 b | 37.87 ± 0.91 a |

Note: Values in a row followed by the same letter are not significantly different at $p < 0.05$ according to LSD test. All data were reported as means ± standard deviation ($n = 3$). $S_{BET}$, $S_{mic}$, $V_T$ and $V_{mic}$ were the BET surface area, micropore surface area, total pore volume and micropore volume, respectively.

The presence of vesicles at BC300 also indicated that volatile components were formed and released (Figure 3a). When the pyrolysis temperature increased from 300 to 400 °C, the rupture of the hemicellulose along with other organic compounds generated more micropores within biochar (Figure 3b) [48], thus, an increase in the surface area and pore volumes was found in BC400. Meanwhile, some amorphous carbon structures were formed during this temperature range due to the degradation of cellulose. Several researchers suggested that the micropores may be formed by amorphous carbon structures [49]. The increase in the biochar porosity was found in BC500 due to the decomposition of lignin and the quick release of $H_2$ and $CH_4$ as the temperature increased up to 500 °C (Figure 3c). This result contributed to a sharp increase of the surface area and pore volumes (Table 2) [18]. Further increases in the temperature to 600 °C resulting in a dramatic increase in $S_{BET}$ and $V_T$, especially for the $S_{mic}$ and $V_{mic}$. This was mostly contributed to the further degradation of lignin and the reaction of aromatic condensation [30], which increased the release of VM (Table 1) and created more pores (Figure 3d) [18]. The enhancement of the pore development at a higher temperature suggesting that the increase of $S_{BET}$ and $V_T$ mostly dependent on the increase of the $S_{mic}$ and $V_{mic}$.

### 3.2.3. Fourier-Transform Infrared Analysis and Functional Groups

The FTIR spectra of biochars produced at four pyrolysis temperatures are presented in Figure 4 and the descriptions for peak assignments are provided in Table 3. As the pyrolysis temperature increased, FTIR spectra of biochars revealed a decrease in the stretching of O–H (3200–3500 cm$^{-1}$) and C–H (2935 cm$^{-1}$) [13], this was attributed to the acceleration of dehydration reaction in biomass [30], which suggested a decrease in the polar functional groups with an increase in pyrolysis temperature [50]. In particular, biochars began to increase aromatic C=C stretching (1440 cm$^{-1}$) [29] and out-of-plane deformation by aromatic C–H groups (885 cm$^{-1}$) [51], while the symmetric C–O stretching (1030–1110 cm$^{-1}$) for the source material began to disappear with the increasing pyrolysis temperature [51].

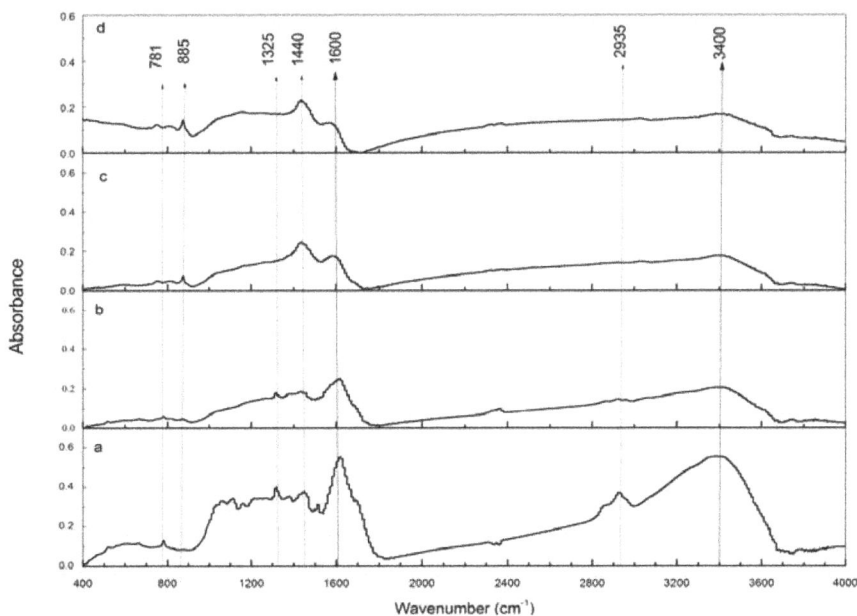

**Figure 4.** Fourier-transform infrared (FTIR) spectra peaks of biochars produced at: (**a**) 300 °C; (**b**) 400 °C; (**c**) 500 °C; and (**d**) 600 °C, respectively.

**Table 3.** Functional groups observed in the Fourier-transform infrared (FTIR) spectra of biocahrs produced at different temperatures.

| Wave Numbers (cm$^{-1}$) | Characteristic Vibrations (Functionality) |
|:---:|:---:|
| 3500–3200 | O–H stretching (water, hydrogen-bonded hydroxyl) [13] |
| 2935 | C–H stretching (aliphatic CHx; 2935-asymmetric) [13] |
| 1600 | Aromatic C=C and C=O stretching of conjugated ketones and quinones [52] |
| 1440 | C=C stretching (lignin carbohydrate) [29] |
| 1325 | O–H bending (phenols, phenolic; ligneous syringyl) [29] |
| 1100–1030 | Symmetric C–O stretching (cellulose, hemicellulose, and lignin) [51] |
| 885 | C–H bending (aromatic C–H out-of-plane deformation) [51] |
| 781 | Pyridine (pyridine ring vibration and C–H deformation) [53] |

This result is likely due to the degradation and depolymerization of cellulose, hemicelluloses and lignin [29]. Instead, intensities of O–H (1325 cm$^{-1}$) stretching [29] and aromatic C=C and C=O stretching of conjugated ketones and quinones (1600 cm$^{-1}$) [52] decreased with temperature which may suggest that phenolic and carboxylic compounds in lignin had been degraded [54]. There was also pyridine in biochars (781 cm$^{-1}$) [53], which was one of the heterocyclic nitrogen compounds commonly observed during pyrolysis [55]. In general, the maximum loss was obtained in –OH, –CH$_2$, and C–O functional groups in biochars as a function of pyrolysis temperature, which was also apparent from their elemental composition (Table 1). Relatively low values of O, H, and H/C in BC600 than those in BC300 revealed the significant elimination of polar functional groups (–OH and C–O).

Different spectra reflected changes in the surface functional groups of biochars produced at different temperatures. A Boehm titration method was used to quantify the surface functional groups of biochars produced at different temperatures. The results showed that when the pyrolysis temperature increased from 300 to 600 °C the concentration of total acidic functional groups, surface carboxylic functional groups, surface phenolic functional groups and surface lactonic functional groups decreased from 1.16 to 0.44 mmol·g$^{-1}$, 0.49 to 0.23 mmol·g$^{-1}$, 0.33 to 0.09 mmol·g$^{-1}$ and 0.34 to 0.13 mmol·g$^{-1}$, respectively (Figure 5). This result was in agreement with the findings in previous studies [56,57]. During the pyrolysis process, organic materials were converted to water and carbonaceous gas, thus, the higher temperature caused an increased in the quantity of volatile compounds resulting in a decrease in the content of acidic groups [18], which was well supported by our FTIR results. It was interesting that the total acidic and carboxylic functional groups significant decreased with increasing temperatures ($p < 0.05$), whereas the temperature trend for phenolic and lactonic functional groups content were less apparent (Figure 5). This was probably due to the fact carboxylic acids represented 48.63% of total acidic functional groups, on average, and were 1.5–2.5 times more abundant than phenolic or lactonic functional groups. This result is similar to the one reported in the literature [15]. While, the total basic functional groups significant increased from 0.11 to 0.32 mmol·g$^{-1}$ when the pyrolysis temperature increased from 300 to 500 °C and then decreased to 0.22 mmol·g$^{-1}$ at 600 °C. The strong direct linear correlation between ash fraction and basic groups ($R^2 = 0.66$) suggest that the basic functional groups are mainly associated with the ash fraction [15], which also implied the highest content of basic groups in BC500 (Figure 5 and Table 1).

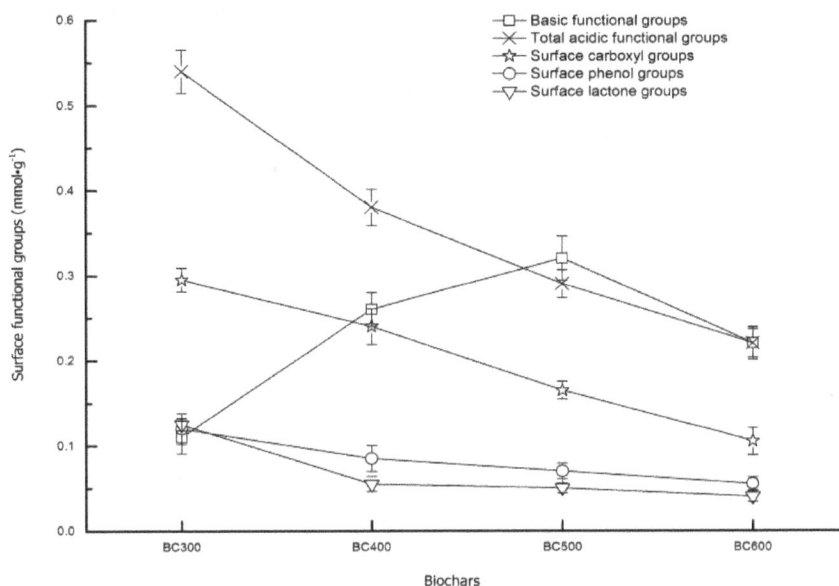

**Figure 5.** Variation in surface functional groups as a function of pyrolysis temperature. Individual points are averages ($n = 3$) and error bars are standard deviations.

### 3.3. Effect of Temperature on the Thermal Stability of Biochars

Representative thermogravimetric (TG) analysis and differential thermogravimetry (DTG) curves for both biochars are presented in Figure 6. During the process, two stages of weight losses were observed (Figure 6a), which was also implied by the DTG curves (Figure 6b). The mass loss occurred slowly under 200 °C and varied from 0.77% to 2.86% (Figure 6a; maximum at around 80 °C, Figure 6b) (under 200 °C), suggesting that this reduction in mass loss commonly associated with the loss of the initial moisture of the sample [58]. Above that, the main mass losses started at around 315 °C, 388 °C, 499 °C and 517 °C for BC300, BC400, BC500 and BC600, respectively. Meanwhile, a sharp weight decrease took place between 313 °C and 500 °C, 388 °C and 644 °C, 499 °C and 738 °C, 517 °C and 753 °C (Figure 6a) in the BC300, BC400, BC500 and BC600, respectively. This was probably due to the fact the generated biochars had undergone a previous heat treatment before the thermal analysis, thus, the tested biochar samples were thermally stable below the temperature which they were produced [59]. For each of the tested biochar, only one peak was found during the temperature range between the temperature they were produced and 1000 °C on the DTG curves (Figure 6b). It is known that secondary pyrolysis reactions could be easily detected and observed if the temperature exceeding the biochars' primary decomposition temperature [58]. Thus, the weight loss over a wider temperature range could be attributed to the degradation and decomposition of organic materials [21]. The maximum weight loss occurred at around 352 °C, 388 °C, 658 °C and 674 °C for BC300, BC400, BC500 and BC600 (Figure 6b), respectively, suggesting that the higher the pyrolysis temperature was, the better thermal stability the biochar showed, which is consistent with the literature [60]. The lower temperature derived biochars were less thermally stable than the higher temperature derived biochars, probably because they were not fully carbonized [21]. Above 600–700 °C, decomposition for all the biochars finished, and the curves became stable. It was interesting that the total weight losses resulting from thermal degradation were 52.19%, 33.32%, 23.80% and 27.27% for BC300, BC400, BC500 and BC600, respectively. This was probably because BC500 had highest ash content.

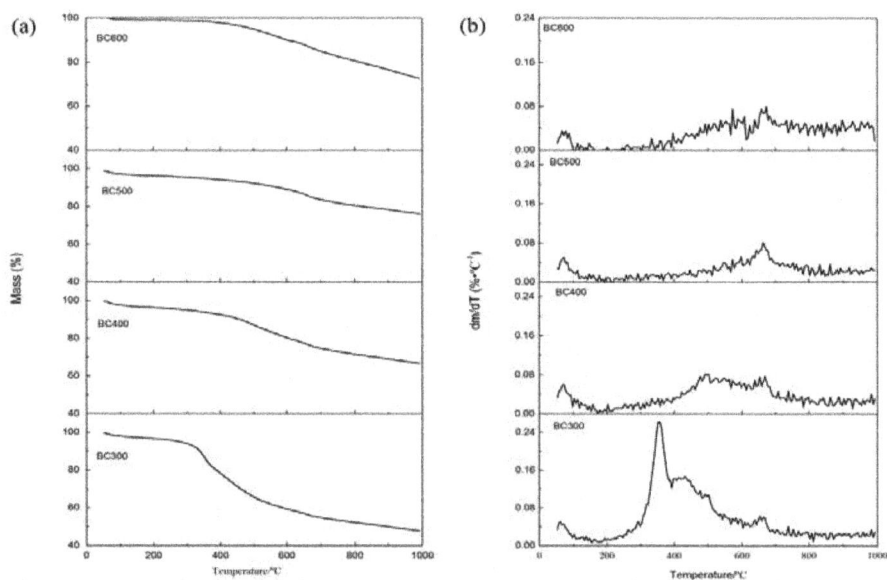

**Figure 6.** Thermogravimetric (TG). (**a**) Differential thermogravimetry (DTG); and (**b**) curves of biochars produced at 300 °C, 400 °C, 500 °C and 600 °C, respectively.

### 3.4. Implications for Environmental and Agronomic Management

Biochar is a carbon-rich product and its addition to soil has been proposed as an effective method for C sequestration [61]. However, the ultimate C sequestration efficiency of biochar mainly depend on its stability, which is mainly effect by the production conditions including pyrolysis temperature [37]. Lehmann et al. suggested that a biochar can be used as an effective C sequestration agents in case its O:C and H:C ratios less than 0.4 and 0.6, respectively [62]. Based on the results, ATB biochars produced at higher temperatures (>400 °C) with H/C of 0.46–0.41 and O/C of 0.10–0.06, may exhibit a high C sequestration potential. Thus, biochars produced at higher temperatures (>400 °C) could be more resistant to mineralization than those pyrolyzed at lower temperature (≤400 °C), thus representing an efficient technique for mitigating greenhouse gas emissions into the environment. In addition, biochars produced at higher temperatures may prove beneficial for use as fertilizer due to their concentrations of minerals like K and Na. However, our results also show that higher pyrolysis temperatures also have the potential of accumulating heavy metals, which can cause soil pollution. Biochars from pyrolysis processes are usually alkaline in nature, especially for the biochars produced at higher pyrolysis temperatures. Therefore, the application of these higher temperature biochars can be useful to increase the pH of acidic soils, which are in risk of aluminum toxicity [63]. In contrast, the application of these higher temperature biochars to arid soils may be critical of concern due to their high salinity and alkalinity. While, compared with the higher temperature biochars, ATB biochars produced at the lower pyrolysis temperature (≤400 °C) have more organic functional groups on their surface, high cation exchange capacity, lower pH values as well as less aromatic content. Thus, the ATB biochars produced at the lower pyrolysis temperature (≤400 °C) may be used to enhance the soil nutrient exchange sites as well as soil cation exchange capacity when they are applied to arid soils.

### 4. Conclusions

The structural and physicochemical properties of biochar derived from ATBs change with pyrolysis temperature. The results show that yield, VM, CEC and H and O were decreased with increasing pyrolysis temperature, whereas total C, fixed C, BET surface area, pore volumes and inorganic minerals (except for Cu and Mn) concentrations increased with the increase in pyrolysis temperature. The pH and ash content increased as temperature increased up to 500 °C and decreased at higher temperature. The increasing temperature also decreased the acidic functional groups, especially for the carboxylic functional groups. While, reverse trend was found for the basic functional groups of biochars. In general, higher temperatures (>400 °C) biochars possessing predominately aromatic carbon structures and highly thermal stability, which can be useful to help mitigate climate change, while, lower temperature (≤400 °C) biochars having more functional groups as well as relatively low pH values may be more suitable for improving the fertility of high pH soils in arid regions. Consequently different ATB biochars can be produced by changing the pyrolysis temperature in order to better meet specific application needs.

**Acknowledgments:** This research was supported by the National Key Technology R&D Program of the Ministry of Science and Technology, China (2012BAD14B11), and the Special Fund for Agro-Scientific Research in the Public Interest of the Ministry of Agriculture, China (201503116).

**Author Contributions:** All authors conceived, designed and performed the experiment. Xu-Dong Wang provided technical and theoretical support; Na Ta contributed to the analysis of experimental results; Shi-Xiang Zhao wrote the paper and all authors read and approved the final version.

**Conflicts of Interest:** The authors declare no conflicts of interest.

## Abbreviations

| | |
|---|---|
| C | Carbon |
| N | Nitrogen |
| H | Hydrogen |
| O | Oxygen |
| CEC | Cation exchange capacity |
| ATB | Apple tree branch |
| VM | Volatile matter |
| SEM | Scanning electron microscopy |
| BET | Brunauer-Emmett-Teller |
| $S_{BET}$ | BET surface area |
| $S_{mic}$ | Micropore surface area; |
| $V_T$ | Total pore volume |
| $V_{mic}$ | Micropore volume |
| FTIR | Fourier-transform infrared |
| TG | Thermogravimetric |
| DTG | Differential thermogravime |

## References

1. Zhao, R.; Coles, N.; Wu, J. Carbon mineralization following additions of fresh and aged biochar to an infertile soil. *Catena* **2015**, *125*, 183–189. [CrossRef]
2. Purakayastha, T.J.; Das, K.C.; Gaskin, J.; Harris, K.; Smith, J.L.; Kumari, S. Effect of pyrolysis temperatures on stability and priming effects of C3 and C4 biochars applied to two different soils. *Soil Tillage Res.* **2016**, *155*, 107–115. [CrossRef]
3. Smith, J.L.; Collins, H.P.; Bailey, V.L. The effect of young biochar on soil respiration. *Soil Biol. Biochem.* **2010**, *42*, 2345–2347. [CrossRef]
4. Abujabhah, I.S.; Bound, S.A.; Doyle, R.; Bowman, J.P. Effects of biochar and compost amendments on soil physico-chemical properties and the total community within a temperate agricultural soil. *Appl. Soil Ecol.* **2016**, *98*, 243–253. [CrossRef]
5. Partey, S.T.; Saito, K.; Preziosi, R.F.; Robson, G.D. Biochar use in a legume–rice rotation system: Effects on soil fertility and crop performance. *Arch. Agron. Soil Sci.* **2015**, *62*, 199–215. [CrossRef]

6. Zhou, J.; Chen, H.; Huang, W.; Arocena, J.M.; Ge, S. Sorption of Atrazine, 17α-Estradiol, and Phenanthrene on Wheat Straw and Peanut Shell Biochars. *Water Air Soil Pollut.* **2016**, *227*, 7. [CrossRef]

7. Venegas, A.; Rigol, A.; Vidal, M. Changes in heavy metal extractability from contaminated soils remediated with organic waste or biochar. *Geoderma* **2016**, *279*, 132–140. [CrossRef]

8. Angin, D.; Sensöz, S. Effect of pyrolysis temperature on chemical and surface properties of biochar of rapeseed (*Brassica napus* L.). *Int. J. Phytoremed.* **2014**, *16*, 684–693. [CrossRef] [PubMed]

9. Roberts, D.A.; De, N.R. The effects of feedstock pre-treatment and pyrolysis temperature on the production of biochar from the green seaweed *Ulva*. *J. Environ. Manag.* **2016**, *169*, 253–260. [CrossRef] [PubMed]

10. Yang, X.; Wang, H.; Strong, P.; Xu, S.; Liu, S.; Lu, K.; Sheng, K.; Guo, J.; Che, L.; He, L. Thermal Properties of Biochars Derived from Waste Biomass Generated by Agricultural and Forestry Sectors. *Energies* **2017**, *10*, 469. [CrossRef]

11. Jouiad, M.; Al-Nofeli, N.; Khalifa, N.; Benyettou, F.; Yousef, L.F. Characteristics of slow pyrolysis biochars produced from rhodes grass and fronds of edible date palm. *J. Anal. Appl. Pyrolysis* **2015**, *111*, 183–190. [CrossRef]

12. Zhang, J.; Liu, J.; Liu, R. Effects of pyrolysis temperature and heating time on biochar obtained from the pyrolysis of straw and lignosulfonate. *Bioresour. Technol.* **2015**, *176*, 288–291. [CrossRef] [PubMed]

13. Claoston, N.; Samsuri, A.W.; Ahmad Husni, M.H.; Mohd Amran, M.S. Effects of pyrolysis temperature on the physicochemical properties of empty fruit bunch and rice husk biochars. *Waste Manag. Res.* **2014**, *32*, 331–339. [CrossRef] [PubMed]

14. Bouraoui, Z.; Jeguirim, M.; Guizani, C.; Limousy, L.; Dupont, C.; Gadiou, R. Thermogravimetric study on the influence of structural, textural and chemical properties of biomass chars on $CO_2$ gasification reactivity. *Energy* **2015**, *88*, 703–710. [CrossRef]

15. Suliman, W.; Harsh, J.B.; Abu-Lail, N.I.; Fortuna, A.M.; Dallmeyer, I.; Garcia-Perez, M. Influence of feedstock source and pyrolysis temperature on biochar bulk and surface properties. *Biomass Bioenergy* **2016**, *84*, 37–48. [CrossRef]

16. Xu, Y.; Chen, B. Investigation of thermodynamic parameters in the pyrolysis conversion of biomass and manure to biochars using thermogravimetric analysis. *Bioresour. Technol.* **2013**, *146*, 485–493. [CrossRef] [PubMed]

17. Guizani, C.; Jeguirim, M.; Valin, S.; Limousy, L.; Salvador, S. Biomass Chars: The Effects of Pyrolysis Conditions on Their Morphology, Structure, Chemical Properties and Reactivity. *Energies* **2017**, *10*, 796. [CrossRef]

18. Shaaban, A.; Se, S.-M.; Dimin, M.F.; Juoi, J.M.; Mohd Husin, M.H.; Mitan, N.M.M. Influence of heating temperature and holding time on biochars derived from rubber wood sawdust via slow pyrolysis. *J. Anal. Appl. Pyrolysis* **2014**, *107*, 31–39. [CrossRef]

19. Liang, C.F.; Gasco, G.; Fu, S.L.; Mendez, A.; Paz-Ferreiro, J. Biochar from pruning residues as a soil amendment: Effects of pyrolysis temperature and particle size. *Soil Tillage Res.* **2016**, *164*, 3–10. [CrossRef]

20. Tag, A.T.; Duman, G.; Ucar, S.; Yanik, J. Effects of feedstock type and pyrolysis temperature on potential applications of biochar. *J. Anal. Appl. Pyrolysis* **2016**, *120*, 200–206. [CrossRef]

21. Sun, Y.N.; Gao, B.; Yao, Y.; Fang, J.N.; Zhang, M.; Zhou, Y.M.; Chen, H.; Yang, L.Y. Effects of feedstock type, production method, and pyrolysis temperature on biochar and hydrochar properties. *Chem. Eng. J.* **2014**, *240*, 574–578. [CrossRef]

22. Brassard, P.; Godbout, S.; Raghavan, V.; Palacios, J.H.; Grenier, M.; Dan, Z. The Production of Engineered Biochars in a Vertical Auger Pyrolysis Reactor for Carbon Sequestration. *Energies* **2017**, *10*, 288. [CrossRef]

23. Colantoni, A.; Zambon, I.; Colosimo, F.; Monarca, D.; Cecchini, M.; Gallucci, F.; Proto, A.R.; Lord, R. An Innovative Agro-Forestry Supply Chain for Residual Biomass: Physicochemical Characterisation of Biochar from Olive and Hazelnut Pellets. *Energies* **2016**, *9*, 526. [CrossRef]

24. Fang, K.; Li, H.; Wang, Z.; Du, Y.; Wang, J. Comparative analysis on spatial variability of soil moisture under different land use types in orchard. *Sci. Hortic.* **2016**, *207*, 65–72. [CrossRef]

25. Bai, S.H.; Reverchon, F.; Xu, C.-Y.; Xu, Z.; Blumfield, T.J.; Zhao, H.; Van Zwieten, L.; Wallace, H.M. Wood biochar increases nitrogen retention in field settings mainly through abiotic processes. *Soil Biol. Biochem.* **2015**, *90*, 232–240. [CrossRef]

26. Eyles, A.; Bound, S.A.; Oliver, G.; Corkrey, R.; Hardie, M.; Green, S.; Close, D.C. Impact of biochar amendment on the growth, physiology and fruit of a young commercial apple orchard. *Trees* **2015**, *29*, 1817–1826. [CrossRef]

27. ASTM International. D5142, Standard Test Methods for Proximate Analysis of the Analysis Sample of Coal and Coke by Instrumental Procedures. In *American Society for Testing and Materials*; ASTM International: West Conshohocken, PA, USA, 2009.

28. Yuan, J.H.; Xu, R.K.; Zhang, H. The forms of alkalis in the biochar produced from crop residues at different temperatures. *Bioresour. Technol.* **2011**, *102*, 3488–3497. [CrossRef] [PubMed]

29. Cantrell, K.B.; Hunt, P.G.; Uchimiya, M.; Novak, J.M.; Ro, K.S. Impact of pyrolysis temperature and manure source on physicochemical characteristics of biochar. *Bioresour. Technol.* **2012**, *107*, 419–428. [CrossRef] [PubMed]

30. Chen, Y.; Yang, H.; Wang, X.; Zhang, S.; Chen, H. Biomass-based pyrolytic polygeneration system on cotton stalk pyrolysis: Influence of temperature. *Bioresour. Technol.* **2012**, *107*, 411–418. [CrossRef] [PubMed]

31. Boehm, H.P. Some aspects of the surface chemistry of carbon blacks and other carbons. *Carbon* **1994**, *32*, 759–769. [CrossRef]

32. Intani, K.; Latif, S.; Kabir, A.K.M.R.; Müller, J. Effect of self-purging pyrolysis on yield of biochar from maize cobs, husks and leaves. *Bioresour. Technol.* **2016**, *218*, 541–551. [CrossRef] [PubMed]

33. Ronsse, F.; Hecke, S.V.; Dickinson, D.; Prins, W. Production and characterization of slow pyrolysis biochar: Influence of feedstock type and pyrolysis conditions. *Glob. Chang. Biol. Bioenergy* **2013**, *5*, 104–115. [CrossRef]

34. Zornoza, R.; Moreno-Barriga, F.; Acosta, J.A.; Muñoz, M.A.; Faz, A. Stability, nutrient availability and hydrophobicity of biochars derived from manure, crop residues, and municipal solid waste for their use as soil amendments. *Chemosphere* **2016**, *144*, 122–130. [CrossRef] [PubMed]

35. Chen, T.; Zhang, Y.; Wang, H.; Lu, W.; Zhou, Z.; Zhang, Y.; Ren, L. Influence of pyrolysis temperature on characteristics and heavy metal adsorptive performance of biochar derived from municipal sewage sludge. *Bioresour. Technol.* **2014**, *164*, 47–54. [CrossRef] [PubMed]

36. Wang, X.; Zhou, W.; Liang, G.; Song, D.; Zhang, X. Characteristics of maize biochar with different pyrolysis temperatures and its effects on organic carbon, nitrogen and enzymatic activities after addition to fluvo-aquic soil. *Sci. Total Environ.* **2015**, *538*, 137–144. [CrossRef] [PubMed]

37. Usman, A.R.A.; Abduljabbar, A.; Vithanage, M.; Ok, Y.S.; Ahmad, M.; Ahmad, M.; Elfaki, J.; Abdulazeem, S.S.; Al-Wabel, M.I. Biochar production from date palm waste: Charring temperature induced changes in composition and surface chemistry. *J. Anal. Appl. Pyrolysis* **2015**, *115*, 392–400. [CrossRef]

38. Zhang, J.N.; Lu, F.; Luo, C.H.; Shao, L.M.; He, P.J. Humification characterization of biochar and its potential as a composting amendment. *J. Environ. Sci.* **2014**, *26*, 390–397. [CrossRef]

39. Li, X.; Shen, Q.; Zhang, D.; Mei, X.; Ran, W.; Xu, Y.; Yu, G. Functional Groups Determine Biochar Properties (pH and EC) as Studied by Two-Dimensional [13]C NMR Correlation Spectroscopy. *PLoS ONE* **2013**, *8*, e65949. [CrossRef] [PubMed]

40. Tan, X.; Liu, Y.; Zeng, G.; Wang, X.; Hu, X.; Gu, Y.; Yang, Z. Application of biochar for the removal of pollutants from aqueous solutions. *Chemosphere* **2015**, *125*, 70–85. [CrossRef] [PubMed]

41. Keiluweit, M.; Nico, P.S.; Johnson, M.G.; Kleber, M. Dynamic molecular structure of plant biomass-derived black carbon (biochar). *Environ. Sci. Technol.* **2010**, *44*, 1247–1253. [CrossRef] [PubMed]

42. Chen, X.; Chen, G.; Chen, L.; Chen, Y.; Lehmann, J.; McBride, M.B.; Hay, A.G. Adsorption of copper and zinc by biochars produced from pyrolysis of hardwood and corn straw in aqueous solution. *Bioresour. Technol.* **2011**, *102*, 8877–8884. [CrossRef] [PubMed]

43. Pituello, C.; Francioso, O.; Simonetti, G.; Pisi, A.; Torreggiani, A.; Berti, A.; Morari, F. Characterization of chemical–physical, structural and morphological properties of biochars from biowastes produced at different temperatures. *J. Soils Sediments* **2015**, *15*, 792–804. [CrossRef]

44. Phuong, H.T.; Uddin, M.A.; Kato, Y. Characterization of Biochar from Pyrolysis of Rice Husk and Rice Straw. *J. Biobased Mater. Bioenergy* **2015**, *9*, 439–446. [CrossRef]

45. Sharma, R.K.; Wooten, J.B.; Baliga, V.L.; Lin, X.; Geoffrey Chan, W.; Hajaligol, M.R. Characterization of chars from pyrolysis of lignin. *Fuel* **2004**, *83*, 1469–1482. [CrossRef]

46. Kim, W.K.; Shim, T.; Kim, Y.S.; Hyun, S.; Ryu, C.; Park, Y.K.; Jung, J. Characterization of cadmium removal from aqueous solution by biochar produced from a giant *Miscanthus* at different pyrolytic temperatures. *Bioresour. Technol.* **2013**, *138*, 266–270. [CrossRef] [PubMed]

47. Li, M.; Liu, Q.; Guo, L.; Zhang, Y.; Lou, Z.; Wang, Y.; Qian, G. Cu(II) removal from aqueous solution by *Spartina alterniflora* derived biochar. *Bioresour. Technol.* **2013**, *141*, 83–88. [CrossRef] [PubMed]
48. Jeong, C.Y.; Dodla, S.K.; Wang, J.J. Fundamental and molecular composition characteristics of biochars produced from sugarcane and rice crop residues and by-products. *Chemosphere* **2016**, *142*, 4–13. [CrossRef] [PubMed]
49. Vamvuka, D.; Sfakiotakis, S. Effects of heating rate and water leaching of perennial energy crops on pyrolysis characteristics and kinetics. *Renew. Energy* **2011**, *36*, 2433–2439. [CrossRef]
50. Zhou, L.; Liu, Y.; Liu, S.; Yin, Y.; Zeng, G.; Tan, X.; Hu, X.; Hu, X.; Jiang, L.; Ding, Y.; et al. Investigation of the adsorption-reduction mechanisms of hexavalent chromium by ramie biochars of different pyrolytic temperatures. *Bioresour. Technol.* **2016**, *218*, 351–359. [CrossRef] [PubMed]
51. Ahmad, M.; Lee, S.S.; Dou, X.; Mohan, D.; Sung, J.K.; Yang, J.E.; Ok, Y.S. Effects of pyrolysis temperature on soybean stover- and peanut shell-derived biochar properties and TCE adsorption in water. *Bioresour. Technol.* **2012**, *118*, 536–544. [CrossRef] [PubMed]
52. Uchimiya, M.; Wartelle, L.H.; Klasson, K.T.; Fortier, C.A.; Lima, I.M. Influence of pyrolysis temperature on biochar property and function as a heavy metal sorbent in soil. *J. Agric. Food Chem.* **2011**, *59*, 2501–2510. [CrossRef] [PubMed]
53. Das, D.D.; Schnitzer, M.I.; Monreal, C.M.; Mayer, P. Chemical composition of acid–base fractions separated from biooil derived by fast pyrolysis of chicken manure. *Bioresour. Technol.* **2009**, *100*, 6524–6532. [CrossRef] [PubMed]
54. Souza, B.S.; Moreira, A.P.D.; Teixeira, A.M.R.F. TG-FTIR coupling to monitor the pyrolysis products from agricultural residues. *J. Therm. Anal. Calorim.* **2009**, *97*, 637–642. [CrossRef]
55. Kazi, Z.H.; Schnitzer, M.I.; Monreal, C.M.; Mayer, P. Separation and identification of heterocyclic nitrogen compounds in biooil derived by fast pyrolysis of chicken manure. *J. Environ. Sci. Health Part B Pestic. Food Contam. Agric. Wastes* **2011**, *46*, 51–61. [CrossRef] [PubMed]
56. Mukherjee, A.; Zimmerman, A.R.; Harris, W. Surface chemistry variations among a series of laboratory-produced biochars. *Geoderma* **2011**, *163*, 247–255. [CrossRef]
57. Doydora, S.A.; Cabrera, M.L.; Das, K.C.; Gaskin, J.W.; Sonon, L.S.; Miller, W.P. Release of Nitrogen and Phosphorus from Poultry Litter Amended with Acidified Biochar. *Int. J. Environ. Res. Publ. Health* **2011**, *8*, 1491–1502. [CrossRef] [PubMed]
58. Santos, L.B.; Striebeck, M.V.; Crespi, M.S.; Ribeiro, C.A.; De Julio, M. Characterization of biochar of pine pellet. *J. Therm. Anal. Calorim.* **2015**, *122*, 21–32. [CrossRef]
59. Wu, H.; Che, X.; Ding, Z.; Hu, X.; Creamer, A.E.; Chen, H.; Gao, B. Release of soluble elements from biochars derived from various biomass feedstocks. *Environ. Sci. Pollut. Res.* **2016**, *23*, 1905–1915. [CrossRef] [PubMed]
60. Bruun, E.W.; Hauggaard-Nielsen, H.; Ibrahim, N.; Egsgaard, H.; Ambus, P.; Jensen, P.A.; Dam-Johansen, K. Influence of fast pyrolysis temperature on biochar labile fraction and short-term carbon loss in a loamy soil. *Biomass Bioenergy* **2011**, *35*, 1182–1189. [CrossRef]
61. Lehmann, J. A handful of carbon. *Nature* **2007**, *447*, 143–144. [CrossRef] [PubMed]
62. Ippolito, J.A.; Laird, D.A.; Busscher, W.J. Environmental benefits of biochar. *J. Environ. Q.* **2012**, *41*, 967–972. [CrossRef] [PubMed]
63. Wan, Q.; Yuan, J.H.; Xu, R.K.; Li, X.H. Pyrolysis temperature influences ameliorating effects of biochars on acidic soil. *Environ. Sci. Pollut. Res. Int.* **2014**, *21*, 2486–2495. [CrossRef] [PubMed]

*energies*

MDPI

*Article*

# Biomass Chars: The Effects of Pyrolysis Conditions on Their Morphology, Structure, Chemical Properties and Reactivity

Chamseddine Guizani [1], Mejdi Jeguirim [2,*], Sylvie Valin [3], Lionel Limousy [2] and Sylvain Salvador [4]

[1] University Grenoble Alpes, CNRS, Grenoble INP, LGP2, F-38000 Grenoble, France; chamseddine.guizani@lgp2.grenoble-inp.fr
[2] Institut de Sciences des Matériaux de Mulhouse, UMR CNRS 7361, 15 rue Jean Starcky, 68057 Mulhouse, France; Lionel.limousy@uha.fr
[3] CEA, LITEN/DTBH/SBRT/LTB, 38054 Grenoble CEDEX 09, France; Sylvie.valin@cea.fr
[4] RAPSODEE, Mines Albi, CNRS UMR 5302, Route de Teillet, 81013 ALBI CT CEDEX 09, France; sylvain.salvador@mines-albi.fr
* Correspondence: mejdi.jeguirim@uha.fr; Tel.: +33-389-608-661

Academic Editor: Shusheng Pang
Received: 12 April 2017; Accepted: 6 June 2017; Published: 11 June 2017

**Abstract:** Solid char is a product of biomass pyrolysis. It contains a high proportion of carbon, and lower contents of H, O and minerals. This char can have different valorization pathways such as combustion for heat and power, gasification for Syngas production, activation for adsorption applications, or use as a soil amendment. The optimal recovery pathway of the char depends highly on its physical and chemical characteristics. In this study, different chars were prepared from beech wood particles under various pyrolysis operating conditions in an entrained flow reactor (500–1400 °C). Their structural, morphological, surface chemistry properties, as well as their chemical compositions, were determined using different analytical techniques, including elementary analysis, Scanning Electronic Microscopy (SEM) coupled with an energy dispersive X-ray spectrometer (EDX), Fourier Transform Infra-Red spectroscopy (FTIR), and Raman Spectroscopy. The biomass char reactivity was evaluated in air using thermogravimetric analysis (TGA). The yield, chemical composition, surface chemistry, structure, morphology and reactivity of the chars were highly affected by the pyrolysis temperature. In addition, some of these properties related to the char structure and chemical composition were found to be correlated to the char reactivity.

**Keywords:** biomass; pyrolysis; entrained flow reactor; char characterization

## 1. Introduction

Economic and environmental issues related to the non-avoidable depletion of fossil fuels are urging governments all over the world to modify their strategies, and shift from a fossil-fuel-based economy to a bio-resources-based one. Hence, bio-refining biomass instead of petroleum to obtain energy and high added-value products represents a major challenge for a sustainable future [1].

In a bio-refinery, the biomass can be transformed into a variety of bio-based products such as biofuels and bio-based chemicals and materials, just as in fossil fuel refineries. Energy can also be obtained from biomass processing for the production of heat and electricity [2].

Different processes exist that allow the transformation of raw biomass into desirable bio-based products and/or energy. For instance, biomass pyrolysis, which is a thermochemical process, has the advantage of transforming the raw material after heating in the absence of oxygen into bio-oil, gas and char that can be valorized separately [3]. The proportion of these three products depends strongly on the pyrolysis conditions; namely, reactor temperature and heating rate. Low temperatures (<600 °C) favor the production of bio-oil and bio-char while high temperatures (>600 °C) maximize the production of gas. At relatively low temperatures, a fast heating rate is desirable to maximize the bio-oil yield.

As mentioned previously, the pyrolysis products may be valorized in different applications. The bio-oil can be used as a fuel for heating, or for the production of chemicals [4]. Gas can be energetically valorized into heat/electricity or further processed to produce biofuels. Depending on its quality, the solid bio-char can be gasified [5], used for the production of activated carbons [6], for the production of graphene [7], or for soil remediation [8].

The reactivity and physicochemical properties of the char are crucial for selecting the suitable valorization pathway. The pyrolysis process highly affects the physicochemical properties of the residual char as well as its reactivity. Indeed, the reactor temperature, the particle heating rate and residence time in the reactor directly influence the char yield and properties. In particular, the char chemical composition depends on the pyrolysis temperature [9]. As temperature increases, O and H atoms are released in the gas phase due to the pyrolysis of biomass components and cracking of the residual char. Moreover, some minerals are volatilized and the solid char becomes richer in carbon [10–12].

The char texture and structural organization change over the pyrolysis reaction due to the thermal degradation of the major biomass components; namely, cellulose, hemicellulose and lignin [13]. The structural features and surface chemistry of the bio-chars are affected by the pyrolysis temperature and residence time. Surface functionality decreases as temperature increases due to biomass pyrolysis and char cracking reactions [14]. Char structure gets more organized when increasing the temperature and residence time in the reactor, and tends towards a graphite-like structure [15–17].

Other studies reported results on the evolution of the yield and above-mentioned physical and chemical characteristics of bio-chars upon fast and slow pyrolysis and correlated them to the pyrolysis experimental conditions such as temperature and heating rate [18,19]. For instance, in [18], the authors found that the char yield decreased when increasing the heating rate up to a value of 600 °C/s, above which the char yield was constant. In [19], the authors analyzed chars coming from slow and fast pyrolysis as well as gasification processes. They observed a pronounced decrease in aromatic C-H functionality between slow pyrolysis and gasification chars using Nuclear Magnetic Resonance (NMR) and Fourier Transform Infrared-Photoacoustic (FTIR-PAS) Spectroscopies. Moreover, NMR estimates of fused aromatic ring cluster size showed fast and slow pyrolysis chars to be similar (7–8 rings per cluster), while higher- temperature gasification char was much more condensed (17 rings per cluster).

Analyzing the char structural evolution upon heat treatment, McDonald-Wharry et al. [15] observed that the $I_V/I_G$ Raman structural ratio was correlated with heat treatment temperature for various types of chars. Morin et al. observed a decrease of beech char reactivity with an increasing pyrolysis temperature [20]. Other authors also correlated several char properties, especially those relating structural parameters of the char and their reactivities [21,22].

In the present work, beech wood pyrolysis experiments were performed in an entrained flow reactor (EFR) at between 500 °C and 1400 °C for different gas residence times. These experimental conditions were wide enough to obtain chars with various properties. The recovered chars were characterized in terms of their morphological, structural, surface chemistry properties as well as their reactivity towards oxygen. Correlations between the different char properties were identified. Such correlations could be helpful for developing a preliminary understanding of char properties, and therefore for identifying their potential valorization pathways.

## 2. Materials and Methods

### 2.1. Raw Biomass

The raw biomass used in this work is beech wood supplied by the Sowood company. The raw wood is finely ground with particles having sizes ranging from tens of microns to a few millimeters. After sieving, particles having a mean size of 370 μm were selected for the pyrolysis experiments. This mean size should be appropriate to avoid heat and mass transfer limitations during the EFR pyrolysis experiments.

The moisture content was determined by drying at 105 °C according to the NF-EN-14774 standard. Ash content was determined by burning the sample in air at 850 °C according to the NF-EN-14775 standard. The oxygen content was determined by difference to the sum of C, H, N and ash contents. The proximate and ultimate analyses are shown in Table 1.

**Table 1.** Proximate and ultimate analyses of beech wood.

| Proximate Analysis | Value |
|---|---|
| Moisture [wt % ar *] | 8.7 |
| Volatile Matter [wt % db **] | 84.3 |
| Fixed carbon [wt % db] | 15.2 |
| Ash (815 °C) [wt % db] | 0.5 |
| Ultimate analysis [wt % db] | |
| C | 49.1 |
| H | 5.7 |
| N | 0.15 |
| S | 0.045 |
| O (by difference) | 44.5 |

* ar: as received; ** db: dry basis.

### 2.2. Experimental EFR Set Up

The EFR experimental set-up was previously described in detail by Billaud et al. [23]. It consists of an alumina tube inserted in a vertical electrical heater with three independent heating zones. The dimensions of the tube are 2.3 m in length and 0.075 m in internal diameter. The heated zone is 1.2 m long. The temperature along the heated zone is homogeneous and verified prior to the experiment by moving a thermocouple vertically along this zone. The temperature profile is assumed to be the same during the experiment, as it is not possible to monitor it along the z axis while injecting biomass and collecting the products. The EFR works at atmospheric pressure and can reach a maximum temperature of 1400 °C. The wood particles are continuously fed into the reactor using a gravimetric feeding system. The main $N_2$ gas stream is electrically pre-heated before entering the reactor. A sampling probe allows collection of the gas produced, which is analyzed by a gas chromatograph. The remaining solid falls by gravity in a settling box. Tars are condensed in a cold trap, while soot is collected by a filter, and is weighed after the reaction to determine its volume.

### 2.3. Pyrolysis Conditions

The pyrolysis experimental conditions are summarized in Table 2. The biomass mass flow rate (1 g/min) and particle size (370 μm) were the same for all experiments. The gas residence time was controlled with the nitrogen flow rate, as the pyrolysis gas flow rate was much lower than that of the $N_2$. The mean gas residence time was determined knowing the nitrogen flowrate, the reactor temperature, length and cross-section.

The chars analyzed in the present work are produced from the pyrolysis of the same biomass and gathered from two experimental campaigns with two different objectives:

- The first is related to medium-temperature pyrolysis (500–600 °C), aiming at maximizing bio-oil production.
- The second concerns high-temperature pyrolysis aiming at maximizing the gas fraction [23].

For the first tests, the gas residence time was fixed at 16.6 s, while in the second tests it was fixed at 4.3 s. At high temperatures, the gas expansion induces an increase in the volumetric flow rate, lowering the residence time of gases. For practical purposes, it is not convenient to operate at lower flow rates (due to oxygen introduction in the reactor). The particle residence time (which is much more important than the gas residence time with regard to the final char characteristics) is calculated considering slip velocity between solid particles and gas. This takes into account the particle density and size, and their evolution as pyrolysis proceeds, between the density and size of wood particles and those of char particles [24]. The robustness and accuracy of the model is validated over a wide range of temperatures, atmospheres and particle sizes [23,25].

**Table 2.** Pyrolysis experimental conditions in the EFR.

| Reactor Temperature $T_{Pyr}$ [°C] | $\dot{m}_{biomass}$ [g/min] | Gas Residence Time [s] | Particle Residence Time $t_{Res}$ [s] |
|---|---|---|---|
| 500 | 1 ± 0.1 | 16.6 ± 0.2 | 2.26 |
| 550 | 1 ± 0.1 | 16.6 ± 0.2 | 4.03 |
| 600 | 1 ± 0.1 | 16.6 ± 0.2 | 5.12 |
| 800 | 1 ± 0.1 | 4.3 ± 0.1 | 2.33 |
| 1000 | 1 ± 0.1 | 4.3 ± 0.1 | 2.40 |
| 1200 | 1 ± 0.1 | 4.3 ± 0.1 | 2.45 |
| 1400 | 1 ± 0.1 | 4.3 ± 0.1 | 2.52 |

The particle residence time varies more at low temperatures, while it remains almost the same for temperatures above 600 °C. This is not obvious at first glance, but it has to do with the change in particle slip velocity during the pyrolysis process. At low temperatures, the pyrolysis reaction rate is not so high, and neither is the particle slip velocity. During the pyrolysis reaction, the particle slip velocity changes from that of the biomass particle to that of the char particle. The latter is much lower than the former. As the pyrolysis kinetics are slower at low temperatures, the change in particle slip velocity (from that of wood to that of char) is also slow, and consequently the residence time is also small. However, when increasing the temperature, the pyrolysis reaction accelerates, the char is formed in a shorter time inside the reactor, the slip velocity drops more rapidly to that of a char particle and, consequently, the residence time becomes higher (see the increase between 500 and 600 °C).

At higher temperatures >800 °C, the pyrolysis reaction rate is quite high and the char is formed in the first centimeters of the reactor. As it has a lower slip velocity than the biomass, it resides inside the reactor for a non-negligible period. One can see that the residence time at high temperatures is similar to that obtained at 500 °C, despite a gas velocity that is, for a time, higher at higher temperatures. As explained, this has to do with the particle slip velocity which is higher at 500 °C than at temperatures above 800 °C.

As for the 500–600 °C experiments, a large sampling probe allowing recovery of all the pyrolysis products is used. The mass of residual char is then directly weighed. The char yield is obtained by the ratio between the total mass of the residual char and the cumulated mass of the injected biomass during the pyrolysis experiment:

$$Y_{char} = \frac{m_{char}}{\int_0^{t_{end}} \dot{m}_{biomass}\, dt}$$

For all the other experiments performed above 800 °C, about 80% of the products were sampled. The char yield is then calculated using the ash tracer method, as described in detail in [24].

*2.4. Characterization Techniques of the Biomass Chars*

2.4.1. Proximate and Ultimate Analysis

Ash content was determined by burning the samples in air at 850 °C according to the NF-EN-14775 standard. The oxygen content was determined by difference.

2.4.2. Morphology

The morphology of the parent wood and solid residues was studied via Scanning Electron Microscopy (Philips model FEI model Quanta 400 SEM). SEM images were obtained in a high vacuum mode for both secondary electrons (topography contrast) and backscattered electrons (chemical contrast). The acceleration voltage and the electron beam spot size were carefully chosen. The solid residues were coated with a thin gold layer by using Agar automatic sputter coater prior to the observation. Energy Dispersive X-Ray analysis was also performed in order to identify the mineral species and analyze their dispersion throughout the char matrix.

2.4.3. Surface Chemistry

The surface functional groups of the solid residues were characterized using a FTIR PerkinElmer 1720 spectrometer. The FTIR spectra were recorded in the spectral range between 4000 and 500 cm$^{-1}$ with a resolution of 4 cm$^{-1}$ and 32 scans in attenuated total reflection (ATR) configuration.

2.4.4. Structure

The structural evolution of the chars was assessed by Raman spectroscopy. Raman spectra of the chars were recorded with a BX40 LabRam, Jobin Yvon/Horiba spectrometer. Several particles were sampled and deposited on a rectangular glass slide for the Raman analysis. Raman spectra were obtained in a backscattered configuration with an excitation laser at 635 nm. The Raman spectra at each position give an average structural information of a large number of carbon micro-crystallites. The Raman spectra were recorded at 6 locations of the char sample. The spectra were afterward analyzed using a home-made software for the background correction, normalization with G band height, average spectrum calculation, as well as different band height and integral areas measurements.

From these spectra, several parameters related to the structure were determined:

- ID: D band intensity height.
- D$_{pos}$: D band position.
- IG: G band intensity height.
- G$_{pos}$: G band position.
- IV: Valley region intensity height.
- The Total Raman Area (TRA):

$$TRA = \int_{800}^{2000} I \, dv \qquad (1)$$

*I* is the intensity and *dv* is the increment of the wave number.

The different band specifications and ratios are further discussed in the results and discussion section.

2.4.5. Reactivity towards Oxygen

Reactivity tests in air atmosphere were performed using the Mettler-Toledo TGA/DSC3+ thermogravimetric analyzer. The air flow rate was set at 50 mL/min. The samples were heated from 30 °C to 700 °C with a heating rate of 10 °C/min and maintained at this temperature during 1 h. The sample mass was below 5 mg to minimize heat and mass transfer limitations.

The TG oxidation experimental data were analyzed with homemade software, for smoothing, derivative calculation, derivative peak positions and values etc.

Hence, the different parameters related to the char reactivities in air were determined and can be summarized as follows:

- $\alpha = \frac{m_{(t)} - m_{(ashes)}}{m_{(0)} - m_{(ashes)}}$ is the conversion level.

- $T_{(2\%)}$ is the temperature at which $\alpha = 2\%$.

- $T_{(50\%)}$ is the temperature at which $\alpha = 50\%$ .

- $T_{(98\%)}$ is the temperature at which $\alpha = 98\%$.

- $\frac{dm}{dt}_{mean} = -\frac{1}{N_{exp}} \sum_{i=1}^{N_{exp}} \left( \frac{dm}{dt}_i \right)$ is the mean reaction rate $-\frac{dm}{dt}_{max}$ is the maximum observed reaction rate.

- $T_{(peak)}$ is the temperature at the maximum reaction rate.

- $R_{mean} = -\frac{1}{N_{exp}} \sum_{i=1}^{N_{exp}} \left( \frac{dm}{dt}_i \frac{1}{m_i} \right)$ is the mean reactivity.

- $R_{index} = -\frac{1}{N_{exp}} \sum_{i=1}^{N_{exp}} \left( \frac{dm}{dt}_i \frac{1}{T_i} \right)$ is the reactivity index.

$T$ is the temperature, $m_{(t)}$ is the char mass at time "$t$", $m_{(0)}$ is the initial char mass, $m_{(ashes)}$ is the mass of the residual ashes, $N_{exp}$ is the number of experimental values.

## 3. Results and Discussion

### 3.1. Pyrolysis Product Yields

Figure 1 shows an area plot of the pyrolysis product yield distribution as a function of the reactor temperature. This area plot is drawn using the discrete values of the different yields obtained from different experimental conditions given in Table 2. Tars include the condensable species in the volatile matter and the initial moisture of the wood. At low temperatures (below 600 °C), tar and solid together represent more than 50 wt % of the products. A temperature of 500 °C maximizes the oil yield, which reaches 62.4 wt %, while the char yield is close to 14 wt %. Several investigations have mentioned 500–550 °C as the optimal temperature for maximizing the bio-oil yield during biomass pyrolysis [3]. The water content of the pyrolysis oil (tars) varied between 30% and 35% at 500 and 600 °C, respectively, which includes both moisture water and pyrolysis reaction water. The water content of the oils obtained at higher temperatures was not evaluated.

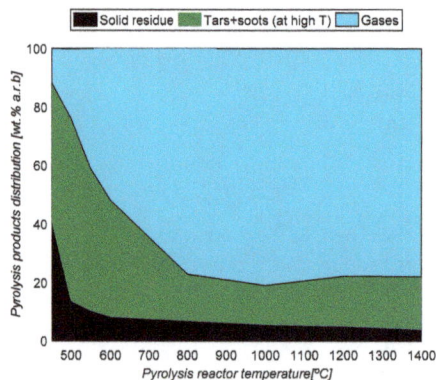

**Figure 1.** Pyrolysis product distribution (as received basis, a.r.b.) versus pyrolysis reactor temperature.

The production of gases increases with temperature, while that of tar and char decreases. For instance, at 800 °C, the gas yield is close to 78 wt %, while the char yield is 7%. Furthermore, at high temperatures, the formation of soot was observed, starting from 1000 °C. A deep analysis of the operating conditions' effect on the pyrolysis product distribution can be found in previous investigations [23,26]. The main focus of the present study was to identify the effect of the operating conditions on the char characteristics.

The evolution of the char yield with the pyrolysis reactor temperature indicates two separate ranges of linear decrease: low temperatures (500–600 °C) and high temperatures (600–1400 °C). The char yield decreased from 14% to 8% between 500 °C and 600 °C (0.05%/°C, $R^2 = 0.991$), while it decreased from 8% to 4% between 600 °C and 1400 °C (0.005%/°C, $R^2 = 0.990$). The relative uncertainties in the char yields are estimated to be lower than 10%. The evolution of the char yield would be thus significant.

The lower char yield at high temperatures may be explained by several factors. Higher heating rates are known to induce lower char yields [18]. Moreover, volatile and char formation are two parallel reactions, and the latter is probably favored at higher temperatures and higher heating rates. Also, above 1000 °C, char gasification by $H_2O$ and $CO_2$ is likely to consume part of the formed solid.

*3.2. Microscopic Analysis of the Biochar Surfaces*

In order to monitor the textural properties and the behavior of minerals on the different chars upon different pyrolysis treatment, we performed SEM analysis coupled with EDX characterization at different enlargements.

Five samples were chosen to observe the effect of the pyrolysis temperature on the surface and the properties of the biomass particles: raw biomass, and samples pyrolysed at low (500, 600 °C) and high (1200, 1400 °C) temperatures.

Figure 2 presents the surface of a particle of beech wood at different enlargements. The cell walls of the raw biomass present a homogeneous surface without any holes; the EDX analysis shows the presence of very small amounts of Ca and K, which are the main minerals present in the biomass. The elemental cartography of the surface shows that K and Ca are homogeneously dispersed inside the biomass.

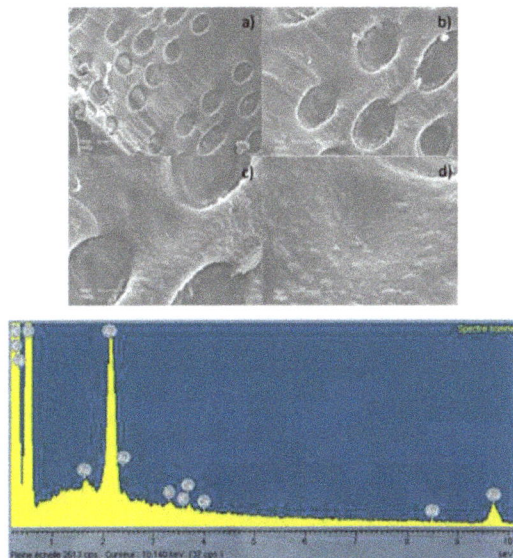

**Figure 2.** SEM photographs and X-ray fluorescence analysis of the surface of the raw wood particles at different enlargements (**a**) ×5000; (**b**) ×10,000; (**c**) ×20,000; and (**d**) ×50,000.

Figures 3 and 4 show the surface of the samples pyrolysed at 500 and 600 °C respectively. Different areas of the particles were observed in order to be sure that the description of these materials was representative of the materials.

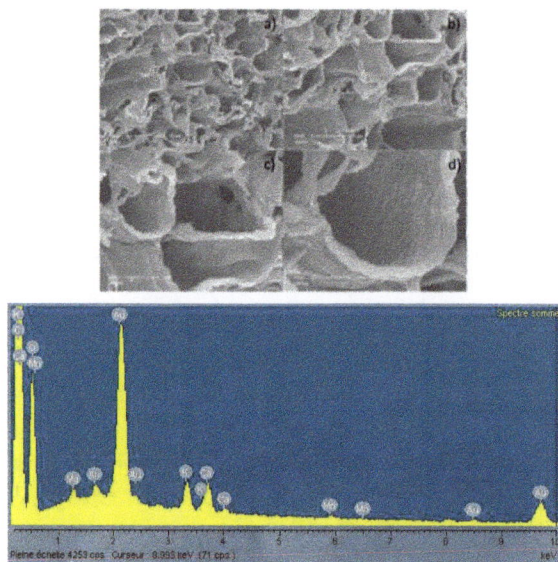

**Figure 3.** SEM photographs and X-ray fluorescence analysis of the surface of the biochar produced after pyrolysis of wood particles at 500 °C and different enlargements (**a**) ×5000; (**b**) ×10,000; (**c**) ×20,000; and (**d**) ×50,000.

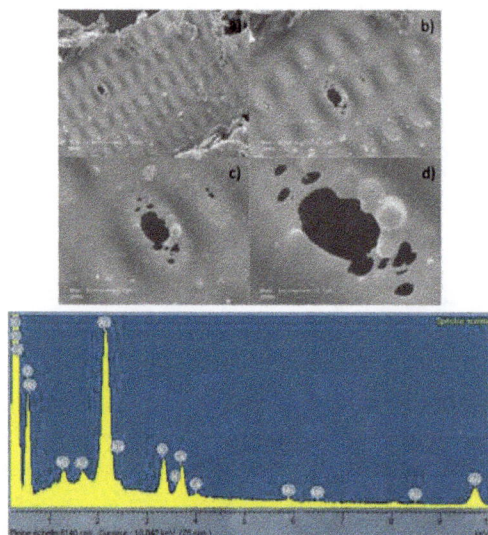

**Figure 4.** SEM photographs and X-ray fluorescence analysis of the surface of the biochar produced after pyrolysis of wood particles at 600 °C and different enlargements (**a**) ×5000; (**b**) ×10,000; (**c**) ×20,000; and (**d**) ×50,000.

The samples Char-500 and Char-600 present a smooth surface, comparable to the one of the raw biomass. They don't present any macropores (pores with diameter higher than 50 nm) and the shape of the particles remains identical to the raw biomass particles. The only difference is the concentrations of Ca and K, which increase with pyrolysis temperature (EDX spectra). The intensities of Ca and K peaks increase greatly with the pyrolysis temperature, bearing comparison to the principal Au peak (close to 2 keV). This phenomenon is due to the loss of volatile compounds from the raw biomass and then to the concentration of Ca and K. It also indicates that Ca and K are not completely emitted in the gas phase during the pyrolysis process when performed at 500 and 600 °C. Small amounts of Mg and Mn were observed in both samples.

Figures 5 and 6 show the surface of the chars obtained at 1200 and 1400 °C respectively. These photographs clearly show the modification of the particle surface shape at these temperatures due to sintering. At high enlargement ($\times$50,000) small particles at the surface of the chars can be observed (Figures 4d and 5d). Analysis of the surface (EDX cartography—not shown) indicated that these particles contain only carbon, and no minerals. For the char obtained at 1200 °C, the amounts of Ca and K increased (with peak intensities comparable to that of Au) significantly, but the surface didn't present cracks or macropores (Figure 5d). Small amounts of Mg and Mn were also observed on these samples.

Figure 6a illustrates the deep modification of the biomass structure after a thermal treatment at 1400 °C. This sample is different from the others, with an important macro/mesoporosity appearing at the surface of the carbon residue (Figure 6d). This modification is also correlated with a decrease of the K content. As can be observed from the EDX analysis presented in Figure 6, the relative intensity of the K peak decreased significantly. This behavior can be explained by the sublimation/boiling of potassium during pyrolysis performed at 1400 °C (boiling point 1420 °C, sublimation at 1500 °C), leading to the appearance of large porosity at the surface of the char particle, or again to the char gasification by $H_2O/CO_2$ formed during the pyrolysis stage. This is consistent with other results obtained under similar conditions [23], which also showed that potassium released from char at 1400 °C was incorporated into soot particles.

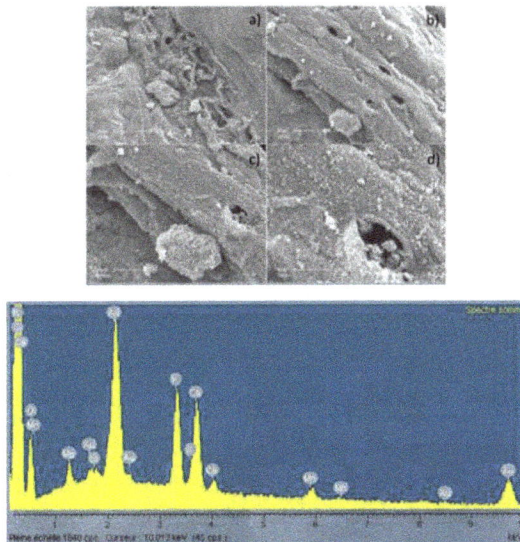

**Figure 5.** SEM photographs and X-ray fluorescence analysis of the surface of the biochar produced after pyrolysis of wood particles at 1200 °C and different enlargements (**a**) $\times$5000; (**b**) $\times$10,000; (**c**) $\times$20,000; and (**d**) $\times$50,000.

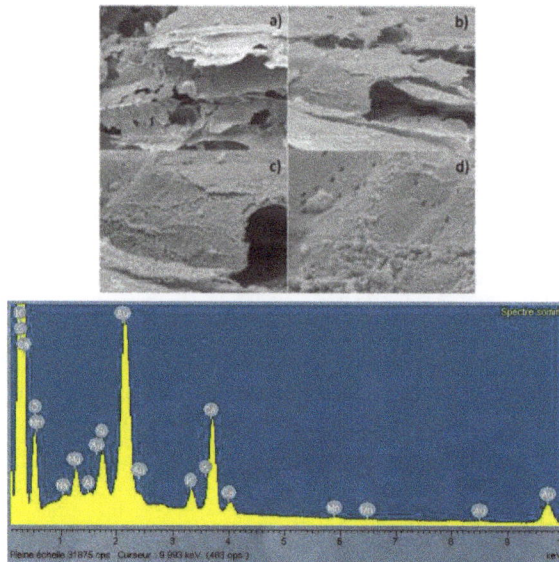

**Figure 6.** SEM photographs and X-ray fluorescence analysis of the surface of the biochar produced after pyrolysis of wood particles at 1400 °C and different enlargements (**a**) ×5000; (**b**) ×10,000; (**c**) ×20,000; and (**d**) ×50,000.

### 3.3. Chemical Composition

The chemical composition of the char samples is given in the Table 3. It shows that the char gets richer in C and poorer in H and O as the pyrolysis temperature increases. The C content increased from 47.1 mol % for the char-500 to 84 mol % for the char-1400. This enrichment in C with increasing temperature is well known in the literature. The O and H contents decreased significantly from 15.4 mol % and 37.5 mol % to 6.2 mol % and 9.9 mol % for the char-500 and the char-1400, respectively. Controlling the chemical composition of chars is relevant to many applications, such as tar removal, which was seen to depend on the C content of chars prepared by pyrolysis of rice husks [26].

Also, the C content of the char was seen to be related to the stability of the char in the soil when focusing on applications aiming at improving the soil quality or sequestrating the C for global warming mitigation [27].

**Table 3.** C, H and O contents of the different chars.

| Sample | C [mol % afb] | H [mol % afb] | O [mol % afb] |
|---|---|---|---|
| char-500 | 47.1 ± 2.3 | 37.5 ± 0.8 | 15.4 ± 1.5 |
| char-550 | 56.1 ± 1.3 | 32.4 ± 0.8 | 11.6 ± 0.5 |
| char-600 | 63.4 ± 2.7 | 26.8 ± 1.1 | 9.9 ± 1.5 |
| char-800 | 64.2 ± 1.5 | 26.9 ± 0.9 | 8.9 ± 0.6 |
| char-1000 | 74.4 ± 3.1 | 15.0 ± 0.2 | 10.6 ± 2.8 |
| char-1200 | 77.6 ± 2.6 | 14.0 ± 0.3 | 8.3 ± 2.2 |
| char-1400 | 84.0 ± 2.7 | 9.9 ± 0.3 | 6.2 ± 2.4 |

### 3.4. FTIR Spectroscopy

FTIR spectra of the fresh wood and pyrolysis chars are shown in Figure 7. The wavenumber range is divided into two sub-ranges for the sake of clarity. The FTIR spectra of the unheated wood sample consist mainly of bands that can be attributed to carbohydrates (cellulose and hemicellulose) and lignin. The most prominent carbohydrate bands in the raw wood can be found between 1000 and 1200 cm$^{-1}$, while those related to lignins were identified at approximately 1221, 1269, 1326, 1367, 1423, 1464, 1510 and 1596 cm$^{-1}$ [11].

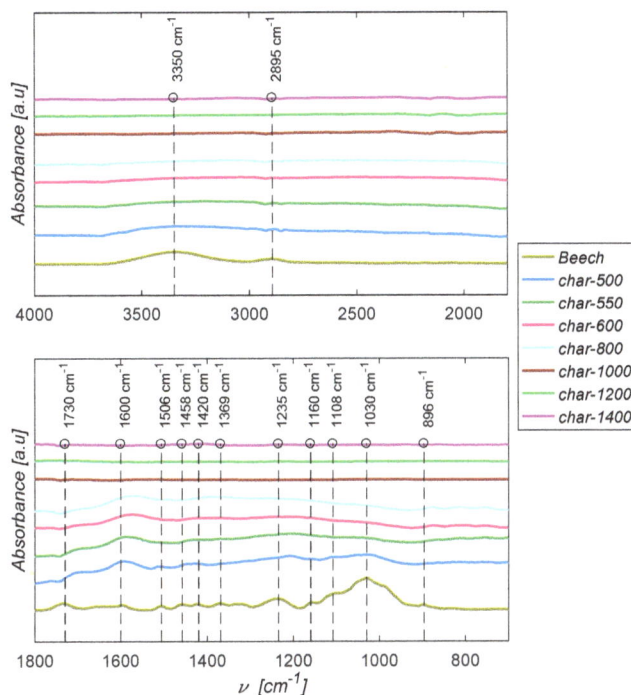

**Figure 7.** FTIR spectra of the pyrolysis solid residues.

Noticeable changes can be observed in the spectra of the chars compared to the raw wood. The OH band (around 3350 cm$^{-1}$) decreased significantly, starting with the char-500. The peak at 1730 cm$^{-1}$, related to the presence of carbonyl groups of esters and uronic acids in the xylan of beech wood progressively disappears as temperature increases. These functions belong to the hemicelluloses of beech wood, which have the lowest thermal stability [19].

Moreover, the two small peaks at 1506 and 1458 cm$^{-1}$, which are related to the Guaiacyl and Syringyl units of the beech wood lignin, are much less pronounced in char-500 than in raw beech, and completely vanish for the char-550 and beyond, confirming the high pyrolysis extent for this experiment. The peak at 1369 cm$^{-1}$ is probably related to the C-H deformation in cellulose and hemicelluloses. This peak is only visible in raw beech, and no more in chars. The signal between 1000 and 1200 cm$^{-1}$ is probably imputable to C-O vibrations in cellulose/hemicelluloses. It shows a high decrease for the char-500 confirming a high cellulose/hemicelluloses pyrolysis extent for this experiment. The signal vanishes completely for the char-550.

As stated previously, from 550 °C, the pyrolysis is nearly completed, which is reflected in the closeness of the char-550, char-600 and char-800 spectra. Beyond 800 °C, the signal intensity is very low through the infrared region of analysis. This denotes a much lower proportion of functional groups on the char surface, which is in agreement with the results of the elemental analysis.

*3.5. Structural Changes of the Biomass Particles as Revealed by Raman Spectroscopy*

The Raman spectra of the different chars are shown in Figure 8. The spectra are normalized according to the G band height. Raman spectra of amorphous biomass chars are known to exhibit two main peaks around 1350–1370 $cm^{-1}$ and 1580–1600 $cm^{-1}$, commonly called the D and G bands.

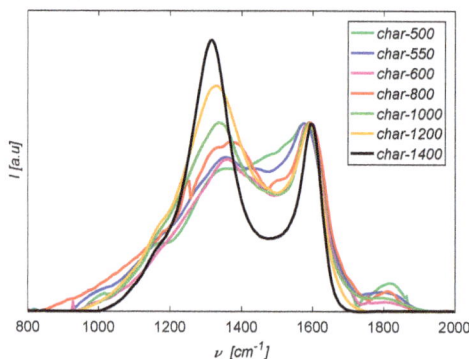

**Figure 8.** Raman spectra of the pyrolysis solid residues.

Increasing the pyrolysis temperature strongly affects the char structure. When increasing the temperature, Raman bands of pyrolysis residues appear as two overlapping but distinguishable peaks at the positions of approximately 1350 and 1600 $cm^{-1}$, which correspond to the in-plane vibrations of $sp^2$-bonded carbon with structural defects D band and the in-plane vibrations of the $sp^2$-bonded graphitic carbon structures G band, respectively [26]. If a high proportion of amorphous carbon structures is present—which is the case for biomass chars—these two bands overlap. This overlapping is associated with hydrogen- and oxygen-rich amorphous carbon structures in the samples. This region (between 1400 and 1550 $cm^{-1}$) is called the valley region "V" [15].

Structural parameters such as the band intensity ratios $\frac{ID}{IG}$, $\frac{IV}{IG}$ or $\frac{IV}{ID}$ are indicators of the char structure. IV represents the valley intensity (taken as the minimum signal intensity between the D and G bands).

These structural parameters are summarized in Table 4. The D band position shifts to lower wavenumbers as pyrolysis temperature increases (from 1363 $cm^{-1}$ at 500 °C to 1327 $cm^{-1}$ at 1400 °C), while the G band position shifts to higher wavenumbers when increasing the pyrolysis temperature (from 1576 $cm^{-1}$ at 500 °C to 1596 $cm^{-1}$ at 1400 °C). This tendency was also observed previously during the characterization of chars prepared by cellulose slow pyrolysis at different temperatures [28]. This "red shift" in the D band peak position is more pronounced for low temperature chars, which have the highest contents of oxygenated defects structures [29].

**Table 4.** Evolution of the Total Raman Area (TRA), D and G band position as well as intensity ratios as a function of the pyrolysis reactor temperature.

| Sample | TRA | Dposition [cm$^{-1}$] | Gposition [cm$^{-1}$] | $\frac{ID}{IG}$ | $\frac{IV}{IG}$ | $\frac{IV}{ID}$ |
|---|---|---|---|---|---|---|
| char-500 | 410.87 ± 10.23 | 1363 ± 4 | 1576 ± 7 | 0.76 ± 0.05 | 0.82 ± 0.05 | 1.08 ± 0.04 |
| char-550 | 418.53 ± 8.01 | 1351 ± 5 | 1576 ± 5 | 0.82 ± 0.03 | 0.74 ± 0.05 | 0.91 ± 0.05 |
| char-600 | 368.25 ± 12.67 | 1357 ± 5 | 1588 ± 6 | 0.81 ± 0.07 | 0.65 ± 0.05 | 0.80 ± 0.03 |
| char-800 | 438.93 ± 15.86 | 1366 ± 6 | 1599 ± 8 | 0.90 ± 0.08 | 0.62 ± 0.05 | 0.69 ± 0.07 |
| char-1000 | 411.66 ± 14.34 | 1336 ± 5 | 1588 ± 6 | 1.01 ± 0.05 | 0.62 ± 0.05 | 0.62 ± 0.05 |
| char-1200 | 362.75 ± 16.22 | 1326 ± 6 | 1590 ± 8 | 1.20 ± 0.06 | 0.63 ± 0.05 | 0.53 ± 0.05 |
| char-1400 | 254.78 ± 12.35 | 1327 ± 7 | 1596 ± 4 | 1.44 ± 0.04 | 0.39 ± 0.05 | 0.27 ± 0.03 |

Furthermore, the intensity in the wavenumber ranges of 800 to 1100 cm$^{-1}$ and 1700 to 1900 cm$^{-1}$ strongly decreases with increasing pyrolysis temperature. This is related to the decrease of the highly reactive structures such as cycloheptane- and cyclooctane-centered ring systems, defective cyclic clusters and aromatic rings with pyrene sizes in the region of 800 to 1100 cm$^{-1}$, and to the pyrolysis of carbonyl bearing structures in the region of 1700 to 1900 cm$^{-1}$ [30]. The decrease of the Raman signal in these regions (together with the valley region which is related to the amorphous carbon structures) is reflected in a decrease of the TRA with temperature, which was also denoted in [16] for cane trash chars, in [30] for mallee wood chars, as well as in [31] for miscanthus chars.

The Raman analysis also shows that the IV/IG ratio decreases sharply between 500 °C and 600 °C (from 0.82 to 0.65), reaches a plateau between 600 °C and 1200 °C, and then decreases sharply between 1200 °C and 1400 °C to reach 0.4. At 1400 °C, the thermal treatment must have been very severe to induce such a brutal change, indicating a probable graphitization in such conditions. McDonald-Wharry et al. [15] observed that the IV/IG height ratios level out at values approaching 0.4 for heat treatment between 700 and 1000 °C. However, the authors used different heat treatment conditions, with a much lower heating rate (7–30 °C/min) and a dwell/hold time of 20 min. Afterwards, the char was allowed to cool in an inert atmosphere to the ambient temperature. The authors think that this value obtained for IV/IG height ratios likely represents the overlap of broad D and G bands in the valley, and may not represent any amorphous carbon content, as might be the case for severe heat treatment.

The ratio $\frac{IV}{ID}$ decreases from 1.08 for the char-500 to 0.27 for the char-1400 which an ordering of the char structure. Moreover, the ID/IG ratio significantly increases with the pyrolysis temperature (from 0.76 at 500 °C to 1.44 at 1400 °C), indicating a higher proportion of condensed aromatic ring structures with defects. These D structures would be formed by the condensation of small aromatic amorphous carbon structures (valley region which intensity highly decreased with temperature). These results are in accordance with the dynamic molecular structure diagram established by Keiluweit et al. [10].

Altogether, an increasing level of order can be noticed in the structure of the char as the pyrolysis temperature increases, with a clearly distinguishable evolution of the Raman spectra of the different chars, especially for the 1400-char sample, the structure of which seems to be highly modified and ordered.

### 3.6. Char Reactivity towards $O_2$

The reactivities of chars towards $O_2$ were examined using thermogravimetric analysis. The obtained results are shown in Figure 9. Plots of the normalized mass ($\frac{m}{m_0} = f(T)$) and negative mass loss rate ($-\frac{dm}{dt} = f(T)$) are presented for raw beech wood and the different prepared chars.

Considering the experiment done on beech wood as reference, the mass loss below 300 °C decreases as pyrolysis temperature increases, indicating lower proportions of low, thermally stable components (reactive carbonaceous structure, incompletely pyrolysed wood) in the chars. The chars prepared at 500 °C and 550 °C (to a lower extent) show peaks (a maximum of degradation rate) at a relatively low temperature of 325 °C, probably indicating the presence of unconverted wood. The final residual mass increases with pyrolysis temperature, which was foreseeable, given that char obtained at higher temperatures has higher mineral content.

The char oxidation characteristics, defined previously in the materials and methods section, are summarized in Table 5. Char-600 unexpectedly shows higher thermal stability compared with char-800 and char-1000. In fact, the temperature corresponding to 50% of conversion is slightly higher for char-600 (405 °C) than for char-800 (379 °C) or char-1000 (389.4 °C). The prolonged char residence time in the reactor after pyrolysis has probably contributes to the thermal stability enhancement. Such behavior is confirmed by examination of the Raman spectra of the different samples. In fact, analysis of the Raman Spectra shows that char-600 has a lower intensity in the regions between 800 and

1200 cm$^{-1}$ and 1700–1900 cm$^{-1}$ when compared to the samples of char-800 and char-1000, respectively. Raman signal in these regions can be associated with the highly reactive structure in the char [29].

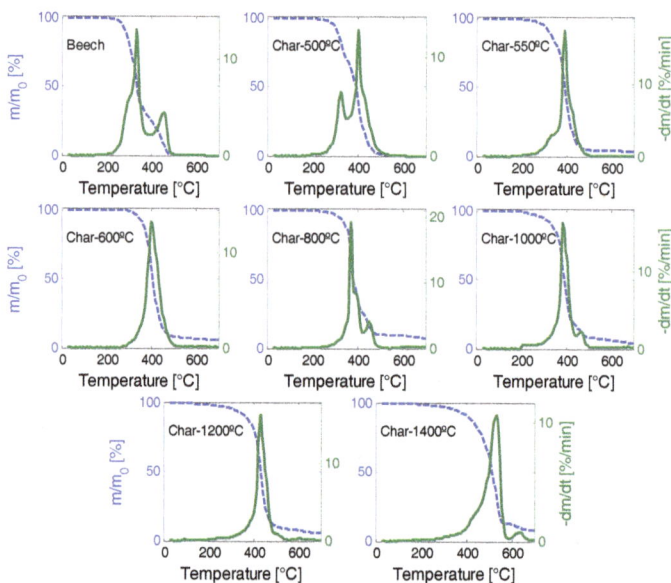

**Figure 9.** Normalized mass as a function of the temperature in the TG air oxidation experiments of the pyrolysis chars.

As shown in Figure 10, although the evolution of $T_{(50\%)}$ is somewhat chaotic between 500 °C and 1000 °C, it increases following a perfect linear relation ($R^2 = 0.9999$) with the char yield for the chars obtained between 1000 °C and 1400 °C. This could be linked to structural ordering and graphitization above 1000 °C.

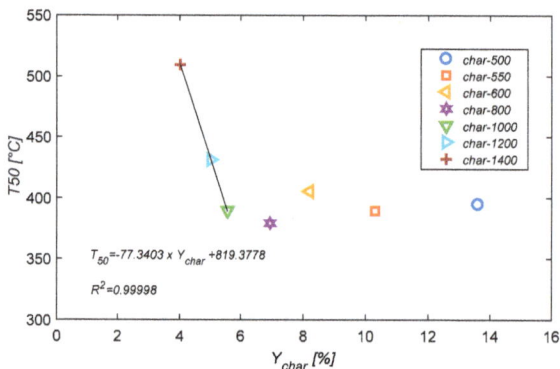

**Figure 10.** Evolution of $T_{50\%}$ with the char yield.

Moreover, the char mean reactivity significantly decreases as pyrolysis temperature increases, going from 0.198 min$^{-1}$ for the char-500 to 0.067 min$^{-1}$ for the char-1400, which represents nearly a threefold decrease.

The reactivity of fast pyrolysis beech bark and beech stick chars obtained between 450 °C and 850 °C were shown to be highly dependent on the pyrolysis temperature [20]. The authors think that at low pyrolysis temperatures, the high H and O contents of the char are associated with the presence of amorphous carbon structures and active sites that increase the char reactivity; whereas, when raising the pyrolysis temperature, the char reactivity decreases due the formation of more aromatic, less functionalized and less reactive structures.

We looked for possible relationships between the different reactivity parameters and the content of heteroatoms in the char (H and O). As shown in Figure 11, we found, for instance, that the mean reaction rate $-\frac{dm}{dt}_{mean}$ is linearly correlated with the O/C ratio. This linear dependence clearly expresses the influence of the surface functional groups containing O atoms on the char oxidation reaction rate.

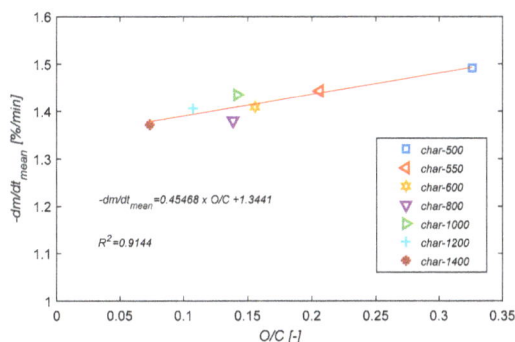

**Figure 11.** Mean mass loss rate of the char in the TG air oxidation experiments as a function of O/C atomic ratio.

Table 5 indicates that Char-1400 clearly presents the highest thermal stability and the lowest reactivity. In particular, Char-1400 has the lowest values for $R_{mean}$ and $R_{index}$ compared to the other samples. This thermal behavior may be linked to a more ordered carbonaceous structure [32], as revealed by the Raman spectroscopy analysis and reflected in the values of ID/IG and IV/IG, which were the highest and lowest, respectively, among all the samples, as well as representing the lowest $\left(\frac{O+H}{C}\right)$ atomic ratio.

**Table 5.** Characteristics of the char TG oxidation experiments.

| Sample | $T_{(2\%)}$ [°C] | $T_{(50\%)}$ [°C] | $T_{(98\%)}$ [°C] | $T_{(-\frac{dm}{dt}\,max)}$ [°C] | $-\frac{dm}{dt}\,max$ [%/min] | $-\frac{dm}{dt}\,mean$ [%/min] | $R_{mean}$ [min$^{-1}$] | $R_{Index}$ [min$^{-1}$] |
|---|---|---|---|---|---|---|---|---|
| Beech | 254.0 | 337.3 | 472.7 | 332.1 | 12.86 | 1.488 | 0.196 | 0.0216 |
| char-500 | 269.6 | 394.6 | 488.3 | 399.8 | 17.53 | 1.488 | 0.198 | 0.0191 |
| char-550 | 264.4 | 389.4 | 483.1 | 389.4 | 13.31 | 1.440 | 0.142 | 0.0265 |
| char-600 | 321.7 | 405.0 | 545.6 | 399.8 | 19.20 | 1.405 | 0.119 | 0.0198 |
| char-800 | 285.2 | 379.0 | 613.3 | 368.5 | 16.27 | 1.378 | 0.074 | 0.0299 |
| char-1000 | 243.5 | 389.4 | 623.8 | 384.2 | 16.35 | 1.430 | 0.077 | 0.0247 |
| char-1200 | 238.3 | 431.0 | 597.7 | 431.0 | 10.62 | 1.402 | 0.074 | 0.0232 |
| char-1400 | 280.0 | 509.2 | 639.4 | 530.0 | 12.86 | 1.369 | 0.067 | 0.0132 |

Moreover, $T_{(50\%)}$ is, remarkably, correlated to the TRA following a linear relation (Figure 12). This relationship, indicating that $T_{(50\%)}$ increases when TRA decreases, makes sense, as the decrease of TRA is mainly due to the decrease of the contribution of the most reactive carbon structures to the Raman signal (amorphous carbon forms in the valley region and highly reactive groups on the two sides of the Raman spectra), which are less present in the char when the pyrolysis treatment is more severe. Other researchers found a linear correlation between $\frac{1}{T_{(50\%)}}$ and the 2490 cm$^{-1}$ band width (second order region in the Raman spectra) for cellulose chars treated between 600 and 2600 °C [19], while others found a linear relationship between $T_{(20\%)}$ and the area ratio of the G band to the TRA [22].

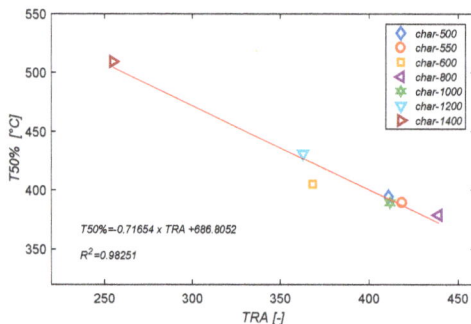

**Figure 12.** Evolution of $T_{(50\%)}$ with the TRA.

Reactivity index can also be correlated to the IV/ID structural ratio, as shown in Figure 13. This evolution here again makes sense, as the more reactive and amorphous small aromatic structures char contains, compared to condensed structures, the higher its reactivity.

Altogether, these oxidation reactivity tests show that the more severe the thermal treatment in the EFR is, the lower the reactivity of the char samples towards O$_2$ will be. The char reactivity, defined through several parameters, is also found to be remarkably correlated with many characteristics of the chars related to their structure and chemical composition.

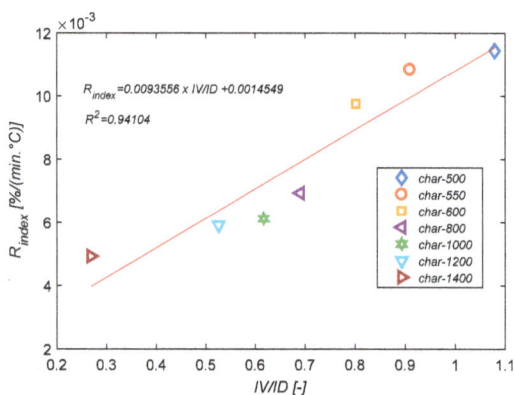

**Figure 13.** Evolution of $R_{(index)}$ with $\frac{IV}{ID}$.

## 4. Conclusions

This study examined the influence of biomass entrained flow pyrolysis conditions on the char properties. It was shown that the pyrolysis reactor temperature highly affects the char yield and properties in terms of morphology, surface chemistry, structure and reactivity. The char yield drops substantially from 14% to 4% when increasing the reactor temperature from 500 °C to 1400 °C. The surface chemistry became poor in surface functional groups beyond 800 °C. The morphology of the chars was highly modified at high temperature with a loss of the initial wood cell architecture, sintering, and macropore formation. Ca was seen to remain on the char while K highly devolatilizes at

1400 °C. Raman analyses showed an increasing ordering of the carbonaceous structures as temperature increases, with a higher content of condensed aromatic structures together with vanishing of the amorphous ones. The char reactivity expressed by means of various parameters equally changed with the pyrolysis temperature. For instance, the mean reactivity decreased nearly threefold for the char-1400 compared to the char-500. Several correlations between reactivity parameters, such as $\frac{dm}{dt}_{mean}$, $T_{(50\%)}$ and $R_{index}$, were seen to be correlated with O/C, TRA and IV/ID, respectively, which evidenced intimate relationships between chemical composition, structure and reactivity. These characterizations also reveal that it is possible to tailor the char properties according to the desired application by controlling the temperature and the particle residence time in an EFR. These two parameters should have the highest impact on the final char properties. We are currently working to further correlate the char properties of these two parameters and to measure their impacts on specific char physical and chemical characteristics.

**Acknowledgments:** The authors thank Jospeh Billaud for providing the char samples (800–1400 °C). They also thank Khouloud Haddad for her help in the acquisition of the Raman spectra.

**Author Contributions:** All authors contributed equally to the work done.

**Conflicts of Interest:** The authors declare no conflict of interest.

## References

1. Markovska, N.; Duić, N.; Vad Mathiesen, B.; Guzović, Z.; Piacentino, A.; Schlör, H.; Lund, H. Addressing the main challenges of energy security in the twenty-first century—Contributions of the conferences on Sustainable Development of Energy, Water and Environment Systems. *Energy* **2016**, *115*, 1504–1512. [CrossRef]
2. Arodudu, O.; Helming, K.; Wiggering, H.; Voinov, A. Bioenergy from low-intensity agricultural systems: An energy efficiency analysis. *Energies* **2017**, *10*, 29. [CrossRef]
3. Chiodo, V.; Zafarana, G.; Maisano, S.; Freni, S.; Urbani, F. Pyrolysis of different biomass: Direct comparison among Posidonia Oceanica, Lacustrine Alga and White-Pine. *Fuel* **2016**, *164*, 220–227. [CrossRef]
4. No, S.-Y. Application of bio-oils from lignocellulosic biomass to transportation, heat and power generation—A review. *Renew. Sustain. Energy Rev.* **2014**, *40*, 1108–1125. [CrossRef]
5. Guizani, C.; Escudero Sanz, F.J.; Salvador, S. Influence of temperature and particle size on the single and mixed atmosphere gasification of biomass char with $H_2O$ and $CO_2$. *Fuel Process. Technol.* **2015**, *134*, 175–188. [CrossRef]
6. Belhachemi, M.; Jeguirim, M.; Limousy, L.; Addoun, F. Comparison of $NO_2$ removal using date pits activated carbon and modified commercialized activated carbon via different preparation methods: Effect of porosity and surface chemistry. *Chem. Eng. J.* **2014**, *253*, 121–129. [CrossRef]
7. Zhu, L.; Shi, T.; Chen, Y. Preparation and characteristics of graphene oxide from the biomass carbon material using fir powder as precursor. *Fuller. Nanotub. Carbon Nanostruct.* **2015**, *23*, 961–967. [CrossRef]
8. Lehmann, J.; Joseph, S. *Biochar for Environmental Management: An Introduction*; Routledge: Abingdon, UK, 2009; pp. 1–14.
9. Brassard, P.; Godbout, S.; Raghavan, V.; Palacios, J.H.; Grenier, M.; Zegan, D. The production of engineered biochars in a vertical auger pyrolysis reactor for carbon sequestration. *Energies* **2017**, *10*, 288. [CrossRef]
10. Keiluweit, M.; Nico, P.S.; Johnson, M.; Kleber, M. Dynamic molecular structure of plant biomass-derived black carbon (biochar). *Environ. Sci. Technol.* **2010**, *44*, 1247–1253. [CrossRef] [PubMed]
11. Rutherford, D.W.; Wershaw, R.L.; Cox, L.G. *Changes in Composition and Porosity Occurring During the Thermal Degradation of Wood and Wood Components*; Scientific Investigations Report 2004-5292; United States Geological Survey: Reston, WV, USA, 2004.
12. Azargohar, R.; Nanda, S.; Kozinski, J.A.; Dalai, A.K.; Sutarto, R. Effects of temperature on the physicochemical characteristics of fast pyrolysis bio-chars derived from Canadian waste biomass. *Fuel* **2014**, *125*, 90–100. [CrossRef]
13. Guerrero, M.; Ruiz, M.P.; Alzueta, M.U.; Bilbao, R.; Millera, A. Pyrolysis of eucalyptus at different heating rates: Studies of char characterization and oxidative reactivity. *J. Anal. Appl. Pyrolysis* **2005**, *74*, 307–314. [CrossRef]

14. Rollinson, A.N. Gasifier reactor engineering approach to understanding the formation of biochar properties. *Proc. A R. Soc.* **2016**, *472*. [CrossRef] [PubMed]

15. McDonald-Wharry, J.; Manley-Harris, M.; Pickering, K. Carbonisation of biomass-derived chars and the thermal reduction of a graphene oxide sample studied using Raman spectroscopy. *Carbon* **2013**, *59*, 383–405. [CrossRef]

16. Keown, D.M.; Li, X.; Hayashi, J.I.; Li, C.Z. Characterization of the structural features of char from the pyrolysis of cane trash using Fourier transform-Raman spectroscopy. *Energy Fuels* **2007**, *21*, 1816–1821. [CrossRef]

17. Guizani, C.; Haddad, K.; Limousy, L.; Jeguirim, M. Document New insights on the structural evolution of biomass char upon pyrolysis as revealed by the Raman spectroscopy and elemental analysis. *Carbon* **2017**, *119*, 519–521. [CrossRef]

18. Trubetskaya, A.; Jensen, P.A.; Jensen, A.D.; Steibel, M.; Spliethoff, H.; Glarborg, P. Influence of fast pyrolysis conditions on yield and structural transformation of biomass chars. *Fuel Process. Technol.* **2015**, *140*, 205–214. [CrossRef]

19. Brewer, C.E.; Schmidt-rohr, K.; Satrio, J.A.; Brown, R.C. Characterization of biochar from fast pyrolysis and gasification systems. *Environ. Prog. Sustain. Energy* **2009**, *28*, 386–396. [CrossRef]

20. Morin, M.; Pécate, S.; Hémati, M.; Kara, Y. Pyrolysis of biomass in a batch fluidized bed reactor: Effect of the pyrolysis conditions and the nature of the biomass on the physicochemical properties and the reactivity of char. *J. Anal. Appl. Pyrolysis* **2016**, *122*, 511–523. [CrossRef]

21. Zaida, A.; Bar-Ziv, E.; Radovic, L.R.; Lee, Y.-J. Further development of Raman Microprobe spectroscopy for characterization of char reactivity. *Proc. Combust. Inst.* **2007**, *31*, 1881–1887. [CrossRef]

22. Sheng, C. Char structure characterised by Raman spectroscopy and its correlations with combustion reactivity. *Fuel* **2007**, *86*, 2316–2324. [CrossRef]

23. Billaud, J.; Valin, S.; Peyrot, M.; Salvador, S. Influence of $H_2O$, $CO_2$ and $O_2$ addition on biomass gasification in entrained flow reactor conditions: Experiments and modelling. *Fuel* **2016**, *166*, 166–178. [CrossRef]

24. Chen, L. Fast Pyrolysis of Millimetric Wood Particles between 800 °C and 1000 °C. Ph.D. Thesis, Claude Bernard Lyon I, Villeurbanne, France, December 2009.

25. Billaud, J.; Valin, S.; Ratel, G.; Thiery, S.; Salvador, S. Biomass gasification between 800 and 1400 °C in the presence of $O_2$: Drop tube reactor experiments and simulation. *Chem. Eng. Trans.* **2014**, *37*, 163–168. [CrossRef]

26. Paethanom, A.; Yoshikawa, K. Influence of pyrolysis temperature on rice husk char characteristics and its tar adsorption capability. *Energies* **2012**, *5*, 4941–4951. [CrossRef]

27. Crombie, K.; Mašek, O.; Sohi, S.P.; Brownsort, P.; Cross, A. The effect of pyrolysis conditions on biochar stability as determined by three methods. *GCB Bioenergy* **2013**, *5*, 122–131. [CrossRef]

28. Guizani, C.; Valin, S.; Billaud, J.; Peyrot, M.; Salvador, S. Biomass fast pyrolysis in a drop tube reactor for bio oil production: Experiments and modeling. *Fuel* **2017**, in press.

29. Smith, M.W.; Dallmeyer, I.; Johnson, T.J.; Brauer, C.S.; McEwen, J.S.; Espinal, J.F.; Garcia-Perez, M. Structural analysis of char by Raman spectroscopy: Improving band assignments through computational calculations from first principles. *Carbon* **2016**, *100*, 678–692. [CrossRef]

30. Asadullah, M.; Zhang, S.; Li, C.Z. Evaluation of structural features of chars from pyrolysis of biomass of different particle sizes. *Fuel Process. Technol.* **2010**, *91*, 877–881. [CrossRef]

31. Elmay, Y.; Le Brech, Y.; Delmotte, L.; Dufour, A.; Brosse, N.; Gadiou, R. Characterization of miscanthus pyrolysis by DRIFTs, UV Raman spectroscopy and mass spectrometry. *J. Anal. Appl. Pyrolysis* **2015**, *113*, 402–411. [CrossRef]

32. Asadullah, M.; Zhang, S.; Min, Z.; Yimsiri, P.; Li, C.Z. Effects of biomass char structure on its gasification reactivity. *Bioresour. Technol.* **2010**, *101*, 7935–7943. [CrossRef] [PubMed]

*Article*

# Volume and Mass Measurement of a Burning Wood Pellet by Image Processing

**Sae Byul Kang [1,\*], Bong Suk Sim [2] and Jong Jin Kim [1]**

[1]   Energy Network Laboratory, Korea Institute of Energy Research, Daejeon 34129, Korea; jjkim@kier.re.kr
[2]   Research and Development Team, Doowon, Choongnam 336864, Korea; bssim@dwdcc.co.kr
\*    Correspondence: byulkang@kier.re.kr; Tel.: +82-42-860-3321

Academic Editors: Mejdi Jeguirim and Lionel Limousy
Received: 2 March 2017; Accepted: 12 April 2017; Published: 1 May 2017

**Abstract:** Wood pellets are a form of solid biomass energy and a renewable energy source. In 2015, the new and renewable energy (NRE) portion of wood pellets was 4.6% of the total primary energy in Korea. Wood pellets account for 6.2% of renewable energy consumption in Korea, the equivalent of 824,000 TOE (ton of oil equivalent, 10 million kcal). The burning phases of a wood pellet can be classified into three modes: (1) gasification; (2) flame burning and (3) charcoal burning. At each wood pellet burning mode, the volume and weight of the burning wood pellet can drastically change; these parameters are important to understand the wood pellet burning mechanism. We developed a new method for measuring the volume of a burning wood pellet that involves no contact. To measure the volume of a wood pellet, we take pictures of the wood pellet in each burning mode. The volume of a burning wood pellet can then be calculated by image processing. The difference between the calculation method using image processing and the direct measurement of a burning wood pellet in gasification mode is less than 8.8%. In gasification mode in this research, mass reduction of the wood pellet is 37% and volume reduction of the wood pellet is 7%. Whereas in charcoal burning mode, mass reduction of the wood pellet is 10% and volume reduction of the wood pellet is 41%. By measuring volume using image processing, continuous and non-interruptive volume measurements for various solid fuels are possible and can provide more detailed information for CFD (computational fluid dynamics) analysis.

**Keywords:** combustion; image processing; renewable; volume measurement; wood pellet

---

## 1. Introduction

With increased fossil fuel prices, the heating costs of households and industry have increased rapidly. Many countries are interested in using renewable energy to decrease consumption of fossil fuel. Wood pellet is a kind of solid biomass energy and a renewable energy source. In 2015, the new and renewable energy portion of wood pellets was 4.6% with respect to the total primary energy in Korea. Most of the renewable energy in Korea is from waste energy, reaching 63.5%. The other renewable energy sources have much smaller contributions, with the bioenergy portion being 20.8%, wind energy 2.1%, solar power 6.4% and solar heat 0.2% [1]. In 2016, the total amount of wood pellet produced in Korea was 52.7 thousand tons and the imported wood pellet amount was 1.72 million tons which was mainly used in power plants for co-firing with coal [2]. Most of Korea's imported wood pellet is from Vietnam, Malaysia and Indonesia. The amount of imported wood pellet into Korea was about 8.5% of the amount of wood pellet consumed in the EU in 2015 (20.3 million tons) [3].

European countries such as Germany, Austria, and Sweden have developed wood pellet boilers and have common norms around the solid biomass boiler, EN 303-5:2012, which classifies and describes the requirements and testing methods of the solid biomass boiler [4]. Fiedler reviewed the state of the art of, small scale pellet-based heating system and relevant regulation in Sweden, Austria and

Germany. He categorized pellet central heating boilers and burners [5]. Recently, lots of studies related to pellet burning use mixed pellet with other solid fuel, such as coal [6], agricultural residue [7,8], bamboo [9,10] and potato pulp [11].

Wood combustion is a complex process that involves many physical and chemical processes. Burning phases of wood pellets can be classified into three modes: (1) gasification; (2) flame burning; and (3) charcoal burning. At each wood pellet burning mode, the volume and weight of the burning wood pellets can drastically change and these parameters are therefore important in understanding the wood pellet burning mechanism. Volume change and burning time at each mode of wood pellet can be important data in designing the combustor of the wood pellet boiler, especially the combustor with the moving stoker type. The main design parameters of moving stoker type wood pellet combustor are (1) shape of grate; (2) number of grate; (3) velocity and moving interval of grate; (4) and combustion air distribution. Volume change and duration of each combustion mode of wood pellet can be useful in determining those design parameters.

Thunman et al. presented a particle model of combustion of wood particles for Eulerian calculations. They performed calculations and compared the results with experimental results for more than 60 samples of particles of different sizes. The model they developed shows the strong influence of shrinkage on devolatilization and char combustion times [12]. Park et al. [13] performed an experimental and theoretical investigation of heat and mass transfer processes during wood pyrolysis. They performed pressure calculations based on a new pyrolysis model and revealed that high pressure is generated inside the biomass particles during pyrolysis. Moreover, sample splitting was observed during the experiments. The splitting is due to both weakening of the structure and the internal pressure generation that takes place during pyrolysis. At low heating rates, structural weakness is the primary factor, whereas at high heating rates, internal pressure is the determining factor [14]. Hshieh and Richards reported the effects of preheating wood on the ignition temperature of wood char. They found that the preheated samples were further heated in air at 5 °C/min to unpiloted ignition. Despite major chemical changes during the various preheating treatments, the ignition temperatures were not significantly affected, except for a slight decrease (11 °C) after five days at 150 °C in air [14]. Ward and Braslaw reported experimental weight loss kinetics of wood pyrolysis under a vacuum. The results showed that kinetic parameters can be used to predict the volatilization rates of wood as a function of temperature in a vacuum. The model can also be used to estimate the quantities of each of the main components initially present in an unknown wood sample [15]. Lautenberger and Fernandez-Pello studied a model for the oxidative pyrolysis of wood. They performed optimized model calculations for mass loss rate, surface temperature, and depth temperatures and reproduced the experimental data well, including the experimentally observed increase in temperature and mass loss rate with increasing oxygen concentration [16]. Kung and Kalelkar determined the heat of the pyrolysis reaction with a mathematical model [17]. Biswas et al. studied on effect of pelletizing conditions on combustion behavior of single wood pellet. They measured a single burning pellet weight by a precision scale during combustion test. They found that time for a single pellet combustion increased with pelletizing temperature and time for flame combustion was about 50 s; time for char oxidation was about 100 s [18]. Fagerström et al. also measured the weight of the wood pellet during the combustion test using a precision scale to determine the release of ash forming elements after the devolatilization phase and the char combustion [19]. Ström and Thunman developed a numerical model for drying and devolatilization of moist wood particles in an inert condition [20]. The results from the CFD calculation were compared with previous experimental data and both were in alignment. Sengupta et al. measured yarn parameters by an image processing technique with a low cost web camera. They measured diameter, diameter variation, and number of thick/thin places [21]. Taghavifar and Mardani measured the contact area of a radial ply tire with the image processing method [22]. They captured RGB data of tire contact area image then converted it to hue, saturation and value (HSV) data to calculate tire contact area.

Arce et al. [23] investigated wood pellet combustion characteristics on a fixed bed reactor (FBR) with various conditions. They concluded that the particle size of the wood pellet and water content are major parameters on combustion and heat transfer. Moradian et al. [24] performed the combustion test on a fluidized bed boiler with fuel of a normal solid waste and compared it to mixing with animal wastes. They found that solid waste comprised of 20–30% animal waste reduced the bed temperature by 70–100 °C and suppressed the deposition growth rate. Chun et al. [25] investigated pyrolysis and gasification characteristics of sewage sludge with a high water and nitrogen content. Their study focused on high quality gas and char production.

Gomez et al. [26] did a simulation study on the effect of water temperature on domestic biomass boiler performance. Their results concluded that low water temperature increased heating performance but also increased CO emission.

The main purpose of the present study was to measure the density and volume of a wood pellet without disturbing combustion condition. Previous studies use a precise electric scale to measure the weight of the burning wood pellet [13,15,18,19]. In this study, we calculate the volume of the burning wood pellet based on the image processing method and we also determine the mass of the burning wood pellet with measured density of the bare wood pellet and the charcoal of the wood pellet. We took combustion state photographs for each mode by recoding video. In order to investigate the volume reduction rate of a wood pellet in combustion, moving images of wood pellet combustion were captured. From the captured moving images, we calculated the volume of the wood pellet at each time point. This research will provide useful background for applying combustor design, analysis, and operation to wood pellet boilers.

## 2. Method

In an image file format, such as JPG, $N \times M \times 3$ matrix data are used to construct an image. Each $N \times M$ matrix has the information of the red, green and blue (RGB) data of the image, the three elements of color. In the matrix, the first $N \times M$ matrix has $x$ and $y$ coordinate data. In the image file, the RGB data can have values from 0 to 255, which is based on 8 bit data.

With the RGB data from an image file, we developed a volume calculation method for an axisymmetric object. If the diameter of the finite volume of an axisymmetric object is known, the volume can be calculated by integration of the finite volume. The radius of an axisymmetric object can be obtained by image processing. With image data, if the border of the axisymmetric object is found, the radius and finite height of a small volume of the axisymmetric object can be obtained. With the obtained border data of the axisymmetric object from the image file, the volume of the axisymmetric object can be calculated.

Figure 1 shows a sample image file of a wood pellet and the lines represent the following: $x = 9$, 150 and 252 pixel. Along the line in the picture, RGB data can be plotted as shown in Figure 2. On the $x = 9$ line in the image data, there is no wood pellet image but only background. Therefore, the RGB data along the line at $x = 9$ should not be changed abruptly. In contrast, on the $x = 150$ line in the image data, both the background and the wood pellet are present. Therefore, if RGB data along the line at $x = 150$ are plotted, there must be two abrupt changes at each border of the wood pellet (Figure 3). On the line at $x = 252$ in the image file, both the background and a thin pin supporting the wood pellet are present. Again, there must be two abrupt changes on the RGB line along the $x = 252$ line (Figure 4). However, the distance between the two changes is shorter than the distance of the $x = 150$ line.

Figure 5 shows a wood pellet in flame burning mode. There is a flame in the middle of the wood pellet. Figure 6 represents RGB data along $x = 150$ of wood pellet of Figure 5. Compared to the RGB data in Figure 3 of the wood pellet in gasification mode, the absolute values of RGB data of the wood pellet in flame burning mode differ from each other, such as at $y = 20$ $R$ value is about 200, $G$ value is about 140 and $B$ value is about 100. However, there are also two abrupt changes of RGB data at the $x = 150$ line, as seen in Figure 6. Thus, we can obtain a border line of the wood pellet in flame burning mode.

If the boundary of the wood pellet is obtained, the volume of the pellet (an axisymmetric object) can be assumed to be an infinitesimal cylinder. However, if the axis of symmetry is not parallel with respect to the *x*-axis, the volume calculation will have error due to an incorrect estimation of the diameter. It is very difficult to align a wood pellet with a vertical line. In Figure 1, the wood pellet is tilted 8.2 degrees from a vertical line. In the error calculation, the diameter and height of the wood pellet are 6 mm and 22 mm, respectively. If the misalignment angle is less than 10 degrees, the maximum volume calculation error will be less than 0.5%. In the volume calculation, coordinate transformation is used as follows:

$$\begin{bmatrix} x' \\ y' \end{bmatrix} = \begin{bmatrix} \cos(\theta) & -\sin(\theta) \\ \sin(\theta) & \cos(\theta) \end{bmatrix} \begin{bmatrix} x \\ y \end{bmatrix} \tag{1}$$

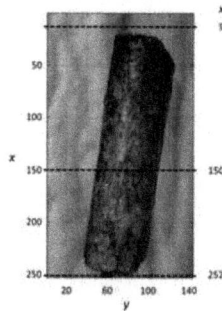

**Figure 1.** Image file of a wood pellet in gasification mode and lines of $x = 9$, 150 and 252.

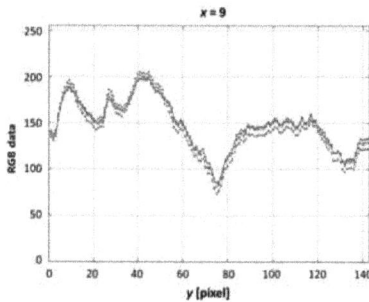

**Figure 2.** Red, Green and Blue (RGB) data of image file of wood pellet (Figure 1) along $y$ pixel at $x = 9$.

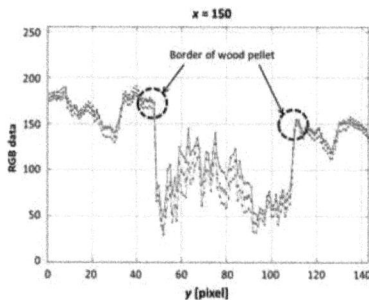

**Figure 3.** RGB data of image file of wood pellet (Figure 1) along $y$ pixel at $x = 150$.

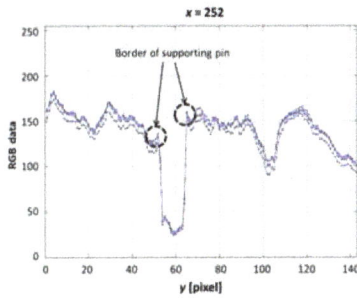

**Figure 4.** RGB data of image file of wood pellet (Figure 1) along $y$ pixel at $x = 252$.

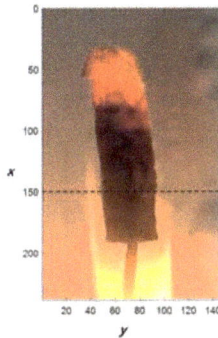

**Figure 5.** Image file of a wood pellet in flame burning mode and line of $x = 150$.

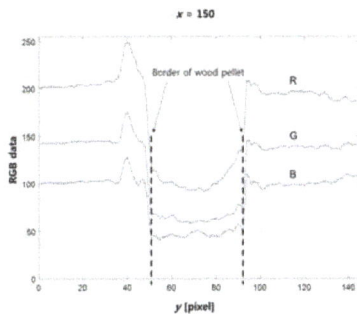

**Figure 6.** RGB data of image file of wood pellet in flame burning mode (Figure 5) along $y$ pixel at $x = 150$.

Figure 7 shows (a) a raw image file of a wood pellet; (b) the border line of the wood pellet and (c) the border line of the wood pellet with rotation. As seen in this figure, the border line of the pellet is precisely found.

To confirm the accuracy of this volume calculation method, we compare the volume of an "AAA" type battery that is manually measured by vernier caliper with one that is measured by the image processing method.

Figure 8 shows a picture of the battery and the captured boundary result by the image processing program. The diameter and length of the "AAA" type battery are 10.27 mm and 42.39 mm, respectively. If the battery is assumed to be asymmetrical, the volume of the battery is 3511.51 mm$^3$. By the image processing method, the volume of the battery is calculated as 3502.37 mm$^3$ with $-0.26\%$ error. With this validation, volume calculation by the image processing method gives very accurate results.

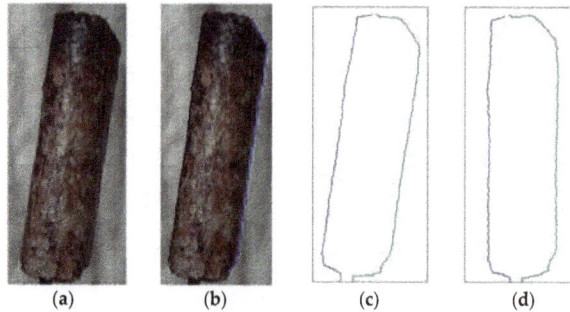

(a)       (b)       (c)       (d)

**Figure 7.** Wood pellet image data and image process results (finding border of wood pellet). (**a**) Raw image file; (**b**) Image file with border line; (**c**) Border line; (**d**) Border line with rotation.

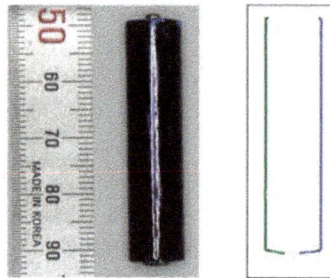

**Figure 8.** Picture of "AAA" type battery and image process result.

## 3. Experimental Results and Discussion

### 3.1. Test Facility

We performed an experimental study to measure the volume and mass of a burning wood pellet and captured images of a burning wood pellet. First, a wood pellet combustion facility was manufactured. A schematic diagram of the combustion facility is shown in Figure 9. To measure the air flow rate, a turbine type gas flow meter was used.

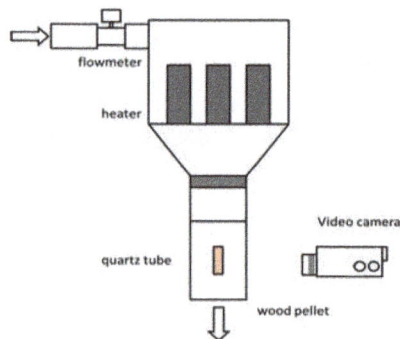

**Figure 9.** Schematic diagram of experimental facility.

The combustion apparatus has $3 \times 1.6$ kW electric heaters with a fan and $2 \times 0.5$ kW bar type heaters to raise the air temperature. Air with a high temperature of up to 600 °C passes through a honeycomb mesh and a transparent quartz tube where a tested wood pellet is placed vertically. Air temperature is measured by a K-type thermocouple. Images of the burning wood pellet are captured by a digital video camera (SONY HDR-CX550). In the image processing program, a video image file is converted into snapshot image files at 0.5 s intervals. The combustion chamber maintained a steady state throughout the test.

## 3.2. Volume Measurement Results

Table 1 shows proximate and ultimate analysis results of the tested wood pellet and Table 2 represents wood pellet size and test condition.

**Table 1.** Proximate and ultimate analysis results of tested wood pellet. (Korea Institute of Energy Research, Testing and Certification Center).

| Higher Heating Value at Dry (MJ/kg, kcal/kg) | | 18.6 (4550) |
|---|---|---|
| Proximate analysis (weight %, as received) | Moisture | 8.73 |
| | Volatile | 73.98 |
| | Ash | 0.40 |
| | Fixed carbon | 16.89 |
| Ultimate analysis (weight %, as dried and ash free) | Carbon | 49.90 |
| | Hydrogen | 6.21 |
| | Nitrogen | 0.38 |
| | Oxygen | 43.10 |
| | Sulfur | 0.01 |

**Table 2.** Wood pellet size and air conditions.

| Wood Pellet | | Air Condition | |
|---|---|---|---|
| mass (g) | 0.8065 | $Q_{air}$ (Nm$^3$/h) | 11.2 |
| length (mm) | 22.01 | $T_0$ (°C) | 20.1 |
| diameter (mm) | 6.12 | $T_h$ (°C) | 544.7 |

In the combustion of the wood pellet, there are three modes: (1) a gasification mode, (2) a flame burning mode, and (3) a charcoal burning mode. When the wood pellet is inserted into the combustion chamber with $T_h > 400$ °C, there is no significant change in its appearance except that the surface color of the pellet becomes dark. This period is the gasification mode. The wood pellet is heated during this period and suddenly a flame starts from the head of the pellet; this flame then expands to the entire wood pellet due to vigorous volatilization. This is the flame burning mode. If there is no flame the charcoal burning mode starts. From the insertion of the wood pellet to the end of the charcoal mode, a video sequence of the burning wood pellet is captured and the volume of the wood pellet is subsequently calculated with an image process. To confirm the image processing result, we also measured nine wood pellet volumes at each mode every 20 s (four wood pellets in gasification mode, one wood pellet in burning mode, and four wood pellets in charcoal mode). Table 3 shows images of wood pellet at each combustion mode and converted images.

Figure 10 shows the volume change results of a burning wood pellet by an off-line and an on-line measurement. The on-line measurement refers to the measurement by the image processing method, which can measure the volume of the burning wood pellet continuously.

The off-line measurement entails measuring the volume of the burning wood pellet by intermittent sampling and it is repeated twice for each off-line measurement. For example, to check the volume of the burning wood pellet at 80 s, we prepare a new wood pellet and put it in the combustion chamber for 80 s, and then remove the burning chamber from the combustion chamber. After cooling the burning wood pellet, we measure the mass of the pellet on a scale and take pictures to measure the volume of the pellet. In the off-line measurement, there is a possibility of error in mass and volume measurements due to mass reduction during cooling of the burning wood pellet and measurement period. There is a difference in the volume between on-line and off-line measurements, especially in charcoal mode; the reason for this is shown in Figure 10.

**Table 3.** Images of wood pellet at each combustion mode.

| Time/Volume | Image | Converted Image |
|---|---|---|
| Gasification<br>7.5 s<br>566.6 mm$^3$ | | |
| Flame burning<br>80 s<br>378.2 mm$^3$ | | |
| Charcoal burning<br>120 s<br>176.9 mm$^3$ | | |

Table 3. *Cont.*

| Time/Volume | Image | Converted Image |
|---|---|---|
| Charcoal burning<br>180 s<br>50.8 mm³ |  |  |

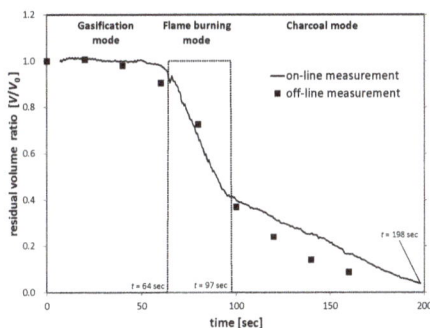

**Figure 10.** Comparison of volume measurement between off-line and on-line measurement.

At the beginning of the gasification mode, there is a volume increase caused by evaporation of water and pyrolysis inside the wood pellet (Figure 10). In the gasification mode, the volume change is very low. However, in the flame burning mode, an abrupt volume change occurs due to high energy emission by flame burning. The charcoal burning mode takes a long time: up to 51.6% of the total burn time. With this volume measurement method, we can easily and continuously measure volume change of the burning wood pellet, without any disturbance of combustion condition, from gasification mode to the end of combustion. At the end of combustion, the wood pellet's volume was 5% of the original pellet. We compare standardized char conversion time (char conversion time divided by density of the wood pellet) to a previous study by Biswas et al. [18]. Char conversion time is defined as the period starting at the moment when the flame around the pellet disappeared until the moment that the glowing of the char stopped [18]. Biswas et al. concluded that standardized char conversion times of the wood pellet with apparent density up to 1200 kg/m³ remain constant about 80–100 cm³·s/g. In the present study, standardized char conversion time is 82.3 cm³·s/g, as shown in Figure 11.

During off-line measurement, we measure volume and mass of the wood pellet at each sampling. From the measurement, we can find the density of the wood pellet at each sampling, i.e., density is mass divided by volume. Figure 12 shows wood pellet density measurement results and data fitting for calculation of the wood pellet. From the beginning of gasification mode to the end of flame burning mode, the density of wood pellet decreases as a quadratic function. On the contrary, density of the wood pellet at charcoal mode is almost constant and density average value is 307.5 kg/m³. So we assume density as a quadratic function that passes through $\rho_{\text{wood pellet}}$ (1213 kg/m³) at $t = 0$ s and $\rho_{\text{charcoal}}$ (307.5 kg/m³) at the end of flame burning mode ($t = 94$ s) as the following equation:

$$\rho = a\,t^2 + \rho_{charcoal} = 0.09627\,t^2 + 1213 \tag{2}$$

Mass measurement by off-line (black square point) and calculation results of the mass of the burning wood pellet based on volume measurement (line) are shown in Figure 13. The mass of the burning wood pellet is calculated by multiplication volume measured by on-line method and density fitted, as shown in Figure 12. The dash line represents the previous research result of residual mass ratio by Biswas et al. [18]. Test conditions between the present study and previous studies differ from each other, especially air temperature. Air temperature in the present study is about 545 °C and 800 °C in a previous study. With two data sets (the result from the present study and the result by Biswas et al.), we can confirm the end time of flame burning. The end time of flame burning of the present study is 97 s, and that of Biswas et al.'s is 50 s. So we add 47 s with time data of Biswas et al. In spite of different test conditions, charcoal mode results are in alignment. However, air temperature of Biswa et al.'s study is higher than that of the present study, so duration of gasification mode of Biswas et al. is less than 5 s. In gasification mode of the present study, residual mass ratio is dropped from 1.0 to 0.63 and in flame burning mode, residual mass ratio is dropped from 0.63 to 0.10. Compared to residual volume ratio result in Figure 10, gasification modes differ from each other. Before flame burning, volume change is less than 7% p (from 100% to 93%). However, mass reduction is about 37% p (from 100% to 63%). Figure 14 shows the volume and mass reduction rate of the burning wood pellet with respect to each burning mode. Volume and mass reduction rates during charcoal burning mode drastically differ from each other.

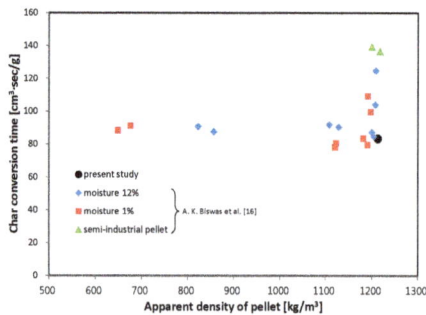

**Figure 11.** Comparison of standardized char conversion time (cm³·s/g) between result of present study and previous research data by Biswas et al. [18].

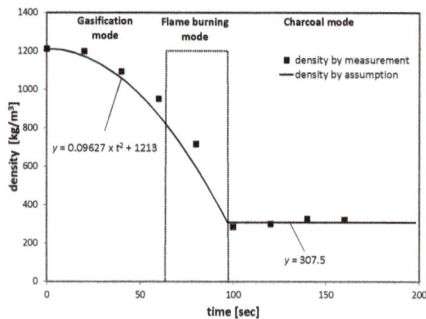

**Figure 12.** Wood pellet density measurement results and data fitting for mass calculation of wood pellet at each time.

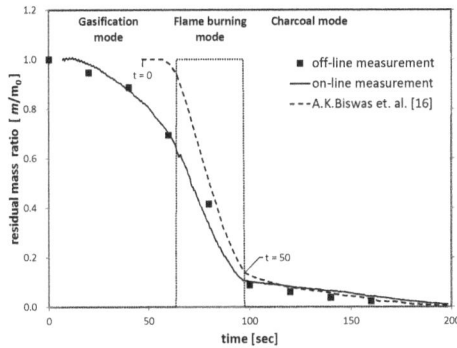

**Figure 13.** Residual mass ratio by off-line and on-line measurement with respect to elapsed time and previous research data by Biswas [18].

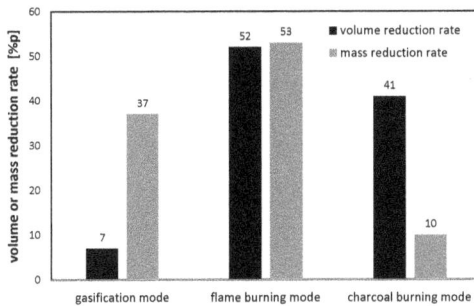

**Figure 14.** Comparison between volume and mass reduction rate of burning wood pellet.

At the end stage of wood pellet burning, 41% of the volume of the original wood pellet remains, however, only 10% of the mass remains. In this study, charcoal burning mode takes 101 s, which is 51% of the whole combustion period. Conversely, flame burning mode takes 33 s, which is 17% of the whole combustion period. Given this result, it is evident that the wood pellet combustor promotes a relatively long period of charcoal burning mode with a bulky volume of charcoal.

## 4. Conclusions

In this research, we developed a volume measurement method for a burning wood pellet. With a video camera, images of a burning wood pellet are continuously captured and then converted to image files to calculate the volume of the wood pellet at each time point. Each of the off-line volume measurements with division into eight parts of complete burning were conducted twice and compared with on-line measurement data. Wood pellets used in the combustion test were 1st grade in Korea. Conclusions and findings of the research are as follows:

1. Compared to the volume measurement taken by a ruler (vernier caliper), the error of the developed volume measurement method is −0.26%.
2. With the image processing method, we can measure the instantaneous volume of the burning wood pellet with high precision without disturbance of the combustion environment.
3. Three combustion modes of the wood pellet can easily be distinguished by the image processing program. Gasification mode takes 32.3% of the total burn time, flame burning mode takes 16.7%, and charcoal burning mode takes 51.0% under the tested conditions.

4. From gasification mode to end of flame burning mode, density of burning wood pellet decreases as a quadratic function. In charcoal mode, the density of the burning wood pellet is constant at about 307.5 kg/m$^3$.

5. In gasification mode in this research, mass reduction of the wood pellet is 37% and volume reduction of the wood pellet is 7%. On the contrary, in charcoal burning mode, mass reduction of the wood pellet is 10% and volume reduction of the wood pellet is 41%. Relatively large volume reduction, small mass reduction and long burning duration in charcoal burning mode should be considered in designing a wood pellet combustor.

**Acknowledgments:** This work was conducted under the framework of Research and Development Program of the Korea Institute of Energy Research (KIER, Korea) (B7-2414-02).

**Author Contributions:** Sae Byul Kang and Jong Jin Kim conceived and designed the experiments. Bong Suk Sim conducted experiment on combustion characteristics and measurement of size of combusted pellets. Calculation of volume of a wood pellet was done by Sae Byul Kang.

**Conflicts of Interest:** The authors declare no conflict of interest.

## References

1. Korea Energy Agency. Overview of New and Renewable Energy in Korea 2015. 2016. Available online: http://www.knrec.or.kr/knrec/14/KNREC140310.asp?idx=80&page=1&num=24&Search=&SearchString=# (accessed on 16 April 2017).

2. Forest Biomass Energy Association (Korea). 2016. Available online: http://www.biomassenergy.kr (accessed on 16 April 2017).

3. European Biomass Association (AEBIOM). *AEBIOM Statistical Report 2016*; European Biomass Association: Brussel, Belgium, 2016.

4. European Committee for Standardization. *EN-303–5:2012 Heating Boilers—Part 5: Heating Boilers for Solid Fuels, Hand and Automatically Stocked, Nominal Heat Output of up to 500 kW—Terminology, Requirements, Testing and Marking*; European Committee for Standardization: Brussel, Belgium, 2012.

5. Fiedler, F. The state of the art of small-scale pellet-based heating systems and relevant regulations in Sweden, Austria and Germany. *Renew. Sustain. Energy Rev.* **2004**, *8*, 201–221. [CrossRef]

6. Finney, K.N.; Sharifi, V.N.; Swithenbank, J. Combustion of spent mushroom compost and coal tailing pellets in a fluidized bed. *Renew. Energy* **2009**, *34*, 860–868. [CrossRef]

7. Niedziołka, I.; Szpryngiel, M.; Kachel-Jakubowska, M.; Kraszkiewicz, A.; Zawislak, K.; Sobczak, P.; Nadulski, R. Assessment of the energetic and mechanical properties of pellets produced from agricultural biomass. *Renew. Energy* **2015**, *76*, 312–317. [CrossRef]

8. Barbanera, M.; Lascaro, E.; Stanzione, V.; Esposito, A.; Altieri, R.; Bufacchi, M. Characterization of pellets from mixing olive pomace and olive tree pruning. *Renew. Energy* **2016**, *88*, 185–191. [CrossRef]

9. Liu, Z.; Liu, X.; Fei, B.; Jiang, Z.; Cai, Z.; Yua, Y. The properties of pellets from mixing bamboo and rice straw. *Renew. Energy* **2013**, *55*, 1–5. [CrossRef]

10. Liu, Z.; Mi, B.; Jiang, Z.; Fei, B.; Cai, Z.; Liu, X. Improved bulk density of bamboo pellets as biomass for energy production. *Renew. Energy* **2016**, *86*, 1–7. [CrossRef]

11. Obidzinski, S.; Piekut, J.; Dec, D. The influence of potato pulp content on the properties of pellets from buckwheat hulls. *Renew. Energy* **2016**, *87*, 289–297. [CrossRef]

12. Thunman, H.; Leckner, B.; Niklasson, F.; Johnsson, F. Combustion of wood particles—A particle model for Eulerian calculations. *Combust. Flame* **2002**, *129*, 30–46. [CrossRef]

13. Park, W.C.; Atreya, A.; Baum, H.R. Experimental and theoretical investigation of heat and mass transfer processes during wood pyrolysis. *Combust. Flame* **2010**, *157*, 481–494. [CrossRef]

14. Hshieh, F.Y.; Richards, G.N. The effect of preheating of wood on ignition temperature of wood char. *Combust. Flame* **1989**, *80*, 395–398. [CrossRef]

15. Ward, S.M.; Braslaw, J. Experimental weight loss kinetics of wood pyrolysis under vacuum. *Combust. Flame* **1985**, *61*, 261–269. [CrossRef]

16. Lautenberger, C.; Fernandez-Pello, C. A model for the oxidative pyrolysis of wood. *Combust. Flame* **2009**, *156*, 1503–1513. [CrossRef]

17. Kung, H.C.; Kalelkar, A.S. On the heat of reaction in wood pyrolysis. *Combust. Flame* **1973**, *20*, 91–103. [CrossRef]

18. Biswas, A.K.; Rudolfsson, M.; Broström, M.; Umeki, K. Effect of pelletizing conditions on combustion behaviour of single wood pellet. *Appl. Energy* **2014**, *119*, 79–84. [CrossRef]

19. Fagerström, J.; Steinvall, E.; Boström, D.; Boman, C. Alkali transformation during single pellet combustion of soft wood and wheat straw. *Fuel Process. Technol.* **2016**, *143*, 204–212. [CrossRef]

20. Ström, H.; Thunman, H. CFD simulations of biofuel bed conversion: A submodel for the drying and devolatilization of thermally thick wood particles. *Combust. Flame* **2013**, *160*, 417–431. [CrossRef]

21. Sengupta, A.; Roy, S.; Sengupta, S. Development of a low cost yarn parameterisation unit by image processing. *Measurement* **2015**, *59*, 96–109. [CrossRef]

22. Taghavifar, H.; Mardani, A. Potential of functional image processing technique for the measurements of contact area and contact pressure of a radial ply tire in a soil bin testing facility. *Measurement* **2013**, *46*, 4038–4044. [CrossRef]

23. Arce, M.E.; Saavedra, Á.; Míguez, J.L.; Granada, E.; Cacabelos, A. Biomass fuel and combustion conditions selection in a fixed bed combustor. *Energies* **2013**, *6*, 5973–5989. [CrossRef]

24. Moradian, F.; Pettersson, A.; Svärd, S.H.; Richards, T. Co-combustion of animal waste in a commercial waste-to-energy BFB boiler. *Energies* **2013**, *6*, 6170–6187. [CrossRef]

25. Chun, Y.N.; Ji, D.W.; Yoshikawa, K. Pyrolysis and gasification characterization of sewage sludge for high quality gas and char production. *J. Mech. Sci. Technol.* **2013**, *27*, 263–272. [CrossRef]

26. Gómez, M.A.; Comesaña, R.; Feijoo, M.A.Á.; Eguía, P. Simulation of the effect of water temperature on domestic biomass boiler performance. *Energies* **2012**, *5*, 1044–1061. [CrossRef]

*energies*

MDPI

*Article*

# Dielectric Properties of Biomass/Biochar Mixtures at Microwave Frequencies

**Candice Ellison [1], Murat Sean McKeown [2], Samir Trabelsi [3] and Dorin Boldor [1,*]**

[1]   Biological and Agricultural Engineering, Louisiana State University, 149 E. B. Doran, Baton Rouge, LA 70803, USA; celli27@lsu.edu

[2]   College of Engineering, University of Georgia, 597 D. W. Brooks Dr., Athens, GA 30602, USA; mckeown@uga.edu

[3]   U. S. Department of Agriculture, Agricultural Research Service, Russell Research Center, 950 College Station Rd., Athens, GA 30605, USA; samir.trabelsi@ars.usda.gov

*   Correspondence: dboldor@agcenter.lsu.edu; Tel.: +1-225-578-7762

Academic Editor: Mejdi Jeguirim
Received: 23 February 2017; Accepted: 4 April 2017; Published: 9 April 2017

**Abstract:** Material dielectric properties are important for understanding their response to microwaves. Carbonaceous materials are considered good microwave absorbers and can be mixed with dry biomasses, which are otherwise low-loss materials, to improve the heating efficiency of biomass feedstocks. In this study, dielectric properties of pulverized biomass and biochar mixtures are presented from 0.5 GHz to 20 GHz at room temperature. An open-ended coaxial-line dielectric probe and vector network analyzer were used to measure dielectric constant and dielectric loss factor. Results show a quadratic increase of dielectric constant and dielectric loss with increasing biochar content. In measurements on biochar, a strong dielectric relaxation is observed at 8 GHz as indicated by a peak in dielectric loss factor at that frequency. Biochar is found to be a good microwave absorber and mixtures of biomass and biochar can be utilized to increase microwave heating rates for high temperature microwave processing of biomass feedstocks. These data can be utilized for design, scale-up and simulation of microwave heating processes of biomass, biochar, and their mixtures.

**Keywords:** biochar; biomass properties; microwave applications; dielectric properties; bioenergy

## 1. Introduction

Biomass resources offer a plentiful, renewable energy alternative to fossil fuels and can reduce $CO_2$ emission due to the potential of net zero emissions [1]. Lignocellulosic biomass materials can be converted to energy-dense products via thermochemical processes (namely pyrolysis and gasification) [2,3]. During these conversion processes, the biomass feedstock is heated to temperatures in the range of 400–700 °C, usually by conventional heating methods, i.e., conduction and convection [4]. In attempts to improve heating efficiency, recent studies have applied dielectric heating to these thermochemical conversion processes [5–7]. Microwave processing has numerous advantages over conventional methods, including no-contact energy transfer, volumetric energy absorption and dissipation, and selective heating in samples composed of two or more materials.

To effectively design microwave reactors for processing biomass feedstocks, an accurate knowledge of the dielectric properties of biomass materials is necessary to evaluate the dielectric response of materials in an applied electric field [8]. Dielectric properties may be determined by the complex relative permittivity expressed by [9]:

$$\varepsilon^* = \varepsilon' - i\varepsilon'' = \varepsilon'(1 - i \times tan\delta) \tag{1}$$

where $\varepsilon'$ is the relative dielectric constant, $\varepsilon''$ is the relative dielectric loss factor, $i = \sqrt{-1}$, and *tan δ* is the loss tangent (*tan δ* = $\varepsilon''/\varepsilon'$). Dielectric constant and dielectric loss factor are dimensionless entities, which are used to measure the ability of a material to store energy and the ability of a material to dissipate energy as heat, respectively. Molecular mechanisms dictate the polarization of molecules in an applied electric field and in the microwave frequency range, dipole rotation and ionic conduction are the dominant mechanisms of molecular polarization [8]. Dielectric properties of a material are dependent on many factors such as measurement frequency, material atomic and molecular composition, and physical characteristics [10]. For materials composed of at least two components, dielectric properties are function of the properties of the bulk material such as bulk density, moisture content, and temperature [11].

Dielectric properties of a variety of lignocellulosic biomasses including woody biomass [11–15], grassy biomass [16–18], oil palm [19], corn stover [20], and pulp mill sludge [21] have been presented in the literature. Due to the low dielectric loss of most lignocellulosic biomasses, these materials alone, especially if dry, require a lot of energy to reach high temperatures by dielectric heating if not aided by a microwave absorbing material [22,23]. To reach the high temperatures required by thermochemical conversion processes, a microwave absorbing material, characterized by high dielectric loss, can be added to biomass feedstocks [6,24,25]. When a high-loss material is added to a low-loss material, the dielectric loss of the overall mixture is increased, resulting in greater heat generation in the bulk material.

Carbonaceous materials, including carbon black, carbon nanotubes, carbon fibers, graphene, activated carbon, SiC, and pyrolytic biochar have been recognized as good microwave absorbers for their potential to convert microwave energy into thermal energy [6,24,26,27]. In this study, biochar was investigated for its microwave absorbing potential since it is a low cost and convenient feedstock additive that is readily available as a byproduct of thermochemical conversion processes. When a mixture of biomass and biochar is irradiated with microwaves, the biochar particles selectively heat, followed by heat transfer to adjacent biomass particles by conduction and by convection. Few studies have investigated the dielectric properties of pyrolytic biochars [17,19,27,28]. Motasemi et al. measured the dielectric properties of hay, switchgrass, and corn stover during pyrolysis [16,17,20]. Low dielectric properties were observed from room temperature to 450 °C during pyrolysis, but a sudden increase in dielectric properties was observed as the feedstock was heated from 450 to 700 °C when the biochar had been formed. Salema et al. [19] and Tripathi et al. [28] measured the dielectric properties of biochar derived from pyrolysis of oil palm shell.

While dielectric properties of biomass and biochar have been investigated separately, dielectric properties of biomass and biochar mixtures have never been measured to our knowledge. Dielectric properties of mixtures of biomass and biochar at room temperature are important for the efficient use of microwaves to initiate dielectric heating for thermochemical conversion processes. This study aims to fill this knowledge gap by characterizing the dielectric properties of biomass/biochar mixtures for four different biomass feedstocks readily available in Louisiana and southeastern United States: energy cane bagasse, pine sawdust (*Pinus* sp.), live oak (*Quercus* sp.), and Chinese tallow tree wood (*Triadica sebifera* (L.)). Each of the biomasses chosen for this study can be sourced from various biomass waste streams, making them viable feedstocks for thermochemical conversion processes. Pine sawdust is a forestry residue from logging operations and a waste from milling processes. Energy cane bagasse is a byproduct of the sugar industry, a residual lignocellulosic material after the juices are pressed from the cane. Chinese tallow tree is an invasive species whose population is controlled to protect native species and wetlands in southeastern United States. Live oak is an urban waste from tree pruning and other tree maintenance services. These lignocellulosic biomass wastes have great potential as feedstocks for biofuel production processes due to their low cost, but the dielectric properties when mixed with biochar as absorbers need to be investigated in order to effectively design microwave-based processes and equipment.

## 2. Materials and Methods

### 2.1. Biochar Preparation

The biochar used in these experiments was obtained from biomass pyrolysis of various feedstocks (mostly pine sawdust). Ground biomass was packed into a stainless-steel tube and heated via induction heating to 400–600 °C under a continuous flow of nitrogen gas. After complete pyrolysis, the tube was cooled to room temperature and the biochar was collected from the tube. The biochar was ground and sieved to obtain particle sizes less than 5 µm.

### 2.2. Biomass Sample Preparation

Energy cane, pine wood, live oak, and Chinese tallow tree samples were obtained by grinding wood chips or shavings using a wood chipper and laboratory blender, and sieved to less than 5 µm particle size. For each biomass, the following biochar mixtures were prepared: 0 wt %, 25 wt %, 50 wt %, 75 wt %, 100 wt % biochar. The moisture contents of each sample were measured on a wet basis using a standardized oven drying method (ASTM E871-82) and are presented in Table 1. Bulk density at the time of measurement was determined gravimetrically by dividing the weight of the sample by the volume of the sample measurement cup (Table 1).

**Table 1.** Moisture content (MC) and density of each of the biomass/biochar samples measured.

| Biochar (% wt) | Energy Cane | | Sawdust | | Live Oak | | Chinese Tallow Tree | |
|---|---|---|---|---|---|---|---|---|
| | MC (% wet basis) | Density (g/cm³) | MC (% wet basis) | Density (g/cm³) | MC (% wet basis) | Density (g/cm³) | MC (% wet basis) | Density (g/cm³) |
| 0 | 10.70 ± 0.13 | 0.21 | 11.09 ± 0.12 | 0.27 | 11.50 ± 0.00 | 0.40 | 11.33 ± 0.10 | 0.35 |
| 25 | 9.32 ± 0.20 | 0.25 | 9.68 ± 0.06 | 0.30 | 10.87 ± 1.00 | 0.41 | 9.00 ± 0.17 | 0.37 |
| 50 | 7.93 ± 0.00 | 0.29 | 8.05 ± 0.02 | 0.37 | 8.57 ± 0.30 | 0.46 | 7.30 ± 0.22 | 0.41 |
| 75 | 6.50 ± 0.01 | 0.32 | 6.57 ± 0.12 | 0.45 | 6.84 ± 0.05 | 0.49 | 6.63 ± 0.08 | 0.48 |
| 100 | 5.00 ± 0.01 | 0.56 | 5.00 ± 0.01 | 0.56 | 5.00 ± 0.01 | 0.56 | 5.00 ± 0.01 | 0.56 |

### 2.3. Measurement Procedure

For this study, the open-ended coaxial-line dielectric probe method was utilized, despite its limitations for measurement of low-loss solid materials [14,29], due to its ease of use and ability to cover a broad range of electromagnetic frequencies. This technique is convenient due to easy sample preparation and small sample size requirement; however, it is sensitive to local inhomogeneities in the test material due to the small measurement region of the probe [30]. If the sample is not homogenous, the resulting measurement is an average value weighted by the intensity of the electric field which is at its highest at the center conductor of the probe tip [31]. Air gaps in the measurement region are another source of error when measuring granular solids [32] and methods to control these errors have been developed and are discussed further [33]. Since the materials being measured in the present study are pulverized to a fine powder consistency, these sources of error can be carefully prevented. Precautions were taken to ensure firm contact of the probe with the sample and to make certain the probe face was presented with a single, smooth, flat surface with gap-free contact. In this study, three to five replicates were measured after agitating and recompressing the sample into the sample holder to verify the consistency of the readings and avoid measurement errors due to sample inhomogeneity.

A schematic of the dielectric measurement setup is shown in Figure 1. The measurement system consisted of an Agilent 85070 high-temperature dielectric probe connected via a coaxial cable to a vector network analyzer (Agilent N5230C PNA-L). The specific sample requirements for the probe are a minimum diameter >20 mm, a granule size <0.3 mm, and a minimum sample thickness given by the following equation [31]:

$$\text{sample thickness} > \frac{20}{\sqrt{|\varepsilon^*|}} \text{ mm} \tag{2}$$

where $|\varepsilon^*|$ is the modulus of the permittivity given by:

$$|\varepsilon^*| = \sqrt{(\varepsilon')^2 + (\varepsilon'')^2}\tag{3}$$

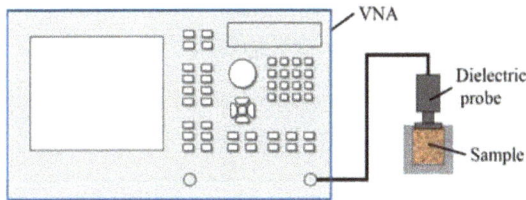

**Figure 1.** Sample measurement set-up consisting of a vector network analyzer (VNA) and high temperature coaxial-line dielectric probe (Agilent 85070) inserted into sample.

The instrument was calibrated using open air, a short, and distilled water at 25 °C as reference standards. The dielectric probe was fixed on a stand and the samples firmly pressed into a cylindrical stainless steel cup (1 cm radius, 2 cm height) by raising an adjustable platform. Measurements consisted of 101-point logarithmic sweep from 0.2 to 20 GHz. Dielectric constant and dielectric loss factor readings were acquired for each biomass/biochar mixture. Due to observed issues with this method at low frequencies, the data measured below 0.5 GHz were discarded. The permittivity measurements at 2.45 GHz were obtained by interpolation of dielectric constant and dielectric loss factor values at their respective neighboring frequencies. This frequency was selected as it is the most commonly used frequency in the industrial, scientific, and medical (ISM) radio bands for microwave heating applications. The accuracy of dielectric properties obtained from the probe is specified by the manufacturer as [31]:

$$\varepsilon' = \varepsilon' \pm 0.05|\varepsilon^*|\tag{4}$$

$$\varepsilon'' = \varepsilon'' \pm 0.05|\varepsilon^*|\tag{5}$$

## 3. Results

In the context of this study, measurements performed at microwave frequencies showed dependence of dielectric constant and dielectric loss factor on frequency, mixture ratio, and biomass type. Figures 2 and 3 depict the frequency dependence of the real ($\varepsilon'$) and imaginary ($\varepsilon''$) parts of permittivity, respectively, for each biomass/biochar mixture. Dielectric constant decreases with increasing frequency for all samples over the measured frequency range. This observed monotonic decrease is due to the decrease in polarization of the dielectric material with increasing frequency, as it is described by Torgovnikov [11]. The phase of the charged particles of the dielectric lags behind the phase of the electric field due to polarization relaxation. With increasing frequency, the number of charged particles that are in phase with the electric field decreases, resulting in a decrease in dielectric constant. Percent decreases of dielectric constant from the maximum at 0.5 GHz to the minimum at 20 GHz are denoted in Table 2. Dielectric loss factor increases to a maximum between 8 and 9 GHz for the biomass/biochar mixtures less than or equal to 75% weight biochar. For the samples with 100% weight biochar, dielectric loss factor decreases between 0.5 and 2 GHz followed by an increase, which peaks at 8 GHz, then decreases from 8 to 20 GHz.

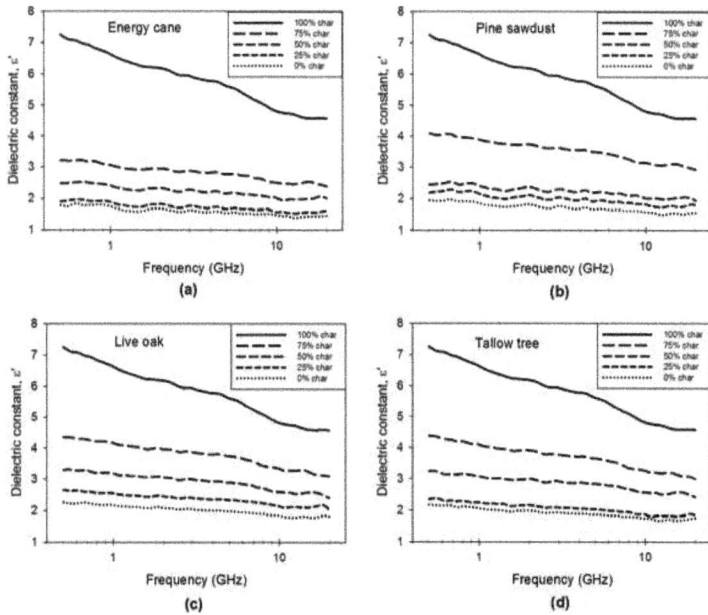

**Figure 2.** Measured dielectric constant from 0.5 to 20 GHz of each biomass sample: (**a**) energy cane bagasse; (**b**) pine sawdust; (**c**) live oak; and (**d**) Chinese tallow tree wood at the indicated biochar contents.

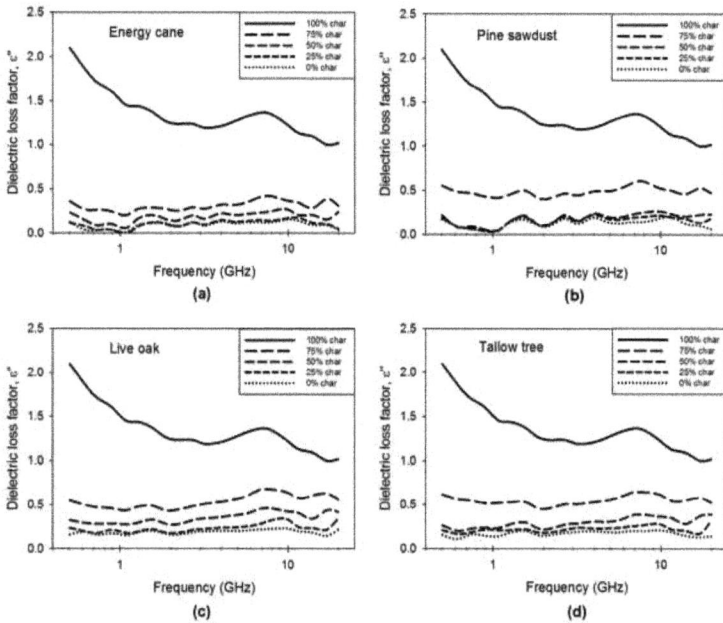

**Figure 3.** Measured dielectric loss constant from 0.5 to 20 GHz of each biomass sample: (**a**) energy cane bagasse; (**b**) pine sawdust; (**c**) live oak; and (**d**) Chinese tallow tree wood) at the indicated biochar contents.

**Table 2.** Percent differences (%) in dielectric constant from 0.5 to 20 GHz.

| Biochar | Energy Cane | Pine Sawdust | Live Oak | Chinese Tallow Tree |
|---------|-------------|--------------|----------|---------------------|
| 0 | 8.83 | 8.98 | 21.48 | 11.61 |
| 25 | 5.06 | 2.35 | 27.04 | 24.37 |
| 50 | 11.86 | 7.64 | 30.81 | 28.09 |
| 75 | 25.49 | 28.91 | 34.98 | 35.02 |
| 100 | 44.41 | 44.41 | 44.41 | 44.41 |

Dielectric properties are dependent on bulk density of air-particle mixtures [34]. The biochar measured in this study is denser than the biomass materials, thus the measured permittivity values of the various biomass/biochar mixtures partially reflect the effect of density on the apparent permittivity. The observed increase in dielectric properties with increasing biochar content cannot be attributed to the biochar alone since with an increase in biochar, there is also an increase in density. To eliminate the effect of density on dielectric properties, the measured permittivity data of each mixture at its respective density was transformed to the corresponding permittivity value at a mean bulk density by the Landeau and Lifshitz, Looyenga equation [10]:

$$\varepsilon_2' = \left[ \frac{\varepsilon_1^{1/3} - 1}{\rho_1} \rho_2 + 1 \right]^3 \tag{6}$$

where $\rho$ is bulk density and $\varepsilon$ represents permittivity. The subscripts 1 and 2 denote the original measured parameter and the transformed parameter, respectively.

Figure 4 illustrates the dielectric constant and dielectric loss factor as a function of biochar content and corrected to a mean density (0.32, 0.39, 0.46 and 0.43 g/cm$^3$, for energy cane, pine sawdust, live oak, and Chinese tallow tree, respectively) using Equation (6). For all biomasses measured, the real and imaginary parts of dielectric properties are shown to increase as biochar content increases. Since biochar has a greater dielectric constant than the biomass samples, addition of biochar to the biomass feedstock increases the overall dielectric constant of the bulk material. At 2.45 GHz, the average dielectric constant of the raw biomasses and biochars are 2.13 and 4.06, respectively, corresponding to twofold increase. A regression analysis was performed and it was determined that the permittivity of the mixture follow a quadratic function of biochar content. The dependency of permittivity on biochar content can be described by a quadratic function of the form $y = Ax^2 + Bx + C$, and the coefficients of the quadratic regression and the regression coefficient are presented in Table 3.

Dielectric loss tangent is the ratio of the loss factor to dielectric constant and its dependency on mixture ratio is depicted in Figure 5a. The loss tangent is an indicator of the ability of a material to dissipate electromagnetic energy. A high dielectric loss factor and moderate dielectric constant would be indicative of a good microwave absorber. Similarly, to dielectric constant and dielectric loss, the regression analysis of the loss tangent follows an increasing quadratic trend with increasing biochar content. Penetration depth is defined as the distance into the material at which the power of an incident electric field has decayed by $1/e$ and is calculated by the following equation (where $\lambda_0$ is the frequency of free space):

$$\delta_p = \frac{\lambda_0}{2\pi\sqrt{2\varepsilon'}} \left( \sqrt{1 + (\varepsilon''/\varepsilon')^2} - 1 \right)^{-\frac{1}{2}} \tag{7}$$

Penetration depth was calculated for each biomass/biochar mixture and is shown to decrease quadratically with increasing biochar content (Figure 5b). Knowledge of penetration depth of a material is important for scale-up of microwave heating systems and is useful for designing reactor geometry and dimensions. The coefficients of quadratic regression analysis for loss tangent and penetration depth as functions of biochar content are presented in Table 4.

**Table 3.** Quadratic regression coefficients ($y = Ax^2 + Bx + C$) for the dependence of dielectric constant ($\varepsilon'$) and dielectric loss factor ($\varepsilon''$) on biochar content for each biomass type at 2.45 GHz.

| Sample | $\varepsilon'$ | | | | $\varepsilon''$ | | | |
|---|---|---|---|---|---|---|---|---|
| | A | B | C | $R^2$ | A | B | C | $R^2$ |
| Energy cane | 0.8187 | 0.5373 | 1.9260 | 0.9604 | 1.9732 | −0.9848 | 0.0615 | 0.9429 |
| Pine sawdust | 2.0488 | −0.2974 | 2.1348 | 0.9711 | 1.6801 | −0.6170 | 0.0639 | 0.9914 |
| Live oak | 2.0412 | 0.3712 | 2.3027 | 0.9947 | 1.5136 | −0.5243 | 0.1488 | 0.9767 |
| Tallow tree | 1.5599 | 0.6214 | 2.1689 | 0.9891 | 1.5184 | −0.4990 | 0.1205 | 0.9896 |

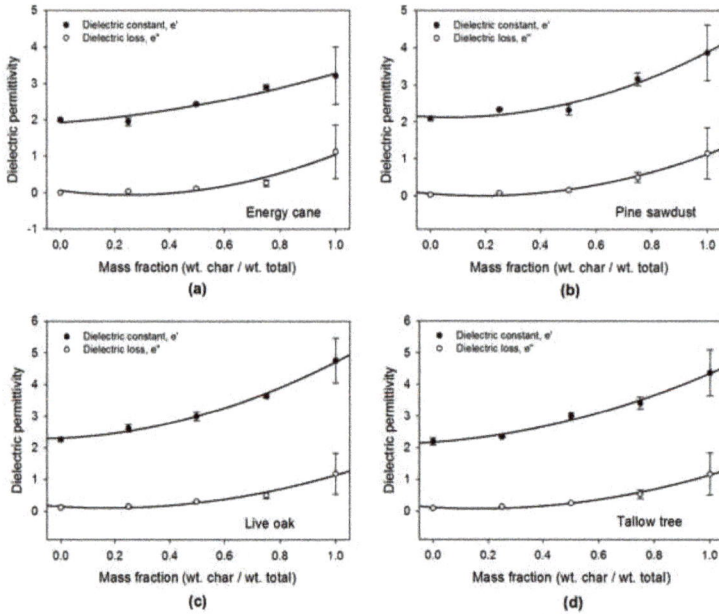

**Figure 4.** Dielectric constant and loss factor measurements as a function of biochar content at 2.45 GHz for each biomass: (**a**) energy cane bagasse; (**b**) pine sawdust; (**c**) live oak; and (**d**) Chinese tallow tree wood). Error bars indicate standard deviations.

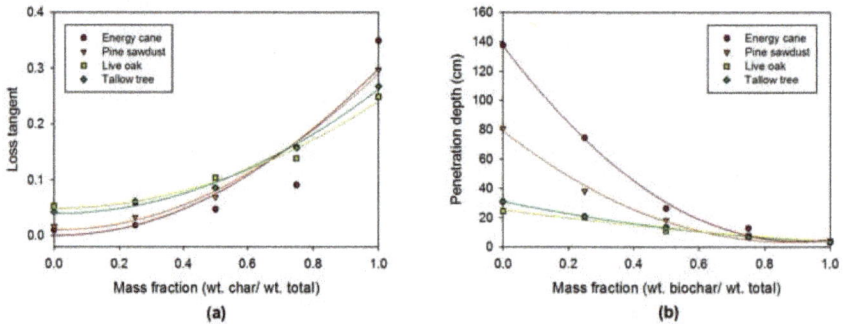

**Figure 5.** (**a**) Dielectric loss factor and (**b**) penetration depth as a function of biochar content at 2.45 GHz frequencies for each biomass.

**Table 4.** Quadratic regression coefficients ($y = Ax^2 + Bx + C$) for loss tan and penetration depth as a function of mass fraction of biochar.

| Sample | Loss Tan | | | | Penetration Depth | | | |
|---|---|---|---|---|---|---|---|---|
| | A | B | C | $R^2$ | A | B | C | $R^2$ |
| Energy cane | 0.2992 | $2.06 \times 10^{-12}$ | $2.06 \times 10^{-12}$ | 0.8827 | 162.4 | −294.7 | 137.3 | 0.9963 |
| Pine sawdust | 0.2802 | $2.52 \times 10^{-12}$ | 0.0101 | 0.9941 | 98.5 | −172.9 | 79.1 | 0.9938 |
| Live oak | 0.1909 | $5.96 \times 10^{-12}$ | 0.0490 | 0.9814 | 7.8 | −29.7 | 25.2 | 0.9789 |
| Tallow tree | 0.2219 | $1.72 \times 10^{-12}$ | 0.0397 | 0.9929 | 17.3 | −44.8 | 31.0 | 0.9996 |

Comparison of the different biomasses studied show that each biomass exhibits similar, but independent dielectric properties. To illustrate the effect of biomass type on dielectric properties, loss tangent was plotted as a function of frequency for 25% biochar (Figure 6a) and 75% biochar (Figure 6b) mixtures. In comparing the measurements on the different biomass at similar biochar contents, it is clear that differences exist between the different materials. By comparing the dielectric properties values of the different biomasses, it is observed that an increased biochar content amplifies the apparent dielectric constant of the mixture.

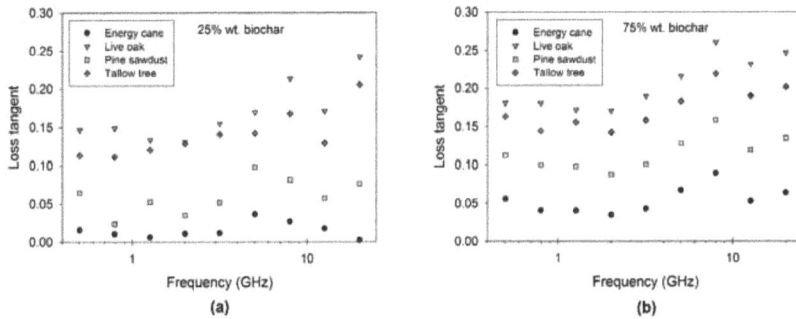

**Figure 6.** Effect of biomass type on the loss tangent of two biomass/biochar mixtures: (**a**) 25% wt biochar and (**b**) 75% wt biochar.

## 4. Discussion

As seen by the quadratic increase in the loss tangent with increasing biochar, mixtures of biomass and biochar are shown to exhibit greater heat generation ability compared to biomass alone. Carbonaceous solid materials are not heated via dipole polarization like water and other polar liquids, but rather via Maxwell-Wagner-Sillars polarization, or interfacial polarization effects. Carbonaceous materials have delocalized $sp^2$ $\pi$-electrons, which move freely within interfacial regions of chars. An applied electric field induces a current in the $\pi$-electrons, which is in phase with the electric field. Over time the $\pi$-electrons become out of phase with the electric field and collisions between electrons generates energy in the form of heat. Thus, the good microwave absorption of biochar reported in literature discussing microwave heating is most likely attributable to the effect of Maxwell-Wagner Sillars polarization [27].

Frequency dependence of biochar is much more pronounced than the frequency dependence of the raw biomasses in this study. An interesting feature of the biochar permittivity data is the dielectric relaxation observed at 8 GHz. This relaxation is seen in the mixtures with high biochar content (75% and 100% wt biochar) and it is indicative of a relaxation frequency of the biochar. It is thought that this frequency could correspond to a wavelength that is equal to the mean free path of the free $\pi$-electrons in the biochar interlayers.

A wide range of dielectric properties values have been presented in the literature for carbonaceous materials (Table 5). Dielectric properties of carbonaceous materials are dependent on the origin of the material. Atwater and Wheeler [32] measured the dielectric properties of three different activated carbons and found the permittivity to be dependent on the origin of the material. In addition, measurement techniques and material conditions at the time of measurement should also be noted as a cause of discrepancy between dielectric measurements on similar materials.

**Table 5.** Dielectric properties of several carbonaceous materials at 2.45 GHz.

| Sample | $\varepsilon'$ | $\varepsilon''$ | Reference |
|---|---|---|---|
| Pyrolysis biochar | 6.00 | 1.22 | This study |
| Oil palm char | 2.83 | 0.23 | [19] |
| Activated carbon | 5.30 | 0.37 | [35] |
| Activated carbon | 14–40 | 4–26 | [36] |
| Graphite | 4.20 | 0.11 | [35] |

Dielectric properties have been found to increase with moisture content due to the high degree of polarization of water molecules under electromagnetic radiation [21]. In this study, moisture contents ranged from $5.00 \pm 0.01\%$ for the biochar samples to $11.16 \pm 0.33\%$ for the average of the biomass samples (Table 1). Despite this decrease in moisture content with increase in biochar, increase of biochar in the mixture increases the dielectric constant and loss factor. If water content was the dominant driving force for variation in dielectric properties, we would expect to see the opposite trend. This confirms that dipole rotation of water molecules is not a dominant mechanism for the dielectric properties of these samples. Permittivity also tends to increase with increasing density [34]. With addition of biochar to biomass, density of the samples increases (Table 1), which could be a cause of the increasing dielectric properties with increasing biochar content. However, as discussed previously, the measured permittivity values were corrected to a medium density using Equation (6) to be able to observe the effect of biochar content without the effect of density.

## 5. Conclusions

The dielectric properties of biomass and biochar mixtures were measured from 0.5 to 20 GHz at room temperature. Results from this study indicate the dependence of permittivity on frequency, biomass type, and mixture ratio. Dielectric properties were found to increase quadratically with increasing biochar content for all biomasses. Dry biomass materials require a considerable amount of microwave energy to reach high processing temperatures by microwave irradiation due to low dielectric properties. Biochar, a byproduct of biomass pyrolysis, was found to be a good microwave absorber and can be used as an additive to biomass feedstocks to increase microwave absorption in the bulk material and accelerate heating rates. The dielectric properties data presented in this study are important for the design, simulation, and scale-up of microwave reactors for high temperature microwave processing of biomass materials. Biomass and biochar mixture ratios can be optimized for a given microwave geometry.

**Acknowledgments:** The authors would like to acknowledge the LSU Agricultural Center, LSU College of Engineering, and LSU Biological and Agricultural Engineering Department for their support of this project. The authors acknowledge NSF CBET (award# 1437810), and USDA Hatch program (LAB #94146) and Louisiana Board of Regents (Graduate Fellowship for Candice Ellison under award #LEQSF(2012-17)-GF-03) for their financial support of this project. A portion of this work was conducted at and using equipment from the USDA Agriculture Research Service lab in Athens, Georgia. The author extends her acknowledgement to Jeff Ortego, McKenna Benbow and Gustavo Aguilar for their technical support. Published with the approval of the Director of the Louisiana Agricultural Experiment Station as manuscript 2017-232-30777.

**Author Contributions:** Candice Ellison and Dorin Boldor conceived and designed the experiments; Candice Ellison, Murat Sean McKeown and Samir Trabelsi performed the experiments; Candice Ellison analyzed the data; Samir Trabelsi and Dorin Boldor contributed reagents/materials/analysis tools; Candice Ellison wrote the paper.

**Conflicts of Interest:** The authors declare no conflict of interest.

# References

1. U.S. Department of Energy. *2016 Billion-Ton Report: Advancing Domestic Resources for a Thriving Bioeconomy*; Langholtz, M.H., Stokes, B.J., Eaton, L.M., Eds.; Oak Ridge National Laboratory: Oak Ridge, TN, USA, 2016; p. 448.
2. Henkel, C.; Muley, P.D.; Abdollahi, K.K.; Marculescu, C.; Boldor, D. Pyrolysis of energy cane bagasse and invasive Chinese tallow tree (*Triadica sebifera* L.) biomass in an inductively heated reactor. *Energy Convers. Manag.* **2016**, *109*, 175–183. [CrossRef]
3. Bridgwater, A.V. Principles and practice of biomass fast pyrolysis processes for liquids. *J. Anal. Appl. Pyrolysis* **1999**, *51*, 3–22. [CrossRef]
4. Bridgwater, A.V. Review of fast pyrolysis of biomass and product upgrading. *Biomass Bioenergy* **2012**, *38*, 68–94. [CrossRef]
5. Salema, A.A.; Ani, F.N. Microwave induced pyrolysis of oil palm biomass. *Bioresour. Technol.* **2011**, *102*, 3388–3395. [CrossRef]
6. Undri, A.; Abou-Zaid, M.; Briens, C.; Berruti, F.; Rosi, L.; Bartoli, M.; Frediani, M.; Frediani, P. Bio-oil from pyrolysis of wood pellets using a microwave multimode oven and different microwave absorbers. *Fuel* **2015**, *153*, 464–482. [CrossRef]
7. Liu, G.; Wright, M.; Zhao, Q.; Brown, R.C. Catalytic fast pyrolysis of duckweed: Effects of pyrolysis parameters and optimization of aromatic production. *J. Anal. Appl. Pyrolysis* **2015**, *112*, 29–36. [CrossRef]
8. Thostenson, E.T.; Chou, T.W. Microwave processing: Fundamentals and applications. *Compos. Part A Appl. Sci. Manuf.* **1999**, *30*, 1055–1071. [CrossRef]
9. Meredith, R.J. *Engineers' Handbook of Industrial Microwave Heating*; Institution of Electrical Engineers: London, UK, 1998.
10. Nelson, S.O. Correlating dielectric properties of solids and particulate samples through mixture relationships. *Trans. ASAE* **1992**, *35*, 625–629. [CrossRef]
11. Torgovnikov, G.I. *Dielectric Properties of Wood and Wood-Based Materials*; Springer: Berlin, Germany, 1993.
12. Olmi, R.; Bini, M.; Ignesti, A.; Riminesi, C. Dielectric Properties of Wood from 2 to 3 GHz. *J. Microw. Power Electromagn. Energy* **2000**, *35*, 135–143. [CrossRef] [PubMed]
13. Ramasamy, S.; Moghtaderi, B. Dielectric Properties of Typical Australian Wood-Based Biomass Materials at Microwave Frequency. *Energy Fuels* **2010**, *24*, 4534–4548. [CrossRef]
14. Paz, A.M.; Trabelsi, S.; Nelson, S.O.; Thorin, E. Measurement of the Dielectric Properties of Sawdust Between 0.5 and 15 GHz. *IEEE Trans. Instrum. Meas.* **2011**, *60*, 3384–3390. [CrossRef]
15. McKeown, M.S.; Trabelsi, S.; Tollner, E.W. Effects of temperature and material on sensing moisture content of pelleted biomass through dielectric properties. *Biosyst. Eng.* **2016**, *149*, 1–10. [CrossRef]
16. Motasemi, F.; Afzal, M.T.; Salema, A.A. Microwave dielectric characterization of hay during pyrolysis. *Ind. Crops Prod.* **2014**, *61*, 492–498. [CrossRef]
17. Motasemi, F.; Afzal, M.T.; Salema, A.A.; Hutcheon, R.M. Microwave dielectric characterization of switchgrass for bioenergy and biofuel. *Fuel* **2014**, *124*, 151–157. [CrossRef]
18. Fennell, L.P.; Boldor, D. Dielectric and Thermal Properties of Sweet Sorghum Biomass. *J. Microw. Power Electromagn Energy* **2014**, *48*, 244–260. [CrossRef]
19. Salema, A.A.; Yeow, Y.K.; Ishaque, K.; Ani, F.N.; Afzal, M.T.; Hassan, A. Dielectric properties and microwave heating of oil palm biomass and biochar. *Ind. Crops Prod.* **2013**, *50*, 366–374. [CrossRef]
20. Motasemi, F.; Salema, A.A.; Afzal, M.T. Dielectric characterization of corn stover for microwave processing technology. *Fuel Process. Technol.* **2015**, *131*, 370–375. [CrossRef]
21. Namazi, A.B.; Allen, D.G.; Jia, C.Q. Probing microwave heating of lignocellulosic biomasses. *J. Anal. Appl. Pyrolysis* **2015**, *112*, 121–128. [CrossRef]
22. Zuo, W.; Tian, Y.; Ren, N. The important role of microwave receptors in bio-fuel production by microwave-induced pyrolysis of sewage sludge. *Waste Manag.* **2011**, *31*, 1321–1326. [CrossRef] [PubMed]
23. Motasemi, F.; Afzal, M.T. A review on the microwave-assisted pyrolysis technique. *Renew. Sustain. Energy Rev.* **2013**, *28*, 317–330. [CrossRef]
24. Shang, H.; Lu, R.R.; Shang, L.; Zhang, W.H. Effect of additives on the microwave-assisted pyrolysis of sawdust. *Fuel Process. Technol.* **2015**, *131*, 167–174. [CrossRef]

25. Zhang, S.; Dong, Q.; Zhang, L.; Xiong, Y. High quality syngas production from microwave pyrolysis of rice husk with char-supported metallic catalysts. *Bioresour. Technol.* **2015**, *191*, 17–23. [CrossRef] [PubMed]
26. Qin, F.; Brosseau, C. A review and analysis of microwave absorption in polymer composites filled with carbonaceous particles. *J. Appl. Phys.* **2012**, *111*, 061301. [CrossRef]
27. Menéndez, J.A.; Arenillas, A.; Fidalgo, B.; Bermúdez, J.M. Review: Microwave heating processes involving carbon materials. *Fuel Process. Technol.* **2010**, *91*, 1–8. [CrossRef]
28. Tripathi, M.; Sahu, J.N.; Ganesan, P.; Monash, P.; Dey, T.K. Effect of microwave frequency on dielectric properties of oil palm shell (OPS) and OPS char synthesized by microwave pyrolysis of OPS. *J. Anal. Appl. Pyrolysis* **2015**, *112*, 306–312. [CrossRef]
29. McKeown, M.S.; Trabelsi, S.; Tollner, E.; Nelson, S.O. Dielectric spectroscopy measurements for moisture prediction in Vidalia onions. *J. Food Eng.* **2012**, *111*, 505–510. [CrossRef]
30. Moreau, J.M.; Aziz, R. Dielectric study of granular media according to the type of measurment device: Coaxial cell or open-ended probe. *Meas. Sci. Technol.* **1993**, *4*, 124–129. [CrossRef]
31. Agilent Technologies. *Agilent 85070E Dielectric Probe Kit Technical Overview*; Agilent Technologies, Inc.: Santa Clara, CA, USA, 2008.
32. Nelson, S.O.; Bartley, P.G. Open-ended coaxial-line permittivity measurements on pulverized materials. *IEEE Trans. Instrum. Meas.* **1998**, *47*, 133–137. [CrossRef]
33. Blackham, D.V.; Pollard, R.D. An improved technique for permittivity measurements using a coaxial probe. *IEEE Trans. Instrum. Meas.* **1997**, *46*, 1093–1099. [CrossRef]
34. Nelson, S.O. Density-Permittivity Relationships for Powdered and Granular Materials. *IEEE Trans. Instrum. Meas.* **2005**, *54*, 2033–2040. [CrossRef]
35. Zhou, F.; Cheng, J.; Liu, J.; Wang, Z.; Zhou, J. Activated carbon and graphite facilitate the upgrading of Indonesian lignite with microwave irradiation for slurryability improvement. *Fuel* **2016**, *170*, 39–48. [CrossRef]
36. Atwater, J.E.; Wheeler, R.R. Complex permittivities and dielectric relaxation of granular activated carbons at microwave frequencies between 0.2 and 26 GHz. *Carbon* **2003**, *41*, 1801–1807. [CrossRef]

*energies*

MDPI

*Article*

# The Potential of Thermal Plasma Gasification of Olive Pomace Charcoal

**Andrius Tamošiūnas [1,\*], Ajmia Chouchène [2], Pranas Valatkevičius [1], Dovilė Gimžauskaitė [1], Mindaugas Aikas [1], Rolandas Uscila [1], Makrem Ghorbel [3] and Mejdi Jeguirim [4]**

[1]  Plasma Processing Laboratory, Lithuanian Energy Institute, Breslaujos str. 3, LT-44403 Kaunas, Lithuania; Pranas.Valatkevicius@lei.lt (P.V.); Dovile.Gimzauskaite@lei.lt (D.G.); Mindaugas.Aikas@lei.lt (M.A.); Rolandas.Uscila@lei.lt (R.U.)

[2]  Institut Supérieur des Sciences et Technologies de l'Environnement, Technopole de Borj-Cedria B.P. 95, 2050 Hammam-Lif, Tunisia; ajmiachouchene@gmail.com

[3]  OliveCoal, ZI El Jem, 5160 Mahdia BP69, Tunisia; makram.ghorbel@olivecoal.net

[4]  Institut de Sciences des Matériaux de Mulhouse, Universite de Haute-Alsace, 15 rue Jean Starcky, 68057 Mulhouse CEDEX, France; mejdi.jeguirim@uha.fr

\*  Correspondence: Andrius.Tamosiunas@lei.lt; Tel.: +370-37-401-999

Academic Editor: Shusheng Pang

Received: 15 March 2017; Accepted: 9 May 2017; Published: 17 May 2017

**Abstract:** Annually, the olive oil industry generates a significant amount of by-products, such as olive pomace, olive husks, tree prunings, leaves, pits, and branches. Therefore, the recovery of these residues has become a major challenge in Mediterranean countries. The utilization of olive industry residues has received much attention in recent years, especially for energy purposes. Accordingly, this primary experimental study aims at investigating the potential of olive biomass waste for energy recovery in terms of synthesis gas (or syngas) production using the thermal arc plasma gasification method. The olive charcoal made from the exhausted olive solid waste (olive pomace) was chosen as a reference material for primary experiments with known composition from the performed proximate and ultimate analysis. The experiments were carried out at various operational parameters: raw biomass and water vapour flow rates and the plasma generator power. The producer gas involved principally CO, $H_2$, and $CO_2$ with the highest concentrations of 41.17%, 13.06%, and 13.48%, respectively. The produced synthesis gas has a lower heating value of 6.09 MJ/nm³ at the $H_2O/C$ ratio of 3.15 and the plasma torch had a power of 52.2 kW.

**Keywords:** biomass; olive pomace; charcoal; thermal plasma; gasification; synthesis gas

## 1. Introduction

Currently, significant quantities of raw biomasses are found in waste streams in the European Union (EU). In 2014, economic activities and households generated approximately 2.6 billion tonnes of wastes in the EU [1]. Recent developments show that additional improvement on resource efficiency is possible, which can lead to significant environmental, economic, and social benefits. Therefore, converting waste into useful products is a key objective to obtain various benefits, including a reduction of greenhouse gas emissions and job creation [2].

The olive oil industry has been mainly concentrated in the Mediterranean region, where a very large amount of waste, such as olive pomace, olive husks, tree prunings, leaves, pits, and branches are being generated annually. Spain, Italy, Greece, Turkey, Tunisia, Portugal, Syria, and Morocco are the major olive oil producers worldwide [3]. The EU produced 69%, and exported 65%, of the world's olive oil in the last five years [4]. Consequently, the estimated quantities of wastes derived from the olive oil industry in the EU accounts for 6.8 million tons/year with a promising energy content of around 18 MJ/kg [5,6].

There are several waste-to-energy conversion pathways into useful products depending on the waste/biomass characteristics and the requirement of the end product and its applications [7–10]. Regarding physicochemical conversion, such as extraction and esterification of biomass to vegetable oil or biodiesel production, biochemical conversion and thermochemical conversion methods have also been extensively applied. Biochemical methods allow the biomass/waste conversion into liquid or gaseous fuels by anaerobic digestion or fermentation with a final primary product of methane or ethanol, respectively. Thermochemical conversion methods include pyrolysis, liquefaction, combustion, and gasification. Although biomass/waste combustion is the main applied process, the overall efficiency of heat production is low. Among all four methods, gasification has been considered to be a more attractive process to exploit the energy from renewable and non-renewable solid biomass with a lower content of moisture when compared to liquefaction at a higher conversion efficiency. The method can be applied not only for direct generation of heat and electricity but also for generation of transportation fuels and chemicals by using low-value feedstocks.

The investigation of olive industry waste feedstock utilization into electrical and thermal power, as well as biogas, biofuels, and synthesis fuels, both by biochemical and thermochemical means, has been extensively studied in [5,6,11–21]. Recently, thermal plasma has attracted the attention as a state of the art waste-to-energy method, showing a better environmental performance over conventional waste treatment technologies in terms of life cycle assessment, as well as process efficiency [22]. Plasma methods can handle not only biomass, but also harmful/toxic wastes, which can be completely converted into products having considerable amounts of useful energy content. In general, plasma, which consists of charged (electrons, ions) and neutral particles, is defined as the fourth state of matter. Depending on the species temperature, plasma can be classified as a high-temperature plasma (fusion plasma) or a low-temperature plasma (gas discharges). The latter group of plasma is the subject of interest of this paper. Low-temperature plasmas, typically related to the pressure, can be classified into thermal plasma, which is in thermal equilibrium (all of the species—electrons, ions, neutrals—are at the same temperature, $T_e = T_i = T_{gas}$), and cold plasma, which is described by a non-equilibrium state (where the electron temperature is much higher than the ion and neutral gas species, $T_e > T_i > T_{gas}$) [23–25].

Thermal plasma gasification process is an allothermal process that requires an external source of power to heat up and sustain high temperatures. However, conventional autothermal gasification has some limitations related to energy efficiency, material yield, syngas purity, compactness, dynamic response and flexibility that might be overcome by plasma utilization [26]. From the chemical point of view, the thermal plasma can significantly contribute to the gasification by enhancing the reaction kinetics due to the generation of active radicals within the plasma medium and improving high-temperature cracking of tars in the generated gas (syngas). From the thermal aspect, enthalpy provided by the plasma can easily be controlled by adjusting the electrical power of the source delivered to the system, thus making the process independent, contrary to the autothermal gasification process. Therefore, numerous investigations have been performed by utilizing thermal plasmas (direct current (DC), microwave (MW), radio frequency (RF)) for biomass/waste treatment to value-added secondary products, such as synthesis gas, hydrogen, biofuels, chemicals, etc. [27–33].

The purpose of the present experimental investigation was to evaluate, for the first time, the potential of olive biomass waste for energy recovery in terms of synthesis gas, or syngas, production using the thermal arc plasma gasification method. The olive charcoal derived from the exhausted olive solid waste (EOSW) (olive pomace) was chosen as a reference material for primary experiments. A DC plasma torch was used as a source for high-temperature, enthalpy, and active radical generation. Water vapour was used as a main gas to form the plasma. The experiments were carried out at the various operational parameters: treated material flow rate, water vapour flow rate, and power of the plasma torch.

## 2. Materials and Methods

### 2.1. Raw Material

The exhausted olive solid waste was used as a feedstock for charcoal production. This residue was provided by the Zouila factory from the region of Mahdia (Tunisia). The exhausted olive solid waste was the solid by-product obtained after the extraction of the residual oil using hexane as a solvent. It includes basically pulp, skin, and stone. The ultimate analyses of olive pomace are shown in Table 1 and compared with available data from the literature for other lignocellulosic biomass.

**Table 1.** Ultimate analysis of different biomasses.

| Biomass Type | C [a] (wt %) | H [a] (wt %) | N [a] (wt %) | S [a] (wt %) |
|---|---|---|---|---|
| Olive pomace (this study) | 47.04 | 5.73 | 0.87 | <0.06 |
| Olive tree wood [34] | 48.20 | 5.30 | 0.70 | 0.03 |
| Pine Sawdust [35] | 51.30 | 6.40 | 0.20 | 0.01 |
| Miscanthus [36] | 47.60 | 6.00 | 0.30 | 0.02 |
| Corn cob [37] | 46.40 | 5.40 | 1.0 | 0.02 |

[a] Weight percentage on dry basis.

#### 2.1.1. Thermogravimetric Study

Thermogravimetric analysis was carried out using a Mettler-Toledo TGA/DSC3+ (Mettler-Toledo Pte Ltd., Cresent, Singapore). Before each test, 10 mg of olive pomace was put in an alumina crucible. TGA experiments were performed under nitrogen atmosphere of 100 mL/min flow rate at heating rates of 5 °C/min from room temperature to 800 °C. Figure 1 shows the thermogravimetry (TG) and derivative thermogravimetry (DTG) curves obtained during the pyrolysis of olive pomace pyrolysis.

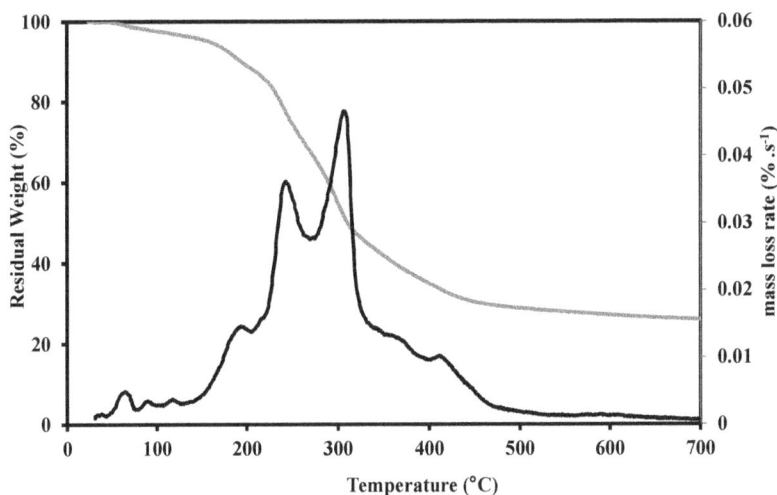

**Figure 1.** TG and DTG curves of olive pomace under inert atmosphere at 5 °C min$^{-1}$.

According to this figure, pyrolysis occurs in two noticeable steps: the first step is attributed to the devolatilization. This took place between 150 °C and 340 °C. This step corresponds to the volatile matter removal and the char formation. In this range, two maximum weight loss rates were observed at 244 °C and 312 °C, respectively. These peaks are attributed to hemicellulose and cellulose degradation. At 190 °C, a slight maximum weight loss rate is shown. This can contribute to the earlier decomposition of lignin. Several researchers have found similar thermal profiles [38–41]. During the first step of pyrolysis, the chemical bonds in the three major constituents of olive pomace, namely cellulose, hemicellulose, and lignin, are thermally cracked. The second pyrolysis step of olive pomace happens in the range of 340–450 °C. During this step, various parallel and serial reactions occur either homogeneously or heterogeneously. These reactions include dehydration, cracking, reforming, condensation, polymerization, oxidation, and gasification reactions [39].

Residual mass at 450 °C is about 30.33% of the initial mass for the olive pomace. Based on the thermogravimetric analysis, pyrolysis experiments at a large scale were conducted at 450 °C in order to prepare the olive pomace charcoal for plasma gasification. The impact of heating rates and particle size were studied by the authors previously in [42].

### 2.1.2. Olive Charcoal Preparation

The experiments of slow pyrolysis of the olive pomace were carried out at an industrial scale in a horizontal multi-stage reactor at the Olive Coal factory (Eljem, Tunisia). The reactor shown in Figure 2 includes four fixed-cylinders. The first cylinder was used essentially for the drying step while the pyrolysis stage occurs slowly in the other cylinders. Each cylinder was comprised of a conveying screw to ensure the progress of the EOSW.

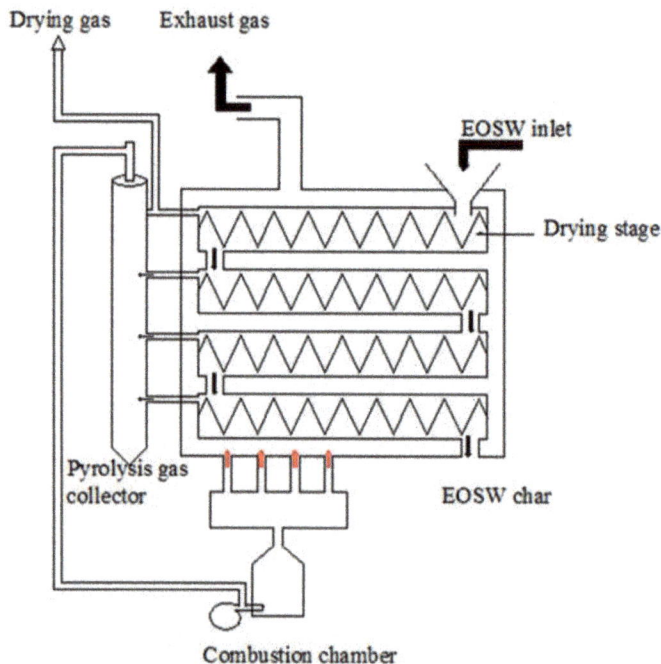

**Figure 2.** Scheme of the pyrolysis reactor (red colour: hot gases).

The hopper is tightly sealed in order to avoid the access of air into the feedstock section. The process is firstly launched with a natural gas burner. The pyrolysis reaction results in fine charcoal as a solid product and combustible pyrolysis gases as a volatile product. The pyrolysis gas was composed of a complex mixture of non-condensable constituents, such as hydrogen, carbon monoxide, carbon dioxide, and methane. Moreover, it comprises condensable constituents, such as water vapour, heavy tars, and other hydrocarbons. Pyrolysis gases, produced by the heated biomass, flowed downwards and went through a collector to be burned in the combustion chamber. The combustion temperature of the gases reached 1200 °C and the generated heat was used for the pyrolysis process.

The input olive pomace had a moisture content between 15% and 20%. It was dried with thermal energy, provided by hot gas, recovered by a heat exchanger from the exhaust gases of the combusted pyrolysis gases. The water evaporated during the drying of the EOSW in the first cylinder was rejected.

The pyrolysis cylinders were heated to different increased temperatures of 450 °C, 550 °C, and 650 °C, which is heated by a hot combustion gas. Therefore, the EOSW is heated only by its contact to the wall of the cylinder. The residence time in each stage was about 20 min. A thermocouple was utilized to determine the temperature of char inside the reactor. The final temperature was about 450 °C. Finally, the obtained charcoal is discharged from the reactor by a rotary valve.

### 2.1.3. Olive Pomace Charcoal Characterization

Tables 2 and 3 show the proximate and the ultimate analysis of the prepared olive pomace char. Hence, moisture content was measured gravimetrically by the oven drying method conforming to the EN 14774-1 standard. Volatile matter was examined using a thermogravimetric method at 900 °C for 7 min according to the NF EN 15148 standard. Ash was determined at 815 °C, conforming to the ISO 1171 standard. The higher heating value (HHV) was determined by employing an adiabatic oxygen calorimeter, according to the NF EN 14918 standard. Ultimate analysis was determined by SOCOR laboratory (France), as reported by the NF EN 15104 standard.

**Table 2.** Proximate analysis of olive pomace charcoal.

| Biomass Type | Moisture (% Dry Basis) | Volatile Matter (% Dry Basis) | Fixed Carbon (% Dry Basis) | Ash (% Dry Basis) |
|---|---|---|---|---|
| Olive pomace char coal | 22 | 17.4 | 77 | 5.6 |

**Table 3.** Energy content and ultimate analysis of different chars.

| Parameters | This Study Olive Pomace Char | Olive Mill Waste Char [43] | Olive Wood Biochar [44] | Hazelnut Wood Biochar [44] |
|---|---|---|---|---|
| Pyrolysis temperature | 450 °C | 480 ± 10 °C | 400–800 °C | 400–800 °C |
| C [a] (wt % ) | 80.4 | 75.3 | 90.1 | 78.1 |
| H [a] (wt % ) | 2.87 | 3.64 | 1.58 | 1.21 |
| N (wt %) | 0.42 | 0.94 | 0.42 | 0.64 |
| S [a] (mg/kg) | 271 | ND | ND | ND |
| HHV [a] (MJ/kg) | 30.89 | 29.21 | 31.71 | 26.62 |
| LHV [a] (MJ/kg) | 30.30 | 28.35 | 30.48 | 25.66 |

[a] On dry basis; ND: not determined.

The obtained results are compared with those found in the literature for different biomass chars. The ultimate analysis revealed that the olive wood char has the highest carbon content, followed by our charcoal. The content of nitrogen found in olive pomace and olive wood chars are of equal amount.

It can also be observed that chars from the olive tree have closer energy content, if compared to hazelnut wood char. Hence, the pyrolysis process can increase the heating value by 75–85% from the initial biomass heating value [44].

### 2.2. Plasma Gasification Setup

The experimental setup used in this research is shown in Figure 3. Its basic parts consists of a DC plasma torch (1); a chemical reactor (2); a charcoal feeding system (3); and a producer gas sampling and analysis system (4).

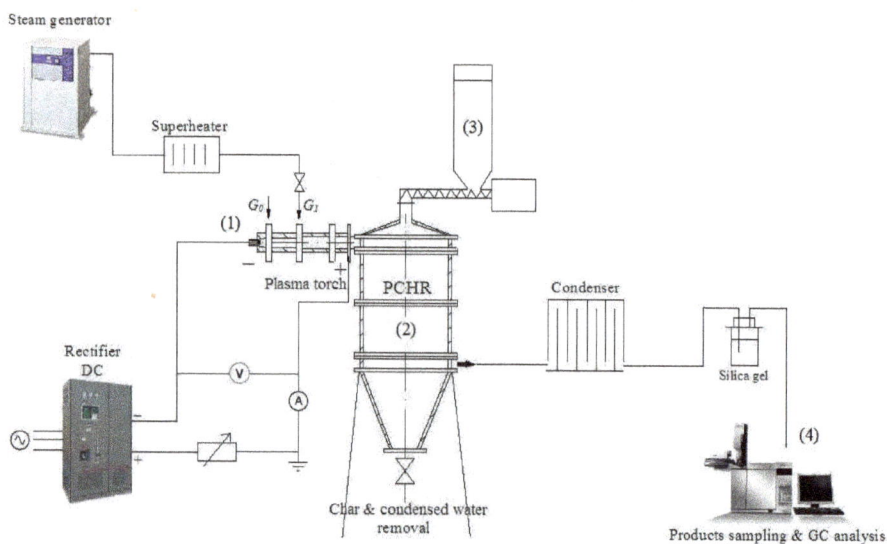

**Figure 3.** Scheme of the plasma gasification setup.

The DC arc plasma torch was used as a source for high temperature, enthalpy, and active radical generation, which operated at atmospheric pressure. Its power ranged from 45 kW to 52.2 kW at a fixed arc current intensity of 180 A and voltage drop of 250–290 V. The plasma torch operated on superheated water vapour at a flow rate in the range of $G_1 = (2.4 - 4.64) \times 10^{-3}$ kg/s. Therefore, the water vapour was superheated to 500 K, serving as the main gas to form the plasma and as a source for active particle/radical (O, H, OH) generation inducing and accelerating the thermochemical reactions. A hafnium cathode of the plasma torch was protected by a small amount of air ($G_0 = 0.62 \times 10^{-3}$ kg/s) to avoid its erosion. The thermal efficiency ($\eta$) of the plasma torch ranged between 0.67 and 0.75 and the mean generated temperature exceeded $2800 \pm 7\%$ K.

The chemical reactor is 1 m long with a 0.4 m inner diameter. Its inner walls are insulated with a refractory material (ceramic coating) to avoid excessive overheating. The EOSW charcoal was fed by a screw feeder from a hopper at an average flow rate of $1.3 \times 10^{-3}$ kg/s. The maximum particle size of charcoal was less than 2 mm. Around 10 kg of the EOSW charcoal was used within the seven individual experimental runs. At the bottom of the plasma-chemical reactor there is a by-product removal section, and in the middle, an outlet chamber for the producer gas sampling and analysis. The producer gas was analysed by means of an Agylent 7890 A gas chromatograph (GC) equipped with dual-channel thermal conductivity detectors (TDCs) and a MRU AIR SWG 300-1 gas analyser (MRU Instruments, Inc., Houston, TX, USA). Each experimental point was measured three times to obtain an average concentration of the produced gas. In each case, the relative deviation was below ±5%.

### 2.3. Main Chemical Reactions

The gasification of the EOSW charcoal to syngas encompasses various complex chemical reactions. Active radicals produced by the plasma torch in the arc discharge chamber can considerably accelerate the reaction kinetics [45,46]. The primary chemical heterogeneous and homogeneous reactions of the charcoal gasification are described in Table 4.

**Table 4.** The main thermo-chemical heterogeneous and homogeneous reactions during solid waste gasification [47].

| | | Oxidation Reactions | |
|---|---|---|---|
| 1. | Carbon partial oxidation | $C + \frac{1}{2}O_2 \rightarrow CO$ | −111 MJ/kmol |
| 2. | Carbon monoxide oxidation | $CO + \frac{1}{2}O_2 \rightarrow CO_2$ | −283 MJ/kmol |
| 3. | Carbon oxidation | $C + O_2 \rightarrow CO_2$ | −394 MJ/kmol |
| 4. | Hydrogen oxidation | $H_2 + \frac{1}{2}O_2 \rightarrow H_2O$ | −242 MJ/kmol |
| | | **Gasification Reactions Involving Steam** | |
| 5. | Water–gas reaction | $C + H_2O \leftrightarrow CO + H_2$ | +131 MJ/kmol |
| 6. | Water–gas shift reaction | $CO + H_2O \leftrightarrow CO_2 + H_2$ | −41 MJ/kmol |
| 7. | Steam methane reforming | $CH_4 + H_2O \leftrightarrow CO + 3H_2$ | +206 MJ/kmol |
| | | **Gasification Reactions Involving Hydrogen** | |
| 8. | Hydrogasification | $C + 2H_2 \leftrightarrow CH_4$ | −75 MJ/kmol |
| 9. | Methanation | $CO + 3H_2 \leftrightarrow CH_4 + H_2O$ | −227 MJ/kmol |
| | | **Gasification Reactions Involving Carbon Monoxide** | |
| 10. | Bouduard reaction | $C + CO_2 \leftrightarrow 2CO$ | +172 MJ/kmol |

## 3. Results and Discussion

### 3.1. Relation between the Water Vapour Flow Rate and the Power of the Plasma Torch

Since the experiments were carried out at the various operational parameters, such as treated material flow rate, water vapour flow rate, and power of the plasma torch, however, there is a direct relation between the water vapour used as a plasma forming gas and the power of the plasma torch. The change in water vapour flow rate changes the power of the plasma torch due to the increase of the voltage drop in the arc at a fixed current intensity. This relation is shown in Figure 4.

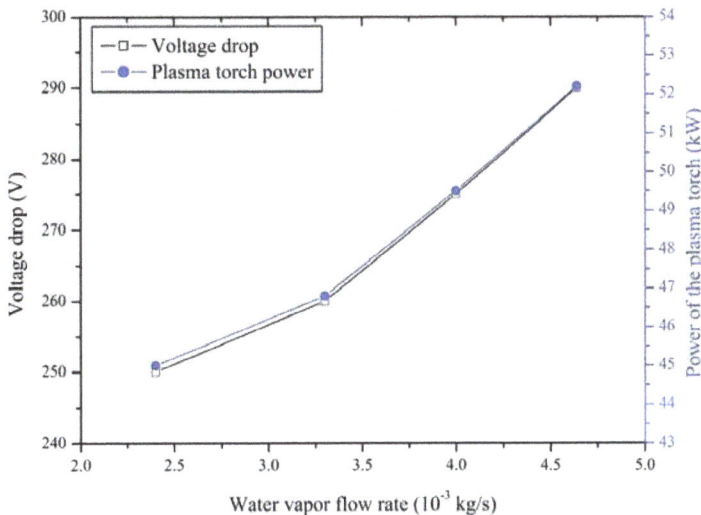

**Figure 4.** Relation between the water vapour flow rate and the power of the plasma torch at the constant arc current intensity of 180 A.

As could be seen from the above figure, as the water vapour flow rate increased from $2.4 \times 10^{-3}$ kg/s to $4.64 \times 10^{-3}$ kg/s, the voltage drop in the arc discharge chamber of the plasma torch also increased from 250 V to 290 V. The increasing water vapour flow rate tangentially supplied through the ring into the arc discharge chamber intensifies the heat transfer mechanism between the arc column and the injected water vapour surrounding the arc and, therefore, the flow changes from laminar to turbulent along the arc discharge chamber. As a result, the electric field of the arc, as well as the arc potential, increases significantly leading to the increased voltage drop [48]. Therefore, this directly influences the plasma torch power, which, during experiments, increased from 44 kW to 52.2 kW, due to the relation $P = VI$, where $P$ is power (W), $V$ is the voltage of the arc (V), and $I$ is the current of the arc (A). Consequently, the operational parameters related to the increased water vapour flow rate at the fixed arc current simultaneously means the increased power of the plasma torch.

### 3.2. Produced Syngas Composition

In this paragraph, the effects of the water vapour to charcoal ($H_2O/C$) ratio, as well as the plasma torch power on the gasification of the EOSW charcoal, are discussed.

Firstly, the experiments were performed in order to determine the reaction time required to reach equilibrium (operating) conditions during charcoal gasification, i.e., an elapsed time to reach the highest concentrations of the formed gases. This is shown in Figure 5.

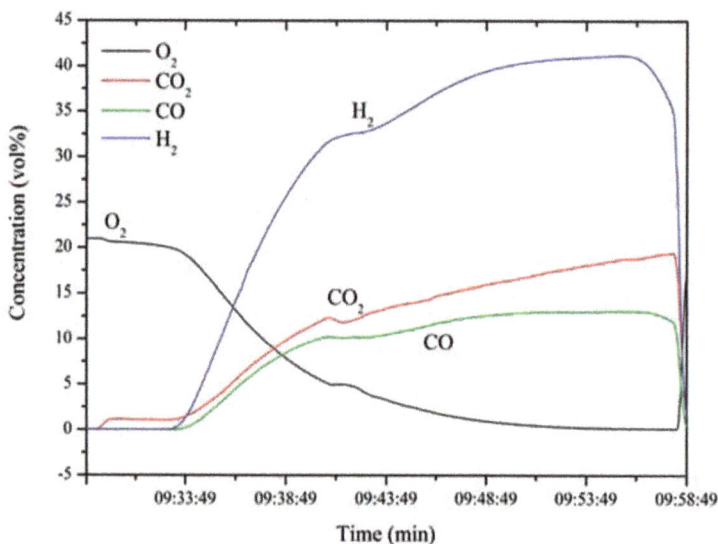

**Figure 5.** Concentration of formed gaseous products during charcoal gasification vs. the time required to reach equilibrium conditions. Arc current $I$ = 180 A, plasma torch power $P$ = 52.2 kW.

It can be seen from the figure that a reaction time required to reach the optimal operating gasification conditions, at which the concentrations of $H_2$ and CO were at the highest level, was approximately 25 min. The same tendency was observed during other experiments at the different regime parameters, such as the $H_2O/C$ ratio and the power of the plasma torch. Additionally, the main produced gases were $H_2$, CO, and $CO_2$, and only traces of other lighter hydrocarbons were detected.

Afterwards, the influence of the $H_2O/C$ ratio on the generated gas concentrations was investigated. The results are shown in Figure 5. As the $H_2O/C$ ratio increased from 1.85 to 3.15, the concentrations of $H_2$ and $CO_2$ increased from 20.3% and 12.3% to 41.17% and 18.68%, respectively, and only at 3.15 to 3.57, decreased to 36.19% and 13.48%, respectively. Meanwhile, the concentration of CO increased all the time, from 8.35% to 14.15%. Since the charcoal steam gasification is an allothermal process requiring an external heat source for initiation of endothermic reactions, it is possible to highlight just three independent main gasification reactions: water–gas reaction (WG) (Equation (5)), water–gas shift (WGS) (Equation (6)), and the Bouduard reaction (Equation (10)). However, it is a simplified explanation, since other chemical elements (H, N, O, S, etc.) and compounds could be involved as reactants and/or products [49]. Therefore, $H_2$, $CO_2$, and CO formation mechanisms and variations of concentrations could be explained by these gasification reactions. In this research, the thermal plasma torch was utilized to provide heat for endothermic reaction initiation. The increase of concentrations of $H_2$ and $CO_2$ at the $H_2O/C$ ratio of 3.15 can be explained by the dominance of WG and WGS reactions, whereas the decrease at the $H_2O/C$ ratio from 3.15 to 3.57, due to the dominance of reversed shift reaction and the Bouduard reaction, as well as less favourable WG reaction.

The figure also indicates the dependence of the $H_2/CO$ ratio, which was in the range of 2.43 to 3.15, on the $H_2O/C$ ratio. The $H_2/CO$ ratio mostly depended on the variation of the concentrations of $H_2$ to CO. At the beginning, as the $H_2O/C$ ratio increased from 1.85 to 3.15, the $H_2/CO$ ratio increased from 2.43 to 3.15. However, between the $H_2O/C$ ratios of 3.15 to 3.57 the latter decreased from 3.15 to 2.56 due to the decrease in $H_2$ and CO concentrations. Generally, a desired $H_2/CO$ ratio for the Fischer–Tropsch synthesis (FTS) of syngas to produce diesel and gasoline is around 2.1:1 [9]. Therefore, the produced syngas required an adjustment to reduce the $H_2$ content or increase the CO content by a WGS reaction (Equation (6)). Nevertheless, the primary experiments showed a promising result for syngas production from the EOSW charcoal.

However, the lower heating value (LHV) of the produced syngas was of low quality. The dependence of the LHV of the syngas on the $H_2O/C$ ratio is shown in Figures 6 and 7. It could be observed that the LHV of the syngas increased from 3.24 MJ/nm$^3$ to 6.09 MJ/nm$^3$ at the $H_2O/C$ ratio in the range of 1.85 to 3.15. However, at the $H_2O/C$ ratio of 3.15 to 3.57, the LHV decreased from 6.09 MJ/nm$^3$ to 5.69 MJ/nm$^3$. This was mostly induced by the decrease of $H_2$ concentration determined by a reversed water–gas shift reaction.

Future prospects will be to assess the performance parameters of the thermal plasma gasification system in terms of cold gas efficiency, carbon conversion efficiency, syngas yield, specific energy requirements, etc.

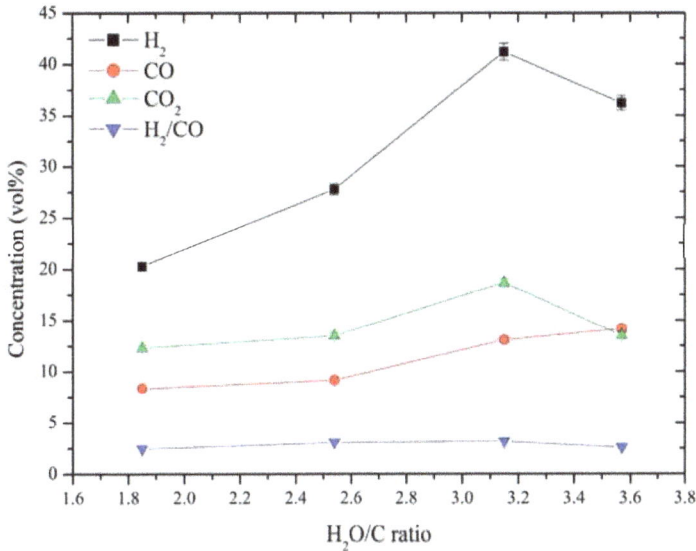

**Figure 6.** Effect of the $H_2O/C$ ratio on the produced gas concentrations.

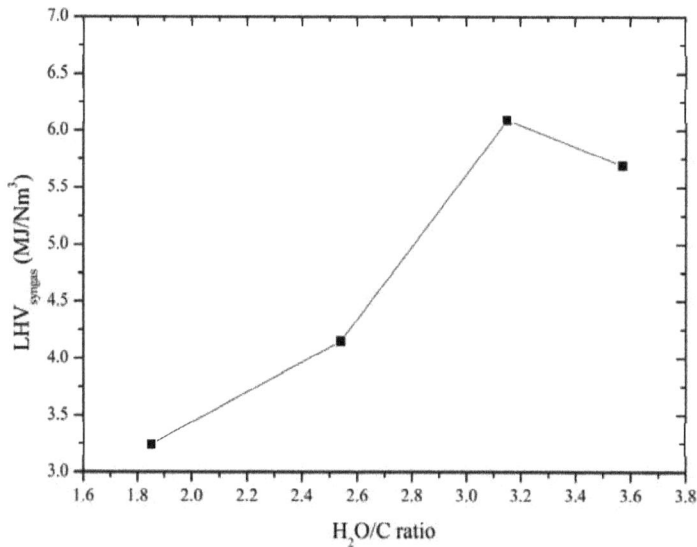

**Figure 7.** Effect of the $H_2O/C$ ratio on the lower heating value of the syngas.

### 3.3. Comparison with Similar Gasification Methods

In this paragraph a comparative study in terms of syngas production with similar gasification methods, both conventional (commercially available and near commercial) and plasma, is discussed. The results are summarized in Table 5.

**Table 5.** Comparison with other similar coal/charcoal gasification methods.

| Method | Gasifying Agent | Material | $H_2$ (vol %) | CO (vol %) | $H_2/CO$ | $LHV_{syngas}$ $(MJ/nm^3)$ | Ref. |
|---|---|---|---|---|---|---|---|
| | Conventional gasification (Lurgi, HTW, Siemens, Shell, etc.) | | | | | | |
| Moving Bed | Oxygen | Coal * | 28.1–42.3 | 15.1–61.2 | 0.5–2.7 | | |
| | Steam/air | | 16.2–23.3 | 17.1–27.8 | 0.69–1.35 | | |
| Fluidized Bed | Oxygen | Coal * | 32.8–40.0 | 31.0–53.0 | 0.71–0.85 | | |
| | Steam/air | | 12.6–28.56 | 12.54–30.7 | 0.56–1.0 | 4–11 | [47,50] |
| Entrained flow | Oxygen | Coal * | 28.1–42.3 | 15.1–61.2 | 0.51–2.72 | | |
| | Steam/air | | 15.7–25.5 | 16.1–31.0 | 0.82–2.00 | | |
| Transport flow | Oxygen | Coal * | 36.2–41.9 | 25.5–39.1 | 0.92–1.64 | | |
| | Steam/air | | 11.8–15.7 | 13.3–23.7 | 0.5–1.18 | | |
| | Plasma gasification (MW, DC) | | | | | | |
| MW | Steam/air | Coal | 48 | 23 | 2.08 | - | [51] |
| | $O_2$/air | Coal | 21.3 | 51.7 | 0.41 | - | |
| MW | Steam | Coal | 62 | 20 | 3.1 | 10.9 | [52] |
| | | Charcoal | 58 | 17 | 3.41 | 10.4 | |
| DC thermal arc plasma | Steam/air | Coal | 58.7 | 35.5 | 1.65 | - | [53] |
| DC thermal arc plasma | Steam/air | Coal | 39 | 34 | 1.14 | - | [31] |
| DC thermal arc plasma | Steam | Charcoal | 41.17 | 13.06 | 3.15 | 6.09 | This work |

* Coal—means various types of coals (brown, lignite, bituminous, sub-bit, anthracite).

There are a number of various gasification technologies. However, according to the configuration of the gasifier, all of the existing gasifiers can basically be divided into four main configurations: moving bed, fluidized bed, entrained flow, and transport flow [50]. This division is suitable for plasma gasification (plasma gasifiers) as well, where torches or arcs are being used as an external heating source. Nevertheless, despite the fact that the end/final product is usually similar, its syngas, conventional gasification, and plasma gasification technologies differ significantly. Traditional gasification methods require high pressures (greater than 30 Bar), long residence times of treated materials (up to several hours), special catalysts and their recovery, higher investment and maintenance costs, etc. [54]. Here, where the thermal plasma comes from with its advantages, such as an easy control of the gasification process, operation at atmospheric pressure, short residence time (up to several minutes), no need of catalysts and their recovery, flexibility and smaller size of the equipment of the same capacity, ability to generate reactive species, etc. Nevertheless, the concentrations of the produced $H_2$ and CO, the $H_2/CO$ and the LHV in both cases are at comparable levels. One of the greatest advantages of the plasma gasification are the low environmental emissions and better environmental impact in terms of life cycle assessment [10,22]. Therefore, this makes plasma an attractive method to use in waste-to-energy applications. However, despite the mentioned advantages, plasma gasification has not been fully commercially proven yet because of the large energy penalty, especially of the DC arc plasma and, consequently, the techno-economic feasibility of such units. High power consumption to operate plasma torches, periodic changes of their electrodes, expensive power supply units, and instability of the plasma flame, especially for MW and RF discharges, makes it challenging for further commercialization.

If compared just between the plasma means, the simplest among these and the most widely used is a DC arc plasma. The power capacity of the MW and RF plasmas can reach several kilowatts (3–5 kW, in some cases up to 10 kW) [51,53], whereas that of the DC arc plasma from hundreds of kilowatts up to several megawatts. This makes the DC arc plasma closer to industrial scale applications, as well as its reliability in operation. The greatest disadvantage of the DC arc plasma is a shorter lifetime of the electrodes due to the electric arc erosion. Moreover, in this particular case, the quality of the produced syngas was lower if compared to Yoon et al. [52], who used MW discharge for coal gasification.

To sum up, the overall situation for the plasmas being applied for syngas production from coal or other carbonaceous materials/wastes is promising in terms of its flexibility, environmental issues, easy process control, generation of active radicals, etc. This primary experiments with the charcoal made from an exhausted olive solid waste showed a promising result for future experiments with the thermal water vapour plasma method for syngas production.

## 4. Conclusions

In the current primary experimental research, the potential of synthesis gas production from the charcoal derived from the exhausted olive solid waste (olive pomace) by utilizing thermal arc plasma was investigated. The experiments were carried out at different $H_2O/C$ ratios, as well as different plasma torch powers. A direct relation between the water vapour flow rate and the plasma torch power at a constant arc current intensity was demonstrated. The main reaction products were hydrogen (41.17%) and carbon monoxide (13.06%), forming syngas, which comprised approximately 55% in the produced gas. The highest LHV of the produced syngas was 6.09 MJ/nm$^3$ and the $H_2/CO$ ratio was 3.15. The performed experiments showed great potential for syngas production from the charcoal derived from olive pomace as a residue/waste from the olive oil industry. Therefore, the future outlook will rely on a more detailed investigation of the performance parameters of the plasma gasification unit and its comparison with similar gasification methods in terms of cold gas efficiency, carbon conversion efficiency, syngas yield, specific energy requirements, etc.

**Author Contributions:** All authors contributed equally to the work done.

**Conflicts of Interest:** The authors declare no conflict of interest.

## References

1. Eurostat. Statistics Explained. Available online: http://ec.europa.eu/eurostat/statistics-explained/index. php/Waste_statistics (accessed on 8 February 2017).
2. Directive of the European Parliament and of the Council amending Directive 2008/98/EC on Waste. Available online: http://eur-lex.europa.eu/legal-content/EN/TXT/?uri=CELEX%3A52015PC0595 (accessed on 12 February 2015).
3. *Food and Agriculture Organization of the United Nations Statistical Databases*; Food and Agriculture Organization of the United Nations: Rome, Italy, 2014; Available online: www.faostat.fao.org/ (accessed on 8 February 2017).
4. European Commission. Agriculture and Rural Development. Available online: http://ec.europa.eu/ agriculture/olive-oil_en (accessed on 8 February 2017).
5. Guizani, C.; Haddad, K.; Jeguirim, M.; Colin, B.; Limousy, L. Combustion characteristics and kinetics of torrefied olive pomace. *Energy* **2016**, *107*, 453–463. [CrossRef]
6. Jeguirim, M.; Elmay, Y.; Limousy, L.; Lajili, M.; Said, R. Devolatilization behaviour and pyrolysis kinetics of potential Tunisian biomass fuels. *Environ. Prog. Sustain. Energy* **2014**, *33*, 1452–1458.
7. Molino, A.; Chianese, S.; Musmarra, D. Biomass gasification technology: The state of the art overview. *J. Energy Chem.* **2016**, *25*, 10–25. [CrossRef]
8. Sansaniwal, S.K.; Pal, K.; Rosen, M.A.; Tyagi, S.K. Recent advances in the development of biomass gasification technology: A comprehensive review. *Renew. Sustain. Energy. Rev.* **2017**, *72*, 363–384. [CrossRef]
9. Kumar, A.; Jones, D.D.; Hanna, M.A. Thermochemical biomass gasification: A review of the current status of the technology. *Energies* **2009**, *2*, 556–581. [CrossRef]
10. Bosmans, A.; Vanderreydt, I.; Geysen, D.; Helsen, L. The crucial role of waste-to-energy technologies in enhanced landfill mining: A technology review. *J. Clean. Prod.* **2013**, *55*, 10–23. [CrossRef]
11. Siciliano, A.; Stillitano, M.A.; De Rosa, S. Biogas production from wet olive mill wastes pretreated with hydrogen peroxide in alkaline conditions. *Renew. Energy* **2016**, *85*, 903–916. [CrossRef]
12. De Rosa, S.; Siciliano, A. A catalytic oxidation process of olive oil mill wastewaters using hydrogen peroxide and copper. *Desalination Water Treat.* **2010**, *23*, 187–193. [CrossRef]
13. Giordano, G.; Perathoner, S.; Centi, G.; De Rosa, S.; Granato, T.; Katovic, A.; Siciliano, A.; Tagarelli, A.; Tripicchio, F. Wet hydrogen peroxide catalytic oxidation of olive oil mill wastewaters using Cu-zeolite and Cu-pillared clay catalysts. *Catal. Today* **2007**, *124*, 240–246. [CrossRef]
14. Al-Mallhi, J.; Furuichi, T.; Ishii, K. Appropriate conditions for applying NaOH-pretreated two-phase olive milling waste for codigestion with food waste to enhance biogas production. *Waste Manag.* **2016**, *48*, 430–439. [CrossRef] [PubMed]
15. Christoforou, E.; Fokaides, P.A. A review of olive solid wastes to energy utilization techniques. *Waste Manag.* **2016**, *49*, 346–363. [CrossRef] [PubMed]
16. Vera, D.; Jurado, F.; Margaritis, N.K.; Grammelis, P. Experimental and economic study of a gasification plant fueled with olive industry wastes. *Energy Sustain. Dev.* **2014**, *23*, 247–257. [CrossRef]
17. Vera, D.; de Mena, B.; Jurado, F.; Schories, G. Study of a downdraft gasifier and gas engine fueled with olive oil industry wastes. *Appl. Therm. Eng.* **2013**, *51*, 119–129. [CrossRef]
18. Kipcak, E.; Akgun, M. Oxidative gasification of olive mill wastewater as a biomass source in supercritical water: Effects on gasification yield and biofuel composition. *J. Supercrit. Fluids* **2012**, *69*, 57–63. [CrossRef]
19. Margaritis, N.; Grammelis, P.; Vera, D.; Jurado, F. Assessment of Operational Results of a Downdraft Biomass Gasifier Coupled with a Gas Engine. *Procedia Soc. Behav. Sci.* **2012**, *48*, 857–867. [CrossRef]
20. Wang, L.; Weller, C.L.; Jones, D.D.; Hanna, M.A. Contemporary issues in thermal gasification of biomass and its application to electricity and fuel production. *Biomass Bioenergy* **2008**, *32*, 573–581. [CrossRef]
21. Vera, D.; Jurado, F.; Panopoulos, K.D.; Grammelis, P. Modelling of biomass gasifier and microturbine for the olive oil industry. *Int. J. Energy Res.* **2012**, *36*, 355–367. [CrossRef]
22. Evangelisti, S.; Tagliaferri, C.; Clift, R.; Lettieri, P.; Taylor, R.; Chapman, C. Life cycle assessment of conventional and two-stage advanced energy-from-waste technologies for municipal solid waste treatment. *J. Clean. Prod.* **2015**, *100*, 212–223. [CrossRef]
23. Bogaerts, A.; Neyts, E.; Gijbels, R.; van der Mullen, J. Gas discharge plasmas and their applications. *Spectrochim. Acta B* **2002**, *57*, 609–658. [CrossRef]

24. Huang, H.; Tang, L. Treatment of organic waste using thermal plasma pyrolysis technology. *Energy Convers. Manag.* **2007**, *48*, 1331–1337. [CrossRef]
25. Tamošiūnas, A.; Valatkevičius, P.; Gimžauskaitė, D.; Jeguirim, M.; Mėčius, V.; Aikas, M. Energy recovery from waste glycerol by utilizing thermal water vapor plasma. *Environ. Sci. Pollut. Res.* **2016**, in press. [CrossRef]
26. Fabry, F.; Rehmet, C.; Rohani, V.; Fulcheri, L. Waste gasification by thermal plasma: A review. *Waste Biomass Valoriz.* **2013**, *4*, 421–439. [CrossRef]
27. Rutberg, G.Ph.; Bratsev, A.N.; Kuznersov, V.A.; Popov, V.E.; Ufimtsev, A.A.; Shtengel, S.V. On efficiency of plasma gasification of wood residues. *Biomass Bioenergy* **2011**, *35*, 495–504. [CrossRef]
28. Arabi, K.; Aubry, O.; Khacef, A.; Cormier, J.M. Syngas production by plasma treatments of alcohols, bio-oils and wood. *J. Phys. Conf. Ser.* **2012**, *406*, 1–8. [CrossRef]
29. Tamošiūnas, A.; Valatkevičius, P.; Valinčius, V.; Levinskas, R. Biomass conversion to hydrogen-rich synthesis fuels using water steam plasma. *CR Chim.* **2016**, *19*, 433–440. [CrossRef]
30. Tamošiūnas, A.; Valatkevičius, P.; Grigaitienė, V.; Valinčius, V.; Striūgas, N. A cleaner production of synthesis gas from glycerol using thermal water steam plasma. *J. Clean. Prod.* **2016**, *130*, 187–194. [CrossRef]
31. Qiu, J.; He, X.; Sun, T.; Zhao, Z.; Zhou, Y.; Guo, Sh.; Zhang, J.; Ma, T. Coal gasification in steam and air medium under plasma conditions: A preliminary study. *Fuel Process. Technol.* **2004**, *85*, 969–982. [CrossRef]
32. Agon, N.; Hrabovsky, M.; Chumak, O.; Hlina, M.; Kopecky, V.; Mašlani, A.; Bosmans, A.; Helsen, L.; Skoblja, S.; Van Oost, G.; et al. Plasma gasification or refuse derived fuel in a single-stage system using different gasifying agents. *Waste Manag.* **2016**, *47*, 246–255. [CrossRef] [PubMed]
33. Du, C.; Wu, J.; Ma, D.; Liu, Y.; Qiu, P.; Qiu, R.; Liao, S.; Gao, D. Gasification of corn cob using non-thermal arc plasma. *Int. J. Hydrogen Energy* **2015**, *40*, 12634–12649. [CrossRef]
34. Vamvuka, D.; Zografos, D. Predicting the behaviour of ash from agricultural wastes during combustion. *Fuel* **2004**, *83*, 2051–2057. [CrossRef]
35. Kraiem, N.; Lajili, M.; Limousy, L.; Said, R.; Jeguirim, M. Energy recovery from Tunisian agri-food wastes: Evaluation of combustion performance and emissions characteristics of green pellets prepared from tomato residues and grape marc. *Energy* **2016**, *107*, 409–418. [CrossRef]
36. Bouraoui, Z.; Jeguirim, M.; Guizani, C.; Limousy, L.; Dupont, C.; Gadiou, R. Thermogravimetric study on the influence of structural, textural and chemical properties of biomass chars on $CO_2$ gasification reactivity. *Energy* **2015**, *88*, 703–710. [CrossRef]
37. Jeguirim, M.; Bikai, J.; Elmay, Y.; Limousy, L.; Njeugna, E. Thermal characterization and pyrolysis kinetics of tropical biomass feedstocks for energy recovery. *Energy Sustain. Dev.* **2014**, *23*, 188–193. [CrossRef]
38. Blanco-López, M.C.; Blanco, C.G.; Martimez-Alonso, A.; Tascou, J.M.D. Composition of gases released during olive stones pyrolysis. *J. Anal. Appl. Pyrolysis* **2002**, *65*, 313–322. [CrossRef]
39. Chouchene, A.; Jeguirim, M.; Favre-Reguillon, A.; Trouvé, G.; Le Buzit, G.; Khiari, B.; Zagrouba, F. Energetic valorisation of olive mill wastewater impregnated on low cost absorbent: Sawdust versus Olive Solid Waste. *Energy* **2012**, *39*, 74–81. [CrossRef]
40. Ozveren, U.; Ozdogan, Z.S. Investigation of the slow pyrolysis kinetics of olive oil pomace using thermo-gravimetric analysis coupled with mass spectrometry. *Biomass Bioenergy* **2013**, *58*, 168–179. [CrossRef]
41. Pantoleontos, G.; Basinas, P.; Skodras, G.; Grammelis, P.; Pintér, J.D.; Topis, S.; Sakellaropoulos, G.P. A global optimization study on the devolatilisation kinetics of coal, biomass and waste fuels. *Fuel Process. Technol.* **2009**, *90*, 762–769. [CrossRef]
42. Chouchene, A.; Jeguirim, M.; Khiari, B.; Trouvé, G.; Zagrouba, F. Thermal degradation behavior of olive solid waste: Influence of the particle size and oxygen atmosphere. *Resour. Conserv. Recycl.* **2010**, *54*, 271–277. [CrossRef]
43. Hmid, A.; Mondelli, D.; Fiore, S.; Fanizzi, F.P.; Al Chami, Z.; Dumontet, S. Production and characterization of biochar from three-phase olive mill waste through slow pyrolysis. *Biomass Bioenergy* **2014**, *71*, 330–339. [CrossRef]
44. Abenavoli, L.M.; Longo, L.; Roto, A.R.; Gallucci, F.; Ghigoli, A.; Zimbalatti, G.; Russo, D.; Colantoni, A. Characterization of biochar obtained from olive and hazelnut prunings and comparison with the standards of European Biochar Certificate (EBC). *Procedia Soc. Behav. Sci.* **2016**, *223*, 698–705. [CrossRef]
45. Kim, Y.; Ferreri, V.W.; Rosocha, L.A.; Anderson, G.K.; Abbate, A.; Kim, K.T. Effect of plasma chemistry on activated propane/air flames. *IEEE Trans. Plasma Sci.* **2006**, *34*, 2532–2536. [CrossRef]

46. Tamošiūnas, A.; Valatkevičius, P.; Grigaitienė, V.; Valinčius, V. Production of synthesis gas from propane using thermal water vapor plasma. *Int. J. Hydrogen Energy* **2014**, *39*, 2078–2086. [CrossRef]

47. Arena, U. Process and technological aspects of municipal solid waste gasification. A review. *Waste Manag.* **2012**, *32*, 626–639. [CrossRef] [PubMed]

48. Tamošiūnas, A.; Valatkevičius, P.; Grigaitienė, V.; Valinčius, V. Water vapor plasma torch: Design, characteristics and applications. *World Acad. Sci. Eng. Technol.* **2012**, *6*, 10–13.

49. De Souza-Santos, M.L. *Solid Fuels Combustion and Gasification: Modeling, Simulation, and Equipment Operations*, 2nd ed.; CRC Press: Boka Raton, FL, USA, 2010; p. 508.

50. Breault, R.W. Gasification processes old and new: A basic review of the major technologies. *Energies* **2010**, *3*, 216–240. [CrossRef]

51. Hong, Y.C.; Lee, S.J.; Shin, D.H.; Kim, Y.J.; Lee, B.J.; Cho, S.Y.; Chang, H.S. Syngas production from gasification of brown coal in a microwave torch plasma. *Energy* **2012**, *47*, 36–40. [CrossRef]

52. Yoon, S.J.; Lee, J.G. Hydrogen-rich syngas production through coal and charcoal gasification using microwave steam and air plasma torch. *Int. J. Hydrogen Energy* **2012**, *37*, 17093–17100. [CrossRef]

53. Messerle, V.E.; Ustimenko, A.B.; Lavrichshev, O.A. Comparative study of coal gasification: Simulation and experiment. *Fuel* **2016**, *164*, 172–179. [CrossRef]

54. Chen, Z.; Dun, Q.; Shi, Y.; Lai, D.; Zhou, Y.; Gao, S.; Xu, G. High quality syngas production from catalytic coal gasification using disposable $Ca(OH)_2$ catalyst. *Chem. Eng. J.* **2017**, *316*, 842–849. [CrossRef]

*energies*

**MDPI**

*Article*

# Optimization of a Bubbling Fluidized Bed Plant for Low-Temperature Gasification of Biomass

**María Pilar González-Vázquez, Roberto García, Covadonga Pevida and Fernando Rubiera \***

Instituto Nacional del Carbón, INCAR-CSIC, Apartado 73, 33080 Oviedo, Spain;
mariapilar.gonzalez@incar.csic.es (M.P.G.-V.); roberto.garcia@incar.csic.es (R.G.); cpevida@incar.csic.es (C.P.)
\* Correspondence: frubiera@incar.csic.es; Tel.: +34-985-119-090

Academic Editor: Shusheng Pang
Received: 8 February 2017; Accepted: 28 February 2017; Published: 4 March 2017

**Abstract:** Investigation into clean energies has been focused on finding an alternative to fossil fuels in order to reduce global warming while at the same time satisfying the world's energy needs. Biomass gasification is seen as a promising thermochemical conversion technology as it allows useful gaseous products to be obtained from low-energy-density solid fuels. Air–steam mixtures are the most commonly used gasification agents. The gasification performances of several biomass samples and their mixtures were compared. One softwood (pine) and one hardwood (chestnut), their torrefied counterparts, and other Spanish-based biomass wastes such as almond shell, olive stone, grape and olive pomaces or cocoa shell were tested, and their behaviors at several different stoichiometric ratios (SR) and steam/air ratios (S/A) were compared. The optimum SR was found to be in the 0.2–0.3 range for S/A = 75/25. At these conditions a syngas stream with 35% of $H_2$ + CO and a gas yield of 2 L gas/g fuel were obtained, which represents a cold-gas efficiency of almost 50%. The torrefaction process does not significantly affect the quality of the product syngas. Some of the obtained chars were analyzed to assess their use as precursors for catalysts, combustion fuel or for agricultural purposes such as soil amendment.

**Keywords:** biomass gasification; bubbling fluidized bed; biomass mixtures; torrefaction; syngas; air–steam oxidation; char reuse

## 1. Introduction

The constant growth of the world's energy demand, combined with the limited reserves of fossil fuels, their fluctuating prices and the environmental damage they cause, has led to the search for a sustainable and environmentally friendly fuel that complements traditional fossil fuels as the main energy source. This aim is supported by several different trans-national policies starting with the Kyoto Protocol of 1998 and most recently the Paris agreement and the European Union's "Horizon 2020" program, which aims to improve efficiency until reaching a 20% reduction in Europe's energy consumption and a 20% increase in the use of renewable energy [1–3]. According to the International Energy Agency (IEA), the use of renewable energy sources in Spain reached 14.9% of overall consumption in 2014, with Spain taking twelfth place among IEA members [4].

Some of the reported advantages of biomass, such as its $CO_2$ life cycle neutrality [5], its moderate $NO_x$ or $SO_2$ emissions and its autonomy as a resource [6], make it a very useful feedstock for achieving these fixed goals. The energy conversion of biomass can be performed in different ways, bio and thermochemical being the most common. Biochemical conversion can be achieved by fermentation or anaerobic digestion [7–9], whilst the thermochemical processes include combustion, pyrolysis and gasification, with the oxidizing agent being the major difference between them, since combustion requires an excess of air while pyrolysis takes place in an inert atmosphere. Of these techniques, gasification is one of the most promising as it allows solid matter with a low energy value to be converted into a clean gaseous fuel that is easy to handle [10] by partially oxidizing carbonaceous fuels from low temperatures (600–650 °C), thereby preventing sintering, agglomeration and other ash-melting-related problems [11,12].

The quality and composition of the outlet gas depends on the selected fuel, oxidizing agent (oxygen, air, steam, carbon dioxide or their mixtures) and its ratio, gasification equipment (fixed, fluidized bed or entrained flow gasifier) and reaction conditions (temperature or bed material) [13]. This product has many possible uses ranging from the direct production of heat and electricity to the production of a wide variety of chemicals [14,15].

The gasification process takes place in five stages: pyrolysis, volatiles and char combustion, char gasification and gas–gas reactions [16–18], the main reactions being:

Water-Gas Reaction I

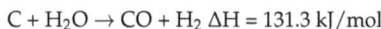

$$C + H_2O \rightarrow CO + H_2 \; \Delta H = 131.3 \; kJ/mol$$

Water-Gas Reaction II

$$C + 2 \, H_2O \rightarrow CO_2 + 2 \, H_2 \; \Delta H = 89.7 \; kJ/mol$$

Water-Gas Shift Reaction III

$$CO + H_2O \rightarrow CO_2 + H_2 \; \Delta H = -41.2 \; kJ/mol$$

Methane-Steam Reforming Reaction IV

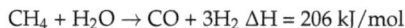

$$CH_4 + H_2O \rightarrow CO + 3H_2 \; \Delta H = 206 \; kJ/mol$$

Water-gas (I), (II) and water-gas shift reactions (III) are the most important when using air/steam mixtures as gasifying agent, as in the present work. The last reaction, methane-steam reforming, is highly useful as it increases the quantity of $H_2$ in the gaseous product [19].

Gasification is quite a well-known energy conversion technique that is commonly applied to biomass, using different types of reactor and gasifying agents. For example, Guizanni et al. [20] used a Macro-TG analyzer to study the influence of the conversion degree (20%, 50% and 70%) and the oxidizing agent used ($CO_2$, $H_2O$ or their mixture) during beech wood gasification on its char structure, surface chemistry and mineral content. Kuo [21] used different air–steam blends to gasify raw bamboo and two of its torrefied products (250, 300 °C) in downdraft fixed bed equipment to study their cold gas efficiency and carbon conversion in each case. Entrained flow gasifiers have also been used in some cases, as in Chen's work [22], where torrefied bamboo was gasified in an $O_2$-impoverished atmosphere, or as in Hernandez's study [16] in which grape marc was oxidized using different steam/$O_2$ mixtures. The latter concluded that the optimum quantity of steam in the oxidizing agent blend ranges between 40% and 70% in mole percentage.

The gasification equipment most widely used is the bubbling fluidized bed, as it has some major advantages like the possibility of using many different fuels and gaseous agents, permitting a wide flexibility of operation [14,23]. Of the works developed at bench-scale those of Makwana [24] and Skoulou [15] both used air to respectively gasify rice husk and olive kernels. Rapagna [25] partially oxidized almond shells in pure steam, whilst Kulkarni [26] used $N_2$-$O_2$ blends to gasify torrefied pine. Zaccariello [23] co-gasified plastic wastes, wood and coal in a pre-pilot plant concluding that the gas yield increases and $H_2$ decreases with plastic content in the feeding mixture. Mohd Salleh [27] treated the biochar of empty fruit bunches in an air atmosphere with the aim of determining the effect of temperature (500–850 °C) on the quality of the product gas, focusing on its composition and HHV. Pinto [28] gasified straw-lignin pellets in steam/oxygen, using stoichiometric ratios (SR) of 0 to 0.3 and a steam-to-biomass ratio (SBR) of 0.7 to 1.2 at different temperatures in the range of 750–900 °C, and achieved conversions of over 65% at 900 °C, for SR = 0.2 and SBR = 1.0. Gil et al. [29] used a pilot scale bubbling fluidized bed to study the influence of the selected gasifying agent (air, steam and steam/$O_2$) and its ratio to biomass (pine chips) on the distribution of the gas product (gas, char and tar yields) and its quality ($H_2$, CO, $CO_2$, $CH_4$). It is also worth noting that if air, $O_2$ or their blends with steam are employed as the gasifying agent, an auto-thermal process is obtained [10,19,30].

As has been widely reported, raw biomass presents certain problems of its own. These are its general heterogeneity, low energy density and highly hygroscopic behavior, which can negatively affect its storage, handling, grinding and transportation properties [31–33]. These limitations can be remedied by torrefaction, which is a thermal pre-treatment at a mild temperature (200–300 °C), for 30–180 min, in an inert ($N_2$) or low reactive atmosphere (3%–6% $O_2$) [34], in which moisture and light volatiles originating from the decomposition of hemicellulose are eliminated [35,36]. As a result, the treated biomass has a much lower moisture content and a higher energy density, qualities that enhance its hydrophobicity and grinding properties [37]. This process can be successfully combined with gasification [28,29,33].

The aim of the present work is to compare the behavior of different biomass samples (wood, torrefied biomass, agricultural and industrial wastes) after gasification in a bubbling fluidized bed gasifier at mild temperatures (600 °C), using an air–steam mixture at different stoichiometric and steam/air ratios (SR, S/A) as oxidizing agent. The quality of the product gas flow was analyzed on the basis of its composition (CO, $H_2$, $CH_4$), high heating value (HHV), gas yield ($\eta_{gas}$) and cold-gas efficiency ($\eta_{cold-gas}$). In addition, the possibility of reusing the carbon-rich partially oxidized chars in agricultural applications, for thermal conversion or as a catalyst-sorbent precursor was evaluated.

## 2. Materials and Methods

### 2.1. Gasification Equipment

A highly versatile gasification pilot plant allowing the use of different fuels and steam and air flows was employed. In this way, a wide range of stoichiometric ratios (SR) [38], steam-to-biomass ratios (SBR) [29] and steam-air ratios (S/A) [16] could be tested. These parameters can be defined as follows:

$$SR = \frac{\left(\frac{\text{Oxygen}}{\text{fuel}}\right)_{used}}{\left(\frac{\text{Oxygen}}{\text{fuel}}\right)_{stoichiometric}} \tag{1}$$

$$SBR = \frac{\text{Total water supplied} \left(\frac{g}{min}\right)}{\text{Fuel supplied to SR} \left(\frac{g}{min}\right)} \tag{2}$$

$$\frac{S}{A} = \frac{\text{Total water supplied} \left(\frac{g}{min}\right)}{\text{Total air supplied} \left(\frac{g}{min}\right)} \tag{3}$$

Depending on the fuel's characteristics the plant can treat up to nearly 10 kg/h, providing a maximum of 55 kWth. The plant consists of the following components: fuel feeding, gas inlet and pre-heating, gasification reactor, outlet gas cleaning system and control and monitorization systems that are shown in Figure 1.

The fuel feeding system includes a 16.7 L storage hopper and an 8 mm-diameter refrigerated worm gear that introduces the sample into the fluidized bed reactor. The auger's rotation speed can be calibrated so that it provides the mass flow of fuel required in each case.

The reactive gases (air, $N_2$ or $O_2$) are supplied by two Bronkhorst High-Tech mass flow controllers that can provide up to 240 NL/min of overall gas, which is enough to satisfy the reaction and fluidization requirements of this equipment. A Wilson 307 piston pump feeds in the selected mass flow of liquid water which is subsequently heated up to 400 °C to ensure a continuous condensate-free flow of steam into the reactor. The mix of gases involved in the reaction process is then pre-heated by a 4.5 kW Watrod SS310 circulation heater.

The SS310 cylindrical gasification reactor is 1 m in length, has an inside diameter of 77 mm and approximately 1.3 kg of coal ash (212–710 μm) was used as reaction bed. The reactor ends in a 529 mm long and 133 mm diameter freeboard. Both the reactor and freeboard are surrounded by two independent ovens that can supply a maximum power of 22 kW, which is sufficient to raise the temperature at the reactor wall to 920 °C.

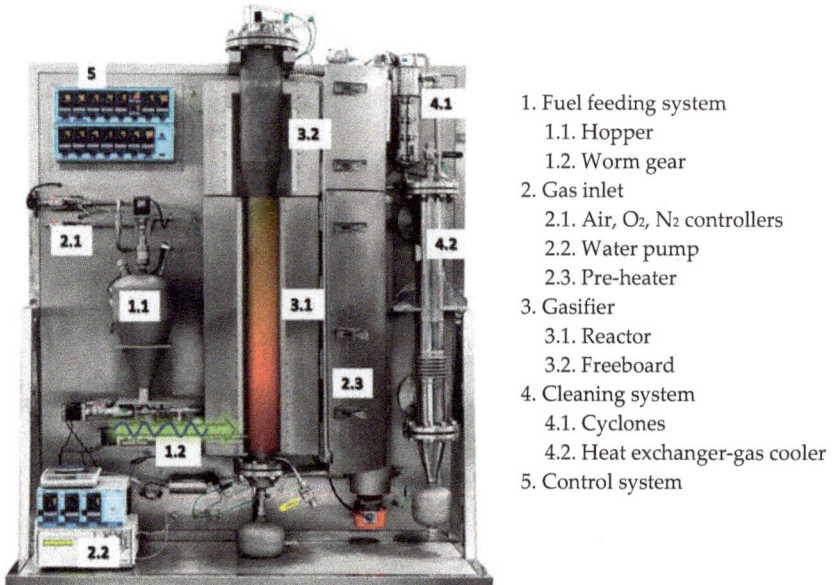

**Figure 1.** Photograph of the gasification plant with its main components.

1. Fuel feeding system
    1.1. Hopper
    1.2. Worm gear
2. Gas inlet
    2.1. Air, $O_2$, $N_2$ controllers
    2.2. Water pump
    2.3. Pre-heater
3. Gasifier
    3.1. Reactor
    3.2. Freeboard
4. Cleaning system
    4.1. Cyclones
    4.2. Heat exchanger-gas cooler
5. Control system

After completing the reaction process, the outgoing gas flow crosses a double cyclone system heated at 400 °C, where particulate matter is eliminated, and then a heat exchanger where condensed water and light tars are separated from gaseous emissions. After this first cleaning step, the remaining flow goes through a cold trap that captures the heavier tars. The cleaned gases are sent to four Rosemount Binos® 100 gas analyzers where the main gases ($CO$, $CO_2$, $H_2$, $CH_4$ and excess $O_2$) are measured in terms of volume percentage.

There is also a control system that measures and continuously monitors every parameter involved in the reaction process, such as the inlet and outlet gas flows, temperatures or pressure drop to assure full control of the whole process.

### 2.2. Biomass Characterization

As previously mentioned, this work focuses on testing the gasification behavior of nine biomass samples of different origin (wood, forest and food industry wastes). One hardwood and two softwood samples (chestnut (CHE) and pine (PIN) sawdust, respectively) together with their torrefied products (CHET, PINT) , obtained by heating at 280 °C for 1 h in a Nabertherm RSR horizontal tubular rotary furnace [37], were studied. In addition, two Spanish seasonal products, almond shells (AS) and olive stones (OS), were tested. Two wastes obtained from well-known Spanish food industries, grape (GP) and olive pomaces (OP), and a sample of cocoa shell (CS) were also studied.

Every sample was air-dried at room temperature for several hours to eliminate external moisture and then ground and sieved down to the range 0.1–1 mm. After this, they were fully characterized;

fixed carbon (FC) and oxygen (O) contents were calculated by difference in mass percentage in dry basis [39,40]:

$$FC\ (\%)\ =\ 100 - (VM + Ash) \tag{4}$$

$$O\ (\%) = 100 - (Ash + C + N + H + S) \tag{5}$$

Bulk density (BD) was determined using a commercial device (Quantachrome Instruments Autotap-tapped density analyzer), by measuring the mass and volume of each sample in a 250 mL test tube after 300 hits.

Gaseous emissions were measured and recorded by means of gas analyzers and suitable software, and the results were used to calculate the gas yield ($\eta_{gas}$) [16] by applying a balance to the inert gas ($N_2$) that came into the reactor as part of the air, the HHV of the syngas obtained [41], measured in kJ/Nm$^3$, and the cold-gas efficiency [42]:

$$\eta_{gas} = \frac{\dot{Q}_{outlet-gas}}{\dot{m}_{fuel}} \tag{6}$$

$$HHV_{gas} = \left(x_{CO} \cdot 3018 + x_{H_2} \cdot 3052 + x_{CH_4} \cdot 9500\right) \tag{7}$$

where $\dot{Q}$ and $\dot{m}$ are the volumetric and mass flows, respectively, and $x_i$ is the volume percentage of each gas in the product flow.

$$\eta_{cold-gas} = \eta_{gas} \cdot \frac{HHV_{gas}}{HHV_{fuel}} \tag{8}$$

The samples that presented the best results were mixed in different mass ratios and their thermogravimetric profiles were studied, using a Setaram TAG24 thermogravimetric analyzer (Caluire, France). To this end 5 mg of fuel was subjected to a 15 °C/min temperature ramp from room temperature to 900 °C maintained for 1 h. The gas flow was fixed at 50 mL/min of air impoverished to simulate the oxidizing gas used during the gasification experiments.

In addition, the particle size distribution of the samples used in the biomass mixtures was determined by sieving 80 g of sample for 15 min on a Retsch AS200 sieve shaker (Haan-Gruiten, Germany) using three sieves with mesh sizes of 150, 500 and 710 μm.

*2.3. Char Samples*

As previously mentioned, the gasification conditions selected for this work (a mild temperature and high steam content in the oxidizing gas flow) imply lower carbon conversion levels and an increase in the amount of partially oxidized chars compared to those obtained in more aggressive atmospheres. Such chars could be reused as fertilizer, fuel for pure combustion or as precursor for catalysts or sorbents as suggested by Qian et al. [43].

With this aim in mind, chars obtained from samples CHE, OS and GP were studied using a Quanta FEG 650 scanning electron microscope (SEM) (Eindhoven, The Netherlands) coupled to an Ametek energy dispersive X-ray analyzer (EDX) (Tilburg, The Netherlands), in order to obtain structural and semi-quantitative composition information.

In addition, textural characterization of some of the studied samples was also carried out by applying nitrogen physical adsorption at −196 °C on a Micromeritics ASAP 2010 and their surface area was calculated by means of the Brunauer-Emmet-Teller (BET) equation.

## 3. Results and Discussion

*3.1. Biomass Analysis*

The ultimate and proximate analysis, HHV and bulk density are listed in Table 1.

**Table 1.** Proximate and ultimate analyses, high heating value (HHV) and bulk density of the studied samples.

| Samples | Ultimate Analysis (wt %, db) | | | | | Proximate Analysis (wt %, db) | | | HHV (MJ/kg, db) | BD (kg/m$^3$) |
|---|---|---|---|---|---|---|---|---|---|---|
| | C | N | H | S | O | Ash | VM * | FC | | |
| CHE | 50.2 | 0.3 | 5.6 | 0.01 | 43.4 | 0.5 | 81.2 | 18.3 | 19.1 | 296 |
| CHET | 51.3 | 0.4 | 5.4 | 0.02 | 42.6 | 0.3 | 80.0 | 19.7 | 19.6 | 374 |
| PIN | 51.0 | 0.3 | 6.0 | 0.02 | 42.2 | 0.4 | 85.1 | 14.5 | 19.9 | 226 |
| PINT | 56.1 | 0.4 | 5.7 | 0.01 | 37.4 | 0.5 | 77.1 | 22.4 | 22.0 | 265 |
| AS | 49.4 | 0.3 | 5.9 | 0.05 | 42.9 | 1.5 | 78.9 | 19.6 | 19.6 | 655 |
| OS | 51.2 | 0.3 | 6.0 | 0.03 | 41.9 | 0.6 | 81.5 | 17.9 | 20.5 | 781 |
| GP | 45.5 | 1.8 | 5.1 | 0.17 | 34.7 | 12.7 | 67.6 | 19.7 | 18.7 | 772 |
| OP | 49.4 | 1.6 | 5.4 | 0.12 | 37.4 | 6.2 | 72.5 | 21.3 | 20.3 | 772 |
| CS | 48.0 | 2.7 | 5.9 | 0.21 | 35.3 | 7.9 | 70.4 | 21.7 | 19.1 | 490 |

* VM = volatile matter; FC: fixed carbon; db: dry basis.

### 3.2. Preliminary Selection of Operation Conditions

The aim of this work is to determine the best gasification conditions for the selected biomass samples from the wide range of SR and S/A ratios originally studied. To this end PIN was subjected to reaction atmospheres with different S/A ratios (85/15, 75/25, 50/50, 25/75 and 15/85), combined with a SR from 0.1 to 0.7, implying a SBR range from 0.14 to 8.58. The results obtained shown in Figure 2, prove that too large SR impoverishes the properties of the outlet flow and leads to a decrease in CO, H$_2$ and CH$_4$, and hence HHV. On the other hand, a too low stoichiometric ratio, SR, minimizes the $\eta_{gas}$, and causes a decrease in the energy yield. In addition, a high S/A ratio seems to improve the properties of the outlet gas but slows down the reaction due to the smaller flow of carrier gas that impoverishes the phase contact in the gasifier. Because of this, an SR range between 0.1–0.4 and a medium-high S/A ratio (50/50, 75/25), i.e., a SBR from 0.24 to 4.69, were selected for the rest of the experiments. These ranges are reasonably similar to the ones commonly reported in the literature [12,15,38].

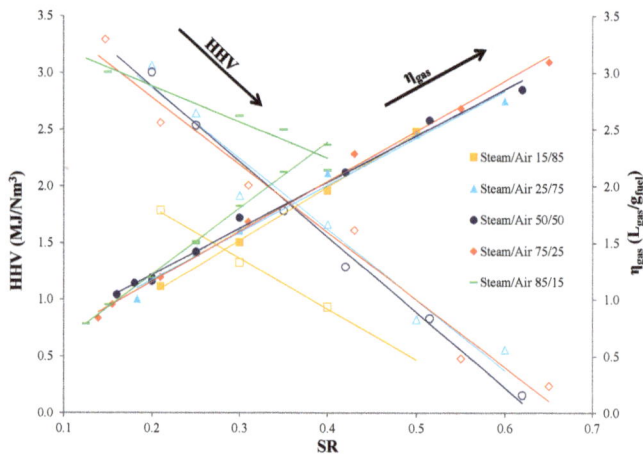

**Figure 2.** Gas yield and HHV of the gas obtained for PIN at different steam/air ratios.

### 3.3. Effect of SR and Steam/Air Ratio on the Product Gas Properties

The sum of gaseous emissions obtained, (CO + H$_2$), $\eta_{gas}$ (L$_{gas}$/g$_{fuel}$) and $\eta_{cold-gas}$ (%) for the previously selected conditions are presented in Figure 3. In addition, details of the gaseous emissions

($CO$, $H_2$, $CH_4$) and their calculated HHV are shown in Table 2. Some general statements can be made on the basis of these results. For example every gaseous emission ($CO$, $H_2$, $CH_4$), and thus HHV, decreases as SR increases for both tested S/A ratios. The gas yield shows the opposite trend (Figure 3), with a linear increase with SR, for both of the tested steam/air ratios. As cold gas efficiency relates these two terms that have conflicting tendencies, an optimum value is achieved, in most cases in the SR range selected, again for both S/A ratios.

A comparison of the biomass samples shows that quite high values were obtained in the total amount of syngas produced from OS, AS, PIN and PINT. By contrast, raw and torrefied chestnut (CHE and CHET), the pomaces (GP and OP) and CS gave poorer yields. This result is much more marked in the experiments carried out at a higher S/A (75/25), where two distinct groups can be observed, with CHET, the best of the second group, yielding 20% syngas at a SR = 0.2, whilst PIN and PINT nearly reach 30% at this point. The highest levels for the combination $CO + H_2$ were obtained for OS at both S/A ratios (50/50 and 75/25), though it was slightly higher in the second case (35% and 30% at SR = 0.2 and 37%–42% at SR = 0.1). From the slight differences between PIN-CHE and their torrefied couples PINT-CHET, it can be concluded that torrefaction does not significantly improve the gasification properties for either hardwood or softwood, as previously reported by Kulkarni et al. [26].

The highest $\eta_{gas}$ values were obtained with the torrefied samples (PINT and CHET), with values close to 2.5 and 2 $L_{gas}/g_{fuel}$, respectively, at SR = 0.4. At SR = 0.2 the values are slightly lower, reaching 1.5 $L_{gas}/g_{fuel}$ in the case of PINT. For all the samples excepting PINT, the $\eta_{gas}$ results obtained at both S/A are similar, so it can be concluded that the steam feeding does not significantly affect the gas-volume production.

It can also be seen that the $\eta_{cold-gas}$ of most samples presents an optimum value in the selected SR range, between 0.1 and 0.4. Important differences between the samples become apparent on considering different steam-air ratios. The best results for the 50/50 air–steam ratio were obtained with PINT and AS, with yields close to 35% at a SR between 0.2 and 0.3 in both cases. OS and CS present a linear increase in the range studied, which means that these samples require a richer air atmosphere to optimize their gasification process at a low S/A ratio. CHET presents its maximum yield at a SR close to 0.25, reaching 30%. All the other samples show levels under 25%.

Things change considerably at the 75/25 steam-air ratio. In this case OS reaches values close to 50% in the SR range between 0.2–0.3, followed by PINT and AS with yields over 40% and 35%, respectively, in the same range as in the previous case, generally considered as the optimum range for biomass gasification in other previous works [26,43,44]. All the other samples reach levels below 30%, with the worst values, just slightly over 10%, corresponding to CHE and CS. A significant difference is observed in the case of the $\eta_{cold-gas}$ of the raw samples and their corresponding torrefied counterparts at both S/A ratios. As can be seen, the values of the torrefied samples are much higher, whilst in the case of the pure gaseous emissions they are quite similar. This can be attributed to the higher density of the samples after the torrefaction process. As a result a smaller mass of sample is required to obtain the same gas flow, which leads to a higher yield.

When analyzing the influence of S/A on the gaseous composition and HHV of the outlet gas, no general conclusion can be drawn. From Table 2 it can be seen that an increase in this ratio from 50/50 to 75/25 does not influence every sample in the same way. In the case of CHE and CHET their behavior is quite similar, as every gaseous compound ($CO$, $CH_4$ and $H_2$), and therefore HHV, decreases as the S/A increases. The only exceptions are the $H_2$ emissions with CHET at a SR of below 0.2 and the CO emissions with CS at a SR of above 0.3 which increase as the S/A ratio increases. The gaseous behavior with PIN, PINT and OS is quite similar, as CO remains almost unvaried as the steam/air increases while $CH_4$, $H_2$ and HHV experience a slight increase in their values as the S/A ratio rises from 50/50 to 75/25. AS and GP also present the same evolution as CO decreases, whereas $CH_4$ remains nearly constant and $H_2$ increases, meaning the overall HHV remains constant or undergoes a slight decrease. OP shows the most changeable behavior with the variation in S/A, as CO, $H_2$ and in turn HHV increase with the steam/air ratios, but only at SR values of below 0.2.

On the other hand $CH_4$ decreases as the steam/air ratio increases from 50/50 to 75/25. From Table 2, it can be seen that the quantity of steam supplied plays a key role in the composition of the outgoing gas, and in the richness of the syngas obtained during the gasification process, but this is very much dependent on the type of fuel and the operation conditions (SR). Consequently, no general statement can be made about the influence of S/A on the gasification process, so each sample must be carefully studied individually and its gasification conditions optimized to maximize the components required in the gaseous stream product.

**Table 2.** Gaseous emissions (vol. %) and HHV (in MJ/Nm$^3$) obtained for the studied samples.

| Samples | 50% Steam-50% Air (v/v) | | | | | | 75% Steam-25% Air (v/v) | | | | | |
|---|---|---|---|---|---|---|---|---|---|---|---|---|
| | SR | SBR | CO | CH$_4$ | H$_2$ | HHV | SR | SBR | CO | CH$_4$ | H$_2$ | HHV |
| CHE | | | | | | | 0.1 | 1.1 | 11.4 | 2.4 | 8.9 | 3.5 |
| | 0.2 | 0.7 | 11.1 | 3.8 | 9.0 | 4.1 | 0.2 | 2.2 | 6.9 | 2.2 | 6.0 | 2.5 |
| | 0.3 | 1.1 | 8.2 | 2.9 | 6.8 | 3.0 | 0.3 | 3.2 | 4.3 | 1.6 | 3.4 | 1.6 |
| | 0.4 | 1.4 | 5.7 | 1.5 | 4.0 | 1.8 | 0.4 | 4.3 | 3.7 | 0.5 | 0.4 | 0.7 |
| CHET | 0.15 | 0.5 | 14.4 | 4.5 | 10.3 | 4.9 | 0.1 | 1.1 | 10.9 | 3.4 | 14.0 | 4.5 |
| | 0.2 | 0.7 | 12.6 | 3.9 | 9.2 | 4.3 | 0.2 | 2.2 | 9.4 | 2.8 | 9.5 | 3.5 |
| | 0.3 | 1.1 | 9.5 | 2.9 | 6.9 | 3.2 | 0.3 | 3.3 | 5.9 | 1.7 | 6.0 | 2.2 |
| | 0.4 | 1.5 | 5.5 | 1.4 | 3.3 | 1.7 | 0.4 | 4.4 | 2.6 | 0.5 | 1.4 | 0.7 |
| PIN | | | | | | | 0.15 | 1.1 | 15.4 | 5.4 | 14.6 | 6.0 |
| | 0.24 | 0.6 | 13.4 | 4.1 | 10.6 | 4.7 | 0.2 | 1.5 | 13.4 | 4.6 | 13.0 | 5.2 |
| | 0.3 | 0.8 | 10.5 | 2.9 | 9.1 | 3.6 | 0.3 | 2.3 | 10.0 | 3.2 | 10.3 | 3.8 |
| | 0.4 | 1.0 | 7.2 | 1.6 | 6.3 | 2.4 | 0.4 | 3.0 | 7.7 | 2.2 | 8.3 | 2.9 |
| PINT | 0.15 | 0.6 | 18.5 | 6.2 | 11.9 | 6.3 | 0.1 | 1.2 | 21.0 | 5.3 | 18.2 | 7.1 |
| | 0.2 | 0.8 | 14.4 | 4.9 | 9.9 | 5.1 | 0.2 | 2.3 | 11.9 | 4.4 | 13.6 | 5.0 |
| | 0.3 | 1.2 | 10.1 | 3.6 | 7.5 | 3.7 | 0.3 | 3.5 | 9.4 | 3.1 | 11.3 | 3.9 |
| | 0.4 | 1.6 | 7.2 | 1.7 | 5.4 | 2.3 | 0.4 | 4.7 | 7.9 | 2.2 | 5.8 | 2.6 |
| AS | 0.1 | 0.2 | 19.7 | 11.1 | 12.5 | 8.5 | 0.1 | 0.7 | 16.1 | 6.1 | 20.1 | 7.0 |
| | 0.2 | 0.5 | 14.3 | 8.6 | 10.3 | 6.5 | 0.2 | 1.4 | 13.3 | 5.1 | 17.4 | 5.9 |
| | 0.3 | 0.7 | 11.2 | 6.7 | 8.5 | 5.2 | 0.3 | 2.2 | 10.4 | 3.5 | 14.2 | 4.5 |
| | 0.4 | 1.0 | 9.3 | 5.0 | 7.1 | 4.1 | 0.4 | 2.9 | 7.6 | 2.5 | 10.8 | 3.3 |
| OS | 0.1 | 0.3 | 25.0 | 5.3 | 12.0 | 6.8 | 0.1 | 0.8 | 23.5 | 8.9 | 18.7 | 8.9 |
| | 0.2 | 0.5 | 18.6 | 5.3 | 9.7 | 5.7 | 0.2 | 1.5 | 17.0 | 6.4 | 16.4 | 6.8 |
| | 0.3 | 0.8 | 14.5 | 4.7 | 8.1 | 4.7 | 0.3 | 2.3 | 14.2 | 5.0 | 14.0 | 5.6 |
| | 0.4 | 1.0 | 12.2 | 4.4 | 7.1 | 4.2 | 0.4 | 3.1 | 10.6 | 4.1 | 11.5 | 4.5 |
| GP | 0.1 | 0.3 | 11.1 | 6.7 | 12.0 | 5.6 | 0.1 | 0.9 | 9.2 | 6.3 | 14.4 | 5.5 |
| | 0.2 | 0.6 | 7.9 | 4.5 | 9.4 | 4.0 | 0.2 | 1.8 | 6.8 | 4.3 | 10.2 | 3.9 |
| | 0.3 | 0.9 | 5.8 | 3.3 | 6.5 | 2.9 | 0.3 | 2.8 | 5.4 | 2.7 | 7.0 | 2.7 |
| | 0.4 | 1.2 | 3.6 | 1.6 | 2.7 | 1.5 | | | | | | |
| OP | 0.1 | 0.3 | 12.5 | 6.3 | 11.9 | 5.6 | 0.1 | 1.0 | 12.0 | 5.2 | 13.9 | 5.4 |
| | 0.2 | 0.7 | 8.6 | 4.5 | 7.7 | 3.9 | 0.15 | 1.6 | 9.9 | 4.0 | 12.9 | 4.5 |
| | 0.3 | 1.1 | 6.4 | 3.2 | 4.8 | 2.7 | 0.3 | 3.2 | 5.0 | 1.2 | 3.2 | 1.5 |
| | 0.4 | 1.4 | 6.2 | 2.3 | 3.6 | 2.1 | | | | | | |
| CS | 0.1 | 0.3 | 6.7 | 3.7 | 5.9 | 3.1 | 0.1 | 1.1 | 8.3 | 5.5 | 14.1 | 5.0 |
| | 0.2 | 0.7 | 6.0 | 2.3 | 4.6 | 2.3 | 0.15 | 1.6 | 6.7 | 4.3 | 12.2 | 4.1 |
| | 0.3 | 1.1 | 5.5 | 1.6 | 3.6 | 1.8 | 0.2 | 2.1 | 4.7 | 3.0 | 8.4 | 2.8 |
| | 0.4 | 1.4 | 4.8 | 1.4 | 2.9 | 1.6 | 0.3 | 3.2 | 2.8 | 2.0 | 4.8 | 1.8 |

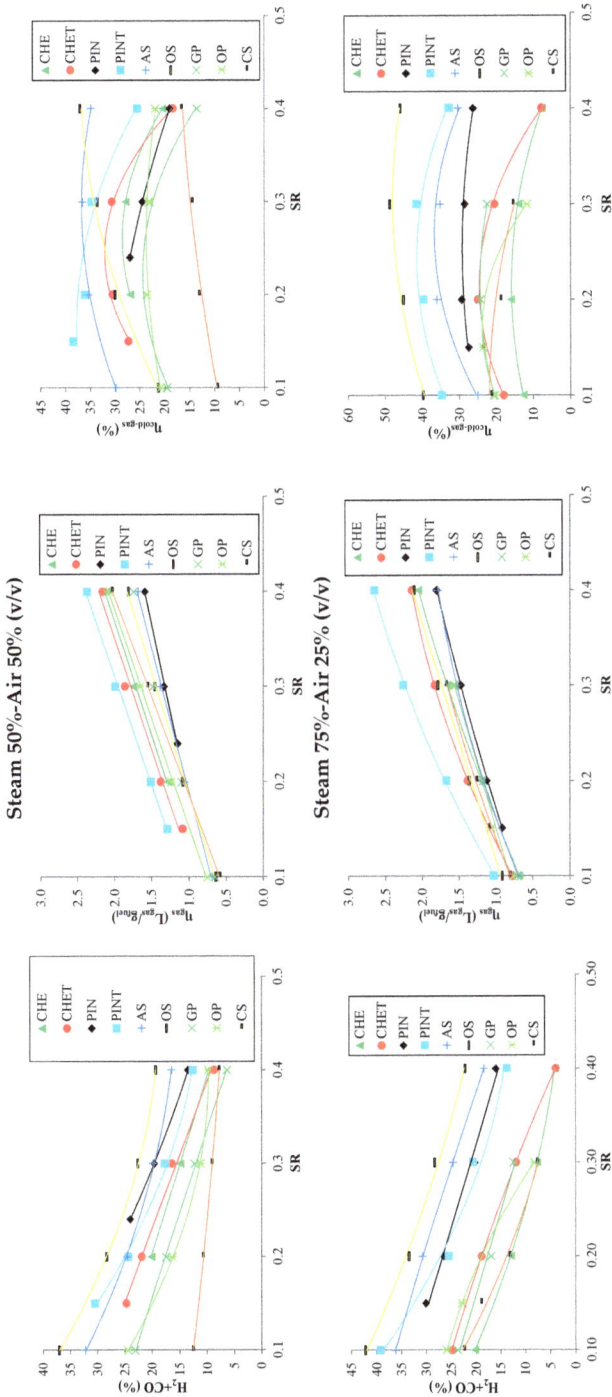

**Figure 3.** Comparison of the obtained results ($H_2$ + CO emissions, gas yield ($L_{gas}/g_{fuel}$) and cold-gas efficiency (%)) at the selected steam/air ratios (50/50 and 75/25) vs. SR.

## 3.4. Selection of Optimum Conditions and Biomass Mixtures

An analysis of the results shows that OS exhibited the best gasification behavior and the highest syngas production whereas the highest gas yield was provided by PINT. The two biomasses were mixed and gasified at 600 °C, at a SR of 0.25, as both of them showed optimum cold gas efficiency in the SR range of 0.2–0.3, at a steam/air ratio of 75/25 in order to maximize syngas production. Three OS/PINT mass ratios (25/75, 50/50, 75/25) were tested and the results were compared with those obtained from the gasification of pure fuels in the same conditions. GP/PINT mixtures were also tested in the same conditions at the same ratios, as GP showed good grinding and handling properties but discrete performance in gasification. All the obtained results are shown in Figure 4.

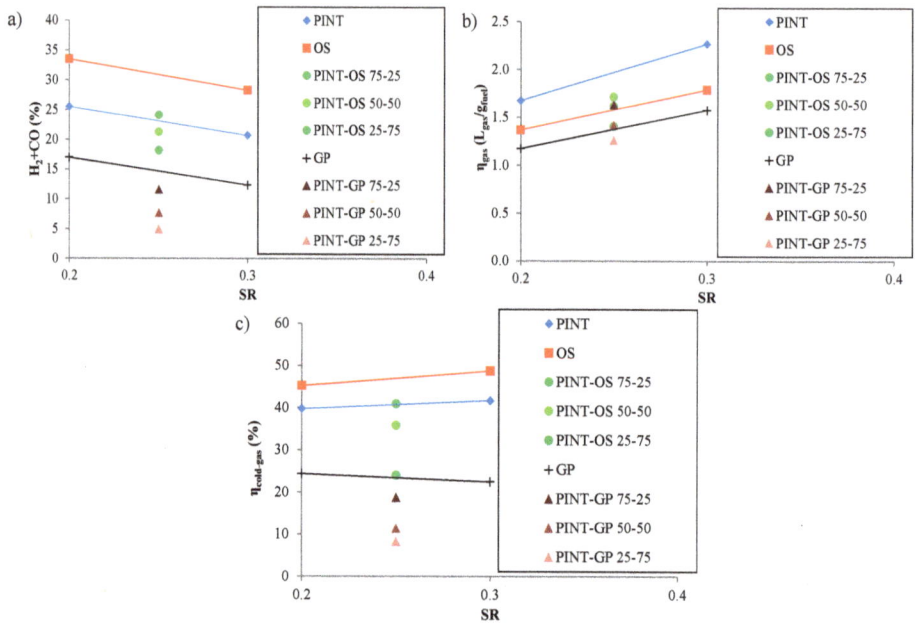

**Figure 4.** Results of the syngas composition (**a**) (CO + $H_2$), (**b**) gas yield and (**c**) cold-gas efficiency of the biomass mixtures.

As can be seen from the results in Figure 4, none of the tested mixtures outperform the best of the single samples, excluding all possibility of synergy. Not only this, but in the OS/PINT series, only the 75/25 steam/air ratio improves the syngas composition yielded by the worst of the single biomasses (PINT) whilst gas and thermal yield results are similar to the worst cases. The results of the GP/PINT series are even poorer, at least at the particle size studied in this work. A possible explanation for this impoverishment is provided by the DTG profiles in Figure 5. As can be seen, both OS and GP have ignition temperatures slightly lower that than of PINT (370, 353 and 389 °C, respectively). In addition to this, whilst the peak temperatures are similar (541, 530 and 542 °C), OS presents a previous peak at approximately 450 °C. This temperature difference may lead, in poor oxygen atmospheres as is the case, to the quasi-total oxidation of one of the fuels, which would remove most of the $O_2$ from the reaction environment ,whilst the other fuel would experience a low reaction rate, a phenomenon known as oxygen starvation [45].

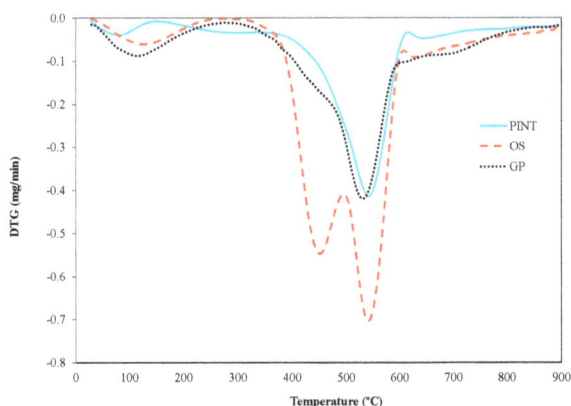

**Figure 5.** Combustion profiles, expressed as the derivative thermogravimetric (DTG) of the samples.

Another interesting feature visible to the naked eye was the heterogeneity of the samples in the size range selected (0.1–1mm). This was especially marked in the case of OS and GP, where two phases, a very thin powdery one and another phase with larger and harder sphere-shaped particles, were observed. An examination of their particle size distribution in Figure 6 shows that the raw OS and GP samples had a higher content of small particles whilst PINT presents a higher percentage of large ones. This lends support to the hypothesis of oxygen starvation as the more reactive fractions of both fuels react sooner, thereby consuming most of the oxidizing agent.

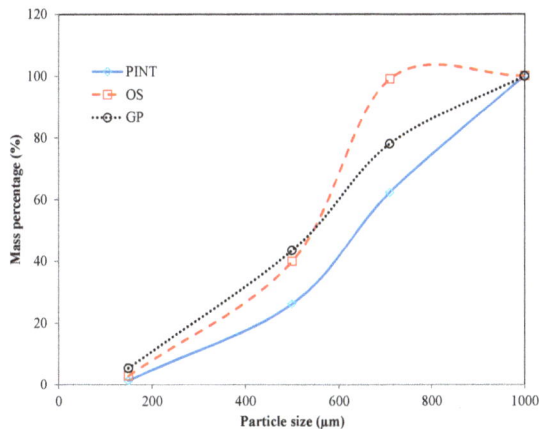

**Figure 6.** Cumulative particle size distribution of the biomass samples used for mixtures.

*3.5. Chars Study*

As previously mentioned, of the potential advantages of carrying out the gasification process at a mild temperature and in a steam-rich environment are the increase in the production of syngas and, at the same time, the possibility of reusing the partially oxidized chars through their thermal conversion, as sorbent or catalyst precursor, or even for soil amendment. To this end SEM and EDX analyses were carried out on three of the studied samples: a lignocellulosic sample, CHET (Figure 7a,b), a hard sample, OS (Figure 7c,d) and an intermediate sample, GP (Figure 7e,f).

**Figure 7.** SEM images of (**a,b**) CHET; (**c,d**) OS and (**e,f**) GP chars at ×80 and ×300 magnifications.

SEM images were obtained at two different magnifications (×80 and ×300) to respectively identify the sample's shape and general structure and to study its surface in more detail. A comparison of the images obtained at ×80 shows that in all three cases there are two kinds of particles: particles that have been almost totally oxidized that appear in clear color and particles that are much darker which are the ones more suitable for thermal conversion. This is confirmed by the EDX results, where it can be seen that the carbon composition ranges from 10%–15% in particles basically formed by elements like Si, Al, Ca, K and O in significant quantities, to 90% in OS, 80% in CHET or close to 75% in GP, all of which are high values that suggest an important remaining HHV. The presence of elements such as Ca and K together with a high N content is highly suitable for agricultural soil improvement [46]. In addition, K is also considered as a natural catalyst for biomass gasification in $CO_2$ atmosphere, controlling it, together with external surface area at low conversion degree. On the other hand, at high conversion levels it is correlated with the catalytic index (CI) = (K + Ca + Mg + Na + Fe)/(Si + Al). [47].

With respect to the particle's shape and surface it can be said that those of CHET present a much more heterogeneous structure with an undefined shape, where both porous and non-porous particles co-exist. This can be seen in detail in the image taken at ×300 (Figure 7b). In contrast, the GP sample consists mainly of spherical particles, with a large quantity of pores (Figure 7f). The OS sample seems to be intermediate between the previous two, with spherical and amorphous particles, with and without pores. To complement the SEM images, textural characterization was carried out and BET surface areas of 2, 27 and 132 $m^2/g$ for OS, CHET and GP, respectively, were determined. Due to this, GP with its high C content, specific surface and porous structure seems to be the most suitable char sample to use as a possible precursor to obtain sorbents or catalysts

## 4. Conclusions

The effect of certain operational variables, SR and S/A, during gasification of different types of biomass on product gas quality, composition and cold-gas-efficiency was evaluated. The results were used for the optimization of the low-temperature gasification of biomass in a purpose-built atmospheric bubbling fluidized bed reactor. It was found that the same gasification conditions do not affect every biomass in the same way, as its performance depends on its particular characteristics. In some cases, richer steam environments maximize the production of the syngas, whilst in other cases it is maximized in poorer environments, so gasification conditions must be carefully tested for each individual biomass. What can be taken as a general rule is that the total richness of syngas decreases as the SR increases.

The torrefaction process does not significantly improve gasification performance, by enhancing the richness of the syngas stream, but it increases the yield. The gas yield linearly increases with SR at approximately the same ratio for all samples. A higher S/A does not apparently affect this in any significant way. Cold gas efficiency ($\eta_{cold-gas}$) exhibits, in most cases, a second-order polynomial behavior in the range studied, with a maximum in the 0.2–0.3 range. This occurs in all cases for the 75/25 steam/air ratio. In a comparison of the fuel samples, OS and AS presented the highest syngas yield in the outlet gas, with values slightly over 35% at SR = 0.2 in both cases when gasified at the highest S/A tested (75/25). On the other hand, PINT cold-gas efficiency reached more than 1.5 $L/g_{fuel}$ at a SR higher than 0.2 for both the S/A tested. The OS sample presented efficiencies close to 50% in the SR range of 0.2–0.3 for a S/A ratio of 75/25. For the 50/50 S/A ratio the results are slightly lower, dropping to 35% in the same range as for AS and PINT. The char study has shown that the OS samples achieve C percentages of up to 90%. The other samples (CHET and GP) present particles with C mass content of over 75%, in addition to a porous structure and substantial surface areas, up to 132 $m^2/g$ in the case of GP. This indicates that they may be effective in other thermal conversion techniques, like combustion, or even as precursors for catalysts or sorbent manufacturing. The presence of nutrients such as N and K suggests that they could also play a positive role in soil amendment.

**Acknowledgments:** This work has received financial support from the Spanish MINECO (ENE2014-53515-P), cofinanced by the European Regional Development Fund (ERDF), and from the Gobierno del Principado de Asturias (PCTI-GRUPIN14-079). María Pilar González-Vázquez also acknowledges a fellowship awarded by the Spanish MINECO (FPI program), cofinanced by the European Social Fund.

**Author Contributions:** María Pilar González-Vázquez and Roberto García conducted experimental work, data analysis and calculations. María Pilar González-Vázquez, Roberto García, Covadonga Pevida and Fernando Rubiera discussed the results and contributed to the writing of the manuscript.

**Conflicts of Interest:** The authors declare no conflict of interest.

## References

1. European Commission. *Communication from the Commission to the European Parliament and the Council;* EC: London, UK, 2015.
2. Pacesila, M.; Burcea, S.G.; Colesca, S.E. Analysis of renewable energies in European Union. *Renew. Sustain. Energy Rev.* **2016**, *56*, 156–170. [CrossRef]
3. Bartolini, F.; Angelini, L.G.; Brunori, G.; Gava, O. Impacts of the CAP 2014–2020 on the agroenergy sector in Tuscany, Italy. *Energies* **2015**, *8*, 1058–1079. [CrossRef]
4. International Energy Agency. Energy Policies of IEA Countries. 2015. Available online: http://www.iea.org/countries/membercountries/spain/ (accessed on 3 February 2017).
5. Siedlecki, M.; de Jong, W.; Verkooijen, A.H.M. Fluidized bed gasification as a mature and reliable technology for the production of bio-syngas and applied in the production of liquid transportation fuels—A review. *Energies* **2011**, *4*, 389–434. [CrossRef]
6. García, R.; Pizarro, C.; Lavín, A.G.; Bueno, J.L. Characterization of Spanish biomass wastes for energy use. *Bioresour. Technol.* **2012**, *103*, 249–258. [CrossRef] [PubMed]
7. Hosseini, S.E.; Wahid, M.A. Hydrogen production from renewable and sustainable energy resources: Promising green energy carrier for clean development. *Renew. Sustain. Energy Rev.* **2016**, *57*, 850–866. [CrossRef]
8. Mudhoo, A.; Kumar, S. Effects of heavy metals as stress factors on anaerobic digestion processes and biogas production from biomass. *Int. J. Environ. Sci. Technol.* **2013**, *10*, 1383–1398. [CrossRef]
9. Wu, T.Y.; Hay, J.X.W.; Kong, L.B.; Juan, J.C.; Jahim, J.M. Recent advances in reuse of waste material as substrate to produce biohydrogen by purple non-sulfur (PNS) bacteria. *Renew. Sustain. Energy Rev.* **2012**, *16*, 3117–3122. [CrossRef]
10. Meng, X.; de Jong, W.; Fu, N.; Verkooijen, A.H.M. Biomass gasification in a 100 kWth steam-oxygen blown circulating fluidized bed gasifier: Effects of operational conditions on product gas distribution and tar formation. *Biomass Bioenergy* **2011**, *35*, 2910–2924. [CrossRef]
11. Kumar, A.; Jones, D.D.; Hanna, M.A. Thermochemical biomass gasification: A review of the current status of the technology. *Energies* **2009**, *2*, 556–581. [CrossRef]
12. Alipour Moghadam, R.; Yusup, S.; Azlina, W.; Nehzati, S.; Tavasoli, A. Investigation on syngas production via biomass conversion through the integration of pyrolysis and air–steam gasification processes. *Energy Convers. Manag.* **2014**, *87*, 670–675. [CrossRef]
13. Dudyński, M.; van Dyk, J.C.; Kwiatkowski, K.; Sosnowska, M. Biomass gasification: Influence of torrefaction on syngas production and tar formation. *Fuel Process. Technol.* **2015**, *131*, 203–212. [CrossRef]
14. Wan Ab Karim Ghani, W.A.; Moghadam, R.A.; Mohd Salleh, M.A.; Alias, A.B. Air gasification of agricultural waste in a fluidized bed gasifier: Hydrogen production performance. *Energies* **2009**, *2*, 258–268. [CrossRef]
15. Skoulou, V.; Koufodimos, G.; Samaras, Z.; Zabaniotou, A. Low temperature gasification of olive kernels in a 5-kW fluidized bed reactor for $H_2$-rich producer gas. *Int. J. Hydrog. Energy* **2008**, *33*, 6515–6524. [CrossRef]
16. Hernández, J.J.; Aranda, G.; Barba, J.; Mendoza, J.M. Effect of steam content in the air–steam flow on biomass entrained flow gasification. *Fuel Process. Technol.* **2012**, *99*, 43–55. [CrossRef]
17. Balat, M.; Balat, M.; Kirtay, E.; Balat, H. Main routes for the thermo-conversion of biomass into fuels and chemicals. Part 2: Gasification systems. *Energy Convers. Manag.* **2009**, *50*, 3158–3168. [CrossRef]
18. Haykiri-Acma, H.; Yaman, S. Interpretation of biomass gasification yields regarding temperature intervals under nitrogen–steam atmosphere. *Fuel Process. Technol.* **2007**, *88*, 417–425. [CrossRef]
19. Sharma, A.M.; Kumar, A.; Huhnke, R.L. Effect of steam injection location on syngas obtained from an air–steam gasifier. *Fuel* **2014**, *116*, 388–394. [CrossRef]
20. Guizani, C.; Jeguirim, M.; Gadiou, R.; Escudero Sanz, F.J.; Salvador, S. Biomass char gasification by $H_2O$, $CO_2$ and their mixture: Evolution of chemical, textural and structural properties of the chars. *Energy* **2016**, *112*, 133–145. [CrossRef]

21. Kuo, P.-C.; Wu, W.; Chen, W.-H. Gasification performances of raw and torrefied biomass in a downdraft fixed bed gasifier using thermodynamic analysis. *Fuel* **2014**, *117*, 1231–1241. [CrossRef]
22. Chen, W.H.; Chen, C.J.; Hung, C.I.; Shen, C.H.; Hsu, H.W. A comparison of gasification phenomena among raw biomass, torrefied biomass and coal in an entrained-flow reactor. *Appl. Energy* **2013**, *112*, 421–430. [CrossRef]
23. Zaccariello, L.; Mastellone, M.L. Fluidized-bed gasification of plastic waste, wood, and their blends with coal. *Energies* **2015**, *8*, 8052–8068. [CrossRef]
24. Makwana, J.P.; Joshi, A.K.; Athawale, G.; Singh, D.; Mohanty, P. Air gasification of rice husk in bubbling fluidized bed reactor with bed heating by conventional charcoal. *Bioresour. Technol.* **2015**, *178*, 45–52. [CrossRef] [PubMed]
25. Rapagnà, S.; Latif, A. Steam gasification of almond shells in a fluidised bed reactor: The influence of temperature and particle size on product yield and distribution. *Biomass Bioenergy* **1997**, *12*, 281–288. [CrossRef]
26. Kulkarni, A.; Baker, R.; Abdoulmomine, N.; Adhikari, S.; Bhavnani, S. Experimental study of torrefied pine as a gasification fuel using a bubbling fluidized bed gasifier. *Renew. Energy* **2016**, *93*, 460–468. [CrossRef]
27. Mohd Salleh, M.A.; Kisiki, N.H.; Yusuf, H.M.; Ab Karim Ghani, W.A.W. Gasification of Biochar from Empty Fruit Bunch in a Fluidized Bed Reactor. *Energies* **2010**, *3*, 1344–1352. [CrossRef]
28. Pinto, F.; André, R.N.; Carolino, C.; Miranda, M.; Abelha, P.; Direito, D.; Dohrup, J.; Sørensen, H.R.; Girio, F. Effects of experimental conditions and of addition of natural minerals on syngas production from lignin by oxy-gasification: Comparison of bench- and pilot scale gasification. *Fuel* **2015**, *140*, 62–72. [CrossRef]
29. Gil, J.; Corella, J.; Aznar, M.P.; Caballero, M.A. Biomass gasification in atmospheric and bubbling fluidized bed: Effect of the type of gasifying agent on the product distribution. *Biomass Bioenergy* **1999**, *17*, 389–403. [CrossRef]
30. Lim, Y.; Lee, U.-D. Quasi-equilibrium thermodynamic model with empirical equations for air–steam biomass gasification in fluidized-beds. *Fuel Process. Technol.* **2014**, *128*, 199–210. [CrossRef]
31. Gucho, E.M.; Shahzad, K.; Bramer, E.A.; Akhtar, N.A.; Brem, G. Experimental study on dry torrefaction of beech wood and miscanthus. *Energies* **2015**, *8*, 3903–3923. [CrossRef]
32. Chen, W.-H.; Lu, K.-M.; Tsai, C.-M. An experimental analysis on property and structure variations of agricultural wastes undergoing torrefaction. *Appl. Energy* **2012**, *100*, 318–325. [CrossRef]
33. Chew, J.J.; Doshi, V. Recent advances in biomass pretreatment—Torrefaction fundamentals and technology. *Renew. Sustain. Energy Rev.* **2011**, *15*, 4212–4222. [CrossRef]
34. Arias, B.; Pevida, C.; Fermoso, J.; Plaza, M.G.; Rubiera, F.; Pis, J.J. Influence of torrefaction on the grindability and reactivity of woody biomass. *Fuel Process. Technol.* **2008**, *89*, 169–175. [CrossRef]
35. Tapasvi, D.; Khalil, R.; Skreiberg, Ø.; Tran, K.Q.; Grønli, M. Torrefaction of Norwegian birch and spruce: An experimental study using macro-TGA. *Energy Fuels* **2012**, *26*, 5232–5240. [CrossRef]
36. Chen, W.-H.; Kuo, P.-C. A study on torrefaction of various biomass materials and its impact on lignocellulosic structure simulated by a thermogravimetry. *Energy* **2010**, *35*, 2580–2586. [CrossRef]
37. Gil, M.V.; García, R.; Pevida, C.; Rubiera, F. Grindability and combustion behavior of coal and torrefied biomass blends. *Bioresour. Technol.* **2015**, *191*, 205–212. [CrossRef] [PubMed]
38. Narváez, I.; Orío, A.; Aznar, M.P.; Corella, J. Biomass gasification with air in an atmospheric bubbling fluidized bed. *Ind. Eng. Chem. Res.* **1996**, *35*, 2110–2120. [CrossRef]
39. Telmo, C.; Lousada, J.; Moreira, N. Proximate analysis, backwards stepwise regression between gross calorific value, ultimate and chemical analysis of wood. *Bioresour. Technol.* **2010**, *101*, 3808–3815. [CrossRef] [PubMed]
40. Obernberger, I.; Brunner, T.; Bärnthaler, G. Chemical properties of solid biofuels—Significance and impact. *Biomass Bioenergy* **2006**, *30*, 973–982. [CrossRef]
41. Xiao, R.; Zhang, M.; Jin, B.; Huang, Y.; Zhou, H. High-temperature air/steam-blown gasification of coal in a pressurized spout-fluid bed. *Energy Fuels* **2006**, *20*, 715–720. [CrossRef]
42. Weiland, F.; Nordwaeger, M.; Olofsson, I.; Wiinikka, H.; Nordin, A. Entrained flow gasification of torrefied wood residues. *Fuel Process. Technol.* **2014**, *125*, 51–58. [CrossRef]
43. Qian, K.; Kumar, A.; Patil, K.; Bellmer, D.; Wang, D.; Yuan, W.; Huhnke, R.L. Effects of biomass feedstocks and gasification conditions on the physiochemical properties of char. *Energies* **2013**, *6*, 3972–3986. [CrossRef]
44. Gómez-Barea, A.; Arjona, R.; Ollero, P. Pilot-plant gasification of olive stone: A technical assessment. *Energy Fuels* **2005**, *19*, 598–605. [CrossRef]

45. De Micco, G.; Fouga, G.G.; Bohé, A.E. Coal gasification studies applied to H$_2$ production. *Int. J. Hydrogen Energy* **2010**, *35*, 6012–6018. [CrossRef]

46. Zambon, I.; Colosimo, F.; Monarca, D.; Cecchini, M.; Gallucci, F.; Proto, A.; Lord, R.; Colantoni, A. An innovative agro-forestry supply chain for residual biomass: Physicochemical characterisation of biochar from olive and hazelnut pellets. *Energies* **2016**, *9*, 526. [CrossRef]

47. Bouraoui, Z.; Jeguirim, M.; Guizani, C.; Limousy, L.; Dupont, C.; Gadiou, R. Thermogravimetric study on the influence of structural, textural and chemical properties of biomass chars on CO$_2$ gasification reactivity. *Energy* **2015**, *88*, 703–710. [CrossRef]

*energies*

MDPI

*Article*

# The Effect of Two Types of Biochars on the Efficacy, Emission, Degradation, and Adsorption of the Fumigant Methyl Isothiocyanate

Wensheng Fang, Aocheng Cao *, Dongdong Yan, Dawei Han, Bin Huang, Jun Li, Xiaoman Liu, Meixia Guo and Qiuxia Wang *

Institute of Plant Protection, Chinese Academy of Agricultural Sciences, State Key Laboratory for Biology of Plant Diseases and Insect Pests, Beijing 100193, China; fws0128@163.com (W.F.); 13260176634@163.com (D.Y.); hdw690464867@163.com (D.H.); huangb1992@163.com (B.H.); 2008evening@163.com (J.L.); lxiaoman163@163.com (X.L.); guomeixia@126.com (M.G.)

* Correspondence: caoac@vip.sina.com (A.C.); qxwang@ippcaas.cn (Q.W.); Tel.: +86-62-815-940 (A.C.); +86-62-894-863 (Q.W.)

Academic Editor: Mejdi Jeguirim
Received: 7 September 2016; Accepted: 19 December 2016; Published: 23 December 2016

**Abstract:** Biochar (BC) is increasingly applied in agriculture; however, due to its adsorption and degradation properties, biochar may also affect the efficacy of fumigant in amended soil. Our research is intended to study the effects of two types of biochars (BC-1 and BC-2) on the efficacy and emission of methyl isothiocyanate (MITC) in biochar amendment soil. Both types of biochars can significantly reduce MITC emission losses, but, at the same time, decrease the concentration of MITC in the soil. The efficacy of MITC for controlling soil-borne pests (*Meloidogyne* spp., *Fusarium* spp. *Phytophthora* spp., *Abutilon theophrasti* and *Digitaria sanguinalis*) was reduced when the biochar (BC-1 and BC-2) was applied at a rate of higher than 1% and 0.5% (on a weight basis) (on a weight basis), respectively. However, increased doses of dazomet (DZ) were able to offset decreases in the efficacy of MITC in soils amended with biochars. Biochars with strong adsorption capacity (such as BC-1) substantially reduced MITC degradation rate by 6.2 times, and increased by 4.1 times following amendment with biochar with high degradability (e.g., BC-2), compared to soil without biochar amendment. This is due to the adsorption and degradation of biochar that reduces MITC emission losses and pest control.

**Keywords:** biochar; methyl isothiocyanate (MITC); dazomet; degradation; adsorption

## 1. Introduction

Soil fumigants are commonly used worldwide to control soil-borne fungal pathogens, nematodes and weeds in high-value crops such as cut flowers and vegetables. Following the phase-out of methyl bromide (MBr), dazomet (DZ) provided a widespread approach for improving the effectiveness of soil disinfestations [1–3]. In moist soil, DZ rapidly decomposes to its active ingredient, methyl isothiocyanate (MITC), which is toxic to soil-borne pests, especially fungi, some soil arthropods, and ectoparasitic nematodes [4,5]. However, the high application rates of DZ (294–450 kg·ha$^{-1}$) and high vapor pressure of MITC (20.7 mmHg at 20 °C) result in the significant volatilization of MITC following fumigation, and this may cause environmental and health problems because of the irritant, lachrymatory and toxic properties of the gas [4,6]. Hence, it is imperative to develop strategies to minimize the emissions of MITC while ensuring high activity against pathogens. Applying biochar (BC) to the soil surface has been demonstrated to reduce the losses of the fumigant caused by volatilization. For example, Wang et al. reported that a small amount of biochar added to the soil surface can reduce 1,3-dichloropropene emissions by more than 92% and reduce chloropicrin losses by 85.7%–97.7% [7,8]. Biochar soil amendments, therefore, offer a possible approach for reducing the emission of MITC [9].

Biochar is a carbon-rich, porous, intentionally produced charcoal made in low oxygen conditions from natural organic materials [10]. Because of its specific properties—high carbon content, rich pore structure, and stable physical and chemical properties—biochar has a potential role in carbon sequestration and the suppression of emissions of greenhouse gases from soil, as well as the improvement of crop productivity and soil health. Consequently, biochar soil amendments are attracting increasing attention from policy makers in China and other countries. However, types of biochar that have an exceptionally high sorption [10] or degradation [7,11] capacity for organic chemicals, especially pesticides, may have both positive and negative impacts on pest management. On the one hand, the high retentive qualities of biochar may prevent or reduce the leaching of soil-applied herbicides and insecticides [12,13] and may decrease the rate at which they are degraded by soil microorganisms. On the other hand, types of biochar that have strong sorption or degradation affinities for pesticides may result in low effective concentrations for pest management, meaning that greater amounts of pesticide must be applied to achieve the same level of pest protection. For example, Kookana et al. demonstrated a decreased efficacy of herbicides and lower bioavailability of pesticides in biochar-amended soils [14,15]. Depending on the sorption strength of the particular biochar, amending the soil with biochar may adversely impact pest control [10].

Obviously, biochar has great potential in terms of reducing emissions of the MITC fumigant, but it also has the potential to adversely impact pest control. Therefore, it is necessary to find a balance between reducing emissions and ensuring adequate agricultural pest control when using biochar and MITC. Previously, our experiments have shown that some biochars could drastically accelerate and some biochars could decelerate the degradation of MITC [16], while both types of biochar accelerated degradation and adsorption potential to reduce MITC emission. However, the impact of biochar on the efficacy and fate of MITC remains poorly understood. Furthermore, it is not clear how MITC fumigation would be impacted by types of biochar that have different physical and chemical characteristics as a result of different feedstock and production parameters. Here, we select two type of biochars: one suppresses MITC degradation and the second accelerates its degradation, in order to investigate the effects of biochar on the efficacy, emission, degradation and adsorption of MITC in biochar-amended soil and to determine the appropriate balance between reducing MITC emissions and ensuring its availability for pest control. The information obtained from this study will be useful for evaluating the effect of biochar on the bioavailability and efficacy of MITC and for evaluating the use of biochar to reduce MITC emissions, as well as identifying the necessary conditions for optimal degradation.

## 2. Results and Discussion

### 2.1. Effects of Biochar Amendment on Methyl Isothiocyanate (MITC) Emissions

MITC emission losses were markedly reduced in soil amended with biochar (Figure 1). The maximum air concentrations of the MITC in the chambers were 22.8, 12.3, and 14.4 $mg \cdot L^{-1}$ in the CK (soil without biochar), BC-1, and BC-2 treatments, respectively. Compared with fumigation without biochar, BC-1 and BC-2 reduced the total fumigant emission losses by 46.1% and 36.8%, respectively. Correspondingly, the emission rates of MITC decreased in soil treated with BC-1 and BC-2. In the three treatments, the emissions flux initially increased with time, exhibiting a peak flux at 1 h after injection, before subsequently declining with time. The maximum emissions fluxes of MITC were 0.17, 0.11, and 0.12 $mg \cdot L^{-1} \cdot min^{-1}$ in the CK, BC-1 and BC-2 treatments, respectively. The results indicated that BC-1 or BC-2 amendments in the soil surface could significantly reduce MITC emissions in the air.

The trend of MITC concentrations in the soil was similar to that in the chamber air, increasing initially and then declining with time (Figure 2). The concentrations of MITC in the soil treated with biochar (BC-1 or BC-2) were significantly lower than in CK. The maximum concentrations of MITC were 21.5, 14.8, and 15.6 $mg \cdot L^{-1}$ in the CK, BC-1, and BC-2 treatments, respectively. These results showed that biochar amendments could reduce the concentrations of MITC in the soil, potentially reducing its efficacy for controlling soil-borne pests.

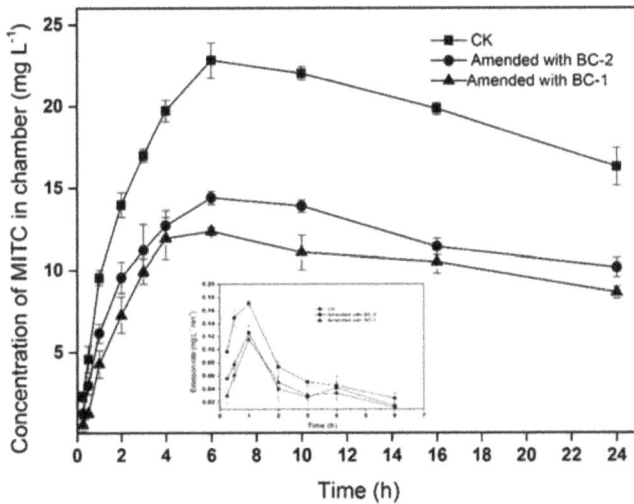

**Figure 1.** Concentration of methyl isothiocyanate (MITC) in chamber treatments and emission rates of MITC. Data points indicate means of three replicate cells, and error bars indicate standard errors of the mean.

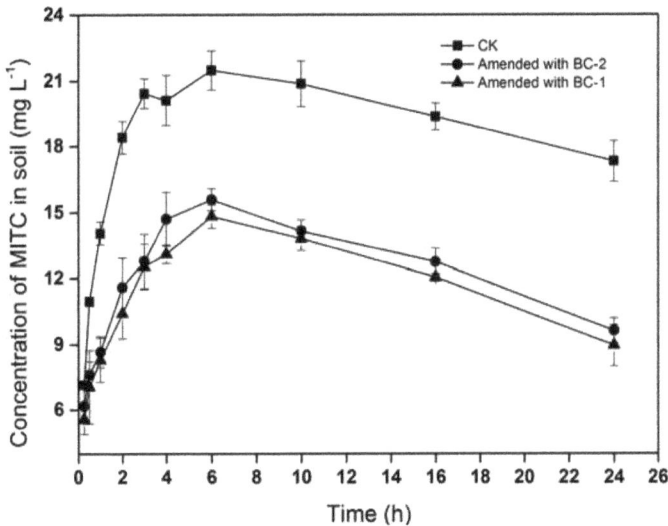

**Figure 2.** Concentration of MITC in soil. Data points indicate means of three replicate cells, and error bars indicate standard error of the mean.

### 2.2. Effects of Biochar Amendment on the Efficacy of MITC against Soil-Borne Pests

The efficacies of MITC with respect to root-knot nematodes (*Meloidogyne* spp.), weed seeds (*Abutilon theophrasti* and *Digitaria sanguinalis*) and key soil-borne fungi (*Fusarium* spp. and *Phytophthora* spp.) in soil amended with BC-1 or BC-2 are listed in Tables 1 and 2, respectively. As shown in Table 1, the corrected mortality of nematodes following fumigation of unamended soil did not differ from that in soil amended with BC-1 at rates of 0.1% to 5%; however, the corrected mortality of nematodes was significantly lower in soil amended with 10% BC-1. When the BC-1 amendment rates were less than 2% and 1%, there was no significant difference in the efficacies of MITC on control of *Digitaria sanguinalis* and *Abutilon theophrasti*, respectively. Compared with samples without biochar, the efficacy of MITC against *Phytophthora* spp. and *Fusarium* spp. was reduced significantly when the BC-1 amendment rates were greater than 2% and 1%, respectively. The above results indicated that BC-1 amendment rates less than or equal to 1% did not have negative effects on MITC's control of soil-borne pests.

**Table 1.** Pest control efficacy of methyl isothiocyanate (MITC) in soil amended with biochar BC-1.

| Treatment | % Corrected Nematode Mortality | % Control of *Digitaria sanguinalis* | % Control of *Abutilon theophrasti* | % Control of *Phytophthora* spp. | % Control of *Fusarium* spp. |
|---|---|---|---|---|---|
| DZ + 0% BC-1 | 75.48 (±1.5) [a] | 68.98 (±2.5) [a] | 92.87 (±1.5) [a] | 69.92 (±2.9) [a,b] | 64.50 (±4.7) [a] |
| DZ + 0.1% BC-1 | 72.88 (±7.7) [a,b] | 69.0 (±1.0) [a] | 90.34 (±2.1) [a] | 72.86 (±4.3) [a] | 64.08 (±11.8) [a] |
| DZ + 0.25%BC-1 | 71.91 (±5.6) [a,b] | 70.17 (±2.4) [a] | 88.17 (±2.6) [a] | 67.94 (±4.5) [a,b] | 62.99 (±15.7) [a] |
| DZ + 0.5% BC-1 | 75.81 (±6.3) [a] | 70.30 (±4.3) [a] | 88.51 (±2.1) [a] | 69.92 (±6.1) [a,b] | 66.93 (±1.4) [a] |
| DZ + 1% BC-1 | 74.66 (±1.2) [a] | 68.10 (±2.1) [a] | 77.66 (±4.1) [b] | 71.03 (±4.4) [a,b] | 57.60 (±5.3) [a,b] |
| DZ + 2% BC-1 | 66.22 (±4.1) [a,b] | 59.78 (±4.0) [b] | 62.76 (±3.7) [c] | 69.94 (±5.1) [a,b] | 42.99 (±3.0) [c] |
| DZ + 5% BC-1 | 65.48 (±8.0) [a,b] | 58.65 (±3.4) [b] | 52.87 (±1.1) [d] | 63.81 (±5.0) [b] | 45.24 (±3.7) [b,c] |
| DZ + 10% BC-1 | 59.51 (±8.3) [b] | 59.52 (±2.4) [b] | 47.56 (±3.0) [e] | 54.29 (±1.4) [c] | 44.20 (±5.2) [b,c] |

Values are means ±SD (*n* = 3) and letters indicate statistical results. Different letters indicate statistical significance at the *p* ≤ 0.05 level using Duncan's multiple range test, the same letter means not different.

**Table 2.** Pest control efficacy of MITC in soil amended with biochar BC-2.

| Treatment | % Corrected Nematode Mortality | % Control of *Digitaria sanguinalis* | % Control of *Abutilon theophrasti* | % Control of *Phytophthora* spp. | % Control of *Fusarium* spp. |
|---|---|---|---|---|---|
| DZ + 0% BC-2 | 100.00 [a] | 100.00 [a] | 100.00 [a] | 100.00 [a] | 99.16 (±1.8) [a] |
| DZ + 0.1% BC-2 | 100.00 [a] | 100.00 [a] | 100.00 [a] | 98.86 (±1.3) [a] | 97.98 (±2.8) [a] |
| DZ + 0.25%BC-2 | 100.00 [a] | 100.00 [a] | 100.00 [a] | 98.44 (±4.5) [a] | 95.29 (±6.7) [a] |
| DZ + 0.5% BC-2 | 90.22 (±14.1) [a,b] | 100.00 [a] | 100.00 [a] | 97.86 (±3.1) [a] | 99.16 (±1.1) [a] |
| DZ + 1% BC-2 | 77.56 (±3.5) [b] | 64.28 (±2.5) [b] | 85.68 (±1.3) [b] | 99.80 (±2.4) [a] | 96.44 (±8.3) [a] |
| DZ + 2% BC-2 | 51.22 (±14.1) [c] | 45.78 (±2.0) [c] | 56.34 (±4.7) [c] | 89.34 (±4.1) [b] | 83.05 (±4.4) [b] |
| DZ + 5% BC-2 | 24.48 (±6.0) [d] | 23.65 (±2.4) [d] | 38.17 (±5.1) [d] | 75.81 (±7.9) [c] | 71.75 (±6.2) [c] |
| DZ + 10% BC-2 | 33.51 (±3.9) [c,d] | 14.42 (±5.4) [e] | 23.56 (±3.8) [e] | 44.39 (±3.5) [d] | 50.20 (±5.2) [d] |

Values are means ±SD (*n* = 3) and letters indicate statistical results. Different letters indicate statistical significance at the *p* ≤ 0.05 level using Duncan's multiple range test, the same letter means not different.

The effects of BC-2 on MITC's efficacy against soil-borne pests were similar to BC-1 (Table 2). The efficacy of MITC against root-knot nematode (*Meloidogyne* spp.) and weed seeds (*Abutilon theophrasti* and *Digitaria sanguinalis*) was inhibited significantly after BC-2 amendment at rates above 1% in soil. However, the efficacy against *Fusarium* spp. and *Phytophthora* spp. was not reduced by BC-2 amendment rates at or below 1% compared with unamended soil. An amendment rate of less than or equal to 0.5% in BC-2 did not have negative effects on MITC's control of soil-borne pests.

To obtain a sufficient control effect, a larger amount of DZ is required in soil amended with biochar. The efficacy of MITC against soil-borne pests was investigated in soil amended with 1% BC-1 or 0.5% BC-2, respectively (Tables 3 and 4). The results indicated that the tested biochars (BC-1 or BC-2) have a negative impact on the biological activity of MITC on most soil-borne pests at a lower application rate of DZ (for example, 25 or 50 mg·kg$^{-1}$). For BC-1, to achieve the comprehensive prevention of all pests in amended soil, an amount of DZ greater than or equal to 125 mg·kg$^{-1}$ was required. For BC-2, a DZ application rate greater than or equal to 100 mg·kg$^{-1}$ did not have negative effects on MITC's control of soil-borne pests. Consequently, increased doses of DZ were able to offset decreases in the efficacy of MITC in soils amended with biochar. The pest control efficacy of MITC in Daxing soil is higher than that in Tongzhou soil (for example, Tables 3 and 4). This may be because Tongzhou soil has a history of previous applications of Metham sodium (MS) or DZ (history), while Daxing soil does not (nonhistory). Many studies show that accelerated MITC degradation in history soils resulted in a significant reduction in *Verticillium dahliae*, *Sclerotium rolfsii* and *Fusarium* spp. mortality compared to exposure in nonhistory soils [17,18].

**Table 3.** Results of MITC against nematodes and fungal pathogens and weeds in soil amended with biochar BC-1.

| Treatment | % Corrected Nematode Mortality | % Control of *Digitaria sanguinalis* | % Control of *Abutilon theophrasti* | % Control of *Phytophthora* spp. | % Control of *Fusarium* spp. |
|---|---|---|---|---|---|
| DZ0 | 0 | 0 | 0 | 0 | 0 |
| DZ0 + 1%BC-1 | 0 | 0 | 7.9 (±3.5) [a] | 0 | 5.9 (±3.0) a |
| DZ25 | 19.62 (±6.2) | 33.45 (±4.5) [a] | 46.23 (±5.5) [a] | 25.87 (±2.0) | 29.97 (±1.2) |
| DZ25 + 1% BC-1 | 17.45 (±5.3) | 12.54 (±2.3) | 21.45 (±2.8) | 32.34 (±1.2) [a] | 24.61 (±2.9) |
| DZ50 | 33.77 (±2.3) [a] | 56.44 (±3.7) [a] | 88.21 (±1.6) [a] | 41.56 (±3.2) [a] | 54.90 (±4.4) [a] |
| DZ50 + 1% BC-1 | 28.93 (±3.4) | 24.78 (±5.7) | 64.78 (±3.2) | 34.48 (±4.4) | 47.92 (±6.0) |
| DZ75 | 45.68 (±5.9) | 67.09 (±4.9) [a] | 94.11 (±2.4) [a] | 58.72 (±6.1) [a] | 66.89 (±3.3) [a] |
| DZ75 + 1% BC-1 | 49.31 (±7.8) [a] | 31.43 (±6.2) | 65.64 (±3.6) | 50.94 (±2.4) | 54.92 (±2.0) |
| DZ100 | 66.94 (±8.9) | 74.33 (±3.2) [a] | 100.00 | 70.34 (±4.9) [a] | 81.61 (±3.3) |
| DZ100 + 1% BC-1 | 65.67 (±4.0) | 61.09 (±8.5) | 95.69 (±4.4) | 64.95 (±7.8) | 82.69 (±2.1) |
| DZ125 | 71.09 (±6.5) | 90.22 (±5.7) | 100.00 | 81.97 (±8.1) | 89.23 (±1.9) |
| DZ125 + 1% BC-1 | 67.77 (±8.4) | 85.38 (±4.3) | 96.13 (±1.1) | 79.81 (±2.1) | 84.09 (±2.2) |
| DZ150 | 76.88 (±4.8) | 98.22 (±1.4) | 100.00 | 95.93 (±4.0) | 95.77 (±6.9) |
| DZ150 + 1% BC-1 | 73.55 (±8.4) | 100.00 | 94.56 (±2.1) | 96.07 (±3.1) | 90.75 (±2.2) |
| DZ175 | 82.53 (±2.4) | 100.00 | 100.00 | 100.00 | 100.00 |
| DZ175 + 1% BC-1 | 79.72 (±1.8) | 100.00 | 100.00 | 100.00 | 100.00 |

[a] $p < 0.05$ (*t*-Test) ($n = 3$) vs. control groups without biochar amendment.

**Table 4.** Results of MITC against nematodes and fungal pathogens and weeds in soil amended with biochar BC-2.

| Treatment | % Corrected Nematode Mortality | % Control of *Digitaria sanguinalis* | % Control of *Abutilon theophrasti* | % Control of *Phytophthora* spp. | % Control of *Fusarium* spp. |
|---|---|---|---|---|---|
| DZ0 | 0 | 0 | 0 | 0 | 0 |
| DZ0 + 0.5%BC-2 | 28.56 (±2.0) [a] | 0 | 8.06 (±1.0) [a] | 12.98 (±8.1) [a] | 17.12 (±7.0) [a] |
| DZ25 | 45.98 (±5.9) | 57.55 (±2.3) [a] | 92.39 (±3.1) [a] | 43.41 (±4.9) | 46.02 (±3.0) |
| DZ25 + 0.5% BC-2 | 53.88 (±6.8) [a] | 29.32 (±4.5) | 43.84 (±6.1) | 62.40 (±5.8) [a] | 64.22 (±2.7) [a] |
| DZ50 | 84.22 (±1.8) | 100.00 [a] | 100.00 [a] | 92.63 (±2.7) [a] | 91.21 (±6.5) [a] |
| DZ50 + 0.5% BC-2 | 80.34 (±1.9) | 62.76 (±2.8) | 76.59 (±3.5) | 84.69 (±4.7) | 83.51 (±7.9) |
| DZ75 | 100.00 | 100.00 [a] | 100.00 | 99.80 (±8.3) | 98.95 (±1.5) |
| DZ75 + 0.5% BC-2 | 100.00 | 96.34 (±2.4) | 100.00 | 93.60 (±5.3) | 97.28 (±2.0) |
| DZ100 | 100.00 | 100.00 | 100.00 | 97.28 (±3.8) | 100.00 |
| DZ100 + 0.5% BC-2 | 100.00 | 100.00 | 100.00 | 94.57 (±2.4) | 100.00 |
| DZ125 | 100.00 | 100.00 | 100.00 | 100.00 | 100.00 |
| DZ125 + 0.5% BC-2 | 100.00 | 100.00 | 100.00 | 100.00 | 100.00 |
| DZ150 | 100.00 | 100.00 | 100.00 | 100.00 | 100.00 |
| DZ150 + 0.5% BC-2 | 100.00 | 100.00 | 100.00 | 100.00 | 100.00 |
| DZ175 | 100.00 | 100.00 | 100.00 | 100.00 | 100.00 |
| DZ175 + 0.5% BC-2 | 100.00 | 100.00 | 100.00 | 100.00 | 100.00 |

[a] $p < 0.05$ (*t*-Test) ($n = 3$) vs. control groups without biochar amendment.

### 2.3. Effects of Biochar Amendment on MITC Degradation and Adsorption

The degradation parameters of MITC in soil, biochar alone, and soil amended with biochar (BC-1 or BC-2) are listed in Table 5. The results show that MITC degradation is slower in BC-1 alone and soil amended with BC-1 (at 1%) than in soil alone. Compared with the control, the degradation rate of MITC was 6.2 times slower in soil amended with 1% BC-1, and the degradation rate was as much as 14 times slower in pure BC-1. It is clear that BC-1 can significantly inhibit the degradation of MITC. In contrast, BC-2 is able to drastically accelerate the degradation of MITC. The MITC degradation rates (k) in soil amended with 1% BC-2 or BC-2 alone increased 4.1 times and 11.3 times over those in soil alone, respectively. We also observed that the MITC degradation was significantly accelerated after amending the soil with biochar that possessed a high degradation capacity. The above results indicated that biochar has the ability to degrade MITC and that the degradation capacity differs with the type of biochar.

**Table 5.** MITC degradation in pure biochar, biochar-amended soil, and unamended soil.

| Treatment | $k$ (Day$^{-1}$) | $t_{1/2}$ (Day) | $r^2$ |
|-----------|------------------|-----------------|-------|
| Soil | 1.23 ± 0.07 [c] | 0.56 | 0.98 |
| BC-1 + soil | 0.17 ± 0.01 [d] | 4.08 | 0.91 |
| BC-2 + soil | 6.23 ± 0.41 [b] | 0.11 | 0.73 |
| BC-1 | 0.08 ± 0.03 [d] | 8.66 | 0.37 |
| BC-2 | 15.19 ± 1.78 [a] | 0.05 | 0.97 |

Values are means ±SD ($n$ = 3) and letters indicate statistical results. Different letters indicate statistical significance at the $p \leq 0.05$ level using Duncan's multiple range test, the same letter means not different.

Previous studies have shown that the major degradation pathway for MITC is reaction with hydroxyl ($\cdot$OH) radicals [19]. In addition, there are many organic radicals existing in biochar [20,21], and these free radicals could induce $\cdot$OH generation. Fang et al. [22] reported that the proposed mechanism of $\cdot$OH generation was for free radicals in biochar to transfer electrons to $O_2$ to produce the superoxide radical anion and hydrogen peroxide, which reacts further with free radicals to produce $\cdot$OH. This indicates that the $\cdot$OH generated on biochar may contribute to MITC degradation. Moreover, Chen et al. suggested that the H/C (hydrogen atom/carbon atom ratio) atomic ratio of organic components in biochar can be used to characterize their aromaticity and polarity [23]: the higher the H/C value, the lower the aromaticity. Therefore, higher H/C values mean that more functional groups are able to generate $\cdot$OH and more MITC could be degraded. For example, BC-2 biochar had a higher H/C value (0.25), and the MITC degradation rate in BC-2 was greater.

Soil amended with BC-1 inhibited MITC degradation, possibly due to BC-1's high adsorption capacity for MITC. To clarify the mechanism by which biochar inhibits MITC degradation, we investigated the biochar adsorption kinetics for MITC at different temperatures. The parameters of MITC adsorption by biochar are listed in Table 6. The $r^2$ values of all treatments were at least 0.82, and the calculated amounts of MITC adsorption at the equilibrium time ($q_e$) approached the values measured in the experiment, indicating that the observed adsorption of MITC to biochar provides a good fit to Bangham models [24]. At 30 °C, the amounts of MITC adsorbed onto BC-1 and BC-2 at the equilibrium time ($q_e$) were 57.87 mg·g$^{-1}$ and 11.97 mg·g$^{-1}$, respectively, indicating that BC-1 has a higher adsorption capacity (for MITC) than BC-2 but much lower than AC (activated charcoal, 211.91 mg·g$^{-1}$), possibly because MITC may be bound to BC-1 and is not available to degrade. In addition, BC-1 has a low H/C value, and chemical degradation was weak, so the MITC degradation rate in BC-1 or soil amended with BC-1 was lower than in unamended soil. As the temperature increased from 15 °C to 45 °C, the adsorption rates of biochar increased; however, the amount of adsorption at the equilibrium time ($q_e$) fell as the temperature increased. Kołohynska et al. note that the sorption capacity of biochar depends mainly on the polarity, aromaticity, surface area, and pore size distribution, etc. [25]. The tested biochars have different surfaces and different pore structures, providing different adsorption rates and adsorption capacities. In general, the adsorption capacity of biochar for MITC increased with the SSA (specific surface area) of the biochar [25]. For example, BC-1 had a larger SSA (382.81 m$^2$·g$^{-1}$) with a higher amount of adsorption (57.87 mg·g$^{-1}$), while BC-2 (SSA 36.14 m$^2$·g$^{-1}$) had a smaller SSA and lower amount of adsorption (11.97 mg·g$^{-1}$).

**Table 6.** The kinetic parameters of biochar (BC-1 and BC-2) adsorption of MITC at 15, 30, and 45 °C.

| Biochar | Temperature | $q_e$ (mg·g$^{-1}$) | $k$ (h$^{-z}$) | $z$ | $r^2$ |
|---------|-------------|---------------------|----------------|-----|-------|
|         | 15 °C       | 87.62               | 0.002          | 1.67| 1.00  |
| BC-1    | 30 °C       | 57.87               | 0.01           | 2.26| 1.00  |
|         | 45 °C       | 50.88               | 0.06           | 2.18| 0.82  |
|         | 15 °C       | 21.08               | 0.02           | 1.38| 0.99  |
| BC-2    | 30 °C       | 11.97               | 0.13           | 1.08| 0.88  |
|         | 45 °C       | 10.80               | 0.15           | 1.95| 0.84  |
| Activated | 15 °C     | 233.58              | 0.0005         | 1.79| 0.98  |
| charcoal (AC) | 30 °C | 211.91              | 0.02           | 1.34| 0.97  |
|         | 45 °C       | 273.39              | 0.01           | 1.68| 0.99  |

As noted above, the degradation rates of MITC were much lower in biochar (BC-1) and biochar-amended soil than in unamended soil. This occurred because MITC was adsorbed onto the biochar and thus degraded much more slowly, decreasing as the adsorption capacity increased. In contrast, BC-2 has a weaker absorbability of MITC but possesses a higher degradability; thus, the degradation rates of MITC were much faster in the biochar (BC-2) and biochar-amended soil than in unamended soil. Therefore, biochar's good absorbability or degradability of MITC were speculated to play an important role in reducing or accelerating MITC's degradation rate in soil amended with these types of biochar. Studies have shown that the degradation of MITC comprises both biological and chemical degradation and that biodegradation accounted for 51%–97% of the total degradation [26,27]. The slow degradation rate of MITC in biochar-amended soil is likely due to reduced microbial degradation. In addition, the surface of biochar contains a large number of chemical functional groups [23], and a vast amount of free radicals may potentially accelerate the degradation of MITC via radical reaction. In summary, the dissipation of MITC in soil amended with biochar depends on the balance between the amount of adsorption and degradation and is positively correlated with the SSA and H/C values, respectively.

However, the effects of biochar on fumigant emissions are known to be complex. Studies have shown that the emission of 1,3-D was reduced after biochar was applied to the surface, due to the enhanced adsorption of 1,3-D onto the biochar [8,28], while the emission of chloropicrin (CP) was reduced by biochar due to the accelerated degradation of CP in soil amended with biochar [7]. We found that biochar used in this experiment can significantly reduce the volatilization of MITC by degradation or adsorption. However, at the same time, biochar amendments also decrease the concentration of MITC in the soil, which potentially reduces its efficacy for controlling soil-borne pests. Through adsorption or degradation, biochar can minimize the concentration of MITC in the soil at different levels. In addition, the amount of reduction varies with different types of biochar. BC-1 has a high SSA (382.2 m$^2$·g$^{-1}$) and a small H/C value (0.01), with a $k$ value of 0.08 d$^{-1}$ and $q_e$ value of 57.87 mg·g$^{-1}$ at 30 °C, meaning that it has a greater absorbability of MITC, resulting in a reduced concentration of MITC in the air and soil. On the contrary, BC-2, with a larger H/C value (0.25) and a smaller SSA (36.1 m$^2$·g$^{-1}$), (the $k$ value was 15.19 d$^{-1}$, $q_e$ value was 11.97 mg·g$^{-1}$ at 30 °C), has a greater degradative effect on MITC and leads to a reduction of MITC both in air and soil. However, the reduction of MITC in the air and soil with the amendment of BC-1 was greater than that in BC-2 (Figures 1 and 2), which may be due to the greater adsorption of MITC by BC-1 rather than the degradation of MITC by BC-2. It is precisely because the concentration of MITC is reduced in soil amended by biochar that its efficacy in controlling soil-borne pests is reduced. For example, 5% BC-1 or 2% BC-2 amendments significantly reduce the efficiency of *Phytophthora* spp. and *Fusarium* spp. Increased doses of DZ were able to offset decreases in the efficacy of MITC in soils amended with biochar. When the DZ rate was only 25 mg·kg$^{-1}$, the efficacy of MITC against *Abutilon theophrasti* was reduced in soil amended with 1% BC-1, compared with unamended soil. However, the above reduction in efficacy was alleviated by increasing the application dose of DZ to 100 mg·kg$^{-1}$ (corresponding to a field rate of 360 kg·ha$^{-1}$).

Wang et al. have indicated that biochar amendment at less than 1% in soil did not have negative effects on the levels of pathogens and nematode control achieved by chloropicrin fumigation [7], and the efficacy of Dimethyl Disulfide (DMDS) for controlling root-knot nematodes and *Fusarium* spp. was not reduced when biochar was applied at a rate less than 2% and 0.5%, respectively [29]. Another study noted that, while nematode control was adequate in the specific system studied, biochar amendment could adversely impact pest control depending on the sorption strength of the particular type of biochar [10]. The actual impact on the efficacy will be a function of the interplay between the application rates of the pesticide and the biochar and the sorption capacity of the specific biochar for the specific pesticide. This issue should be considered when determining the desirable physical and chemical characteristics of biochar for agronomic systems. In the specific experimental system studied here, adequate pest control was achieved at a standard biochar application rate (13 Mg·ha$^{-1}$). However, different source materials and different production processes create types of biochar with different physical and chemical properties, as well as different adsorption properties or degradability. For example, biochars produced at higher temperatures have been shown to have substantially higher sorption capacities than those produced at lower temperatures. It is apparent that the potential detrimental impact of biochar amendments on pest control must be taken into account when considering the use of biochar in agriculture. In summary, the impact of biochar soil amendments on the efficacy of MITC against soil-borne pests depends on the biochar type, amendment rate, and the application dose of DZ. Because biochar can play a significant role in reducing fumigant emissions [7,8], it is important to select an appropriate biochar amendment that does not affect the efficacy of fumigants such as MITC.

## 3. Materials and Methods

### 3.1. Soil, Biochars and Chemicals

Soil samples were collected from the top 20 cm of cucumber greenhouses in Tongzhou and Daxing, Beijing, where the occurrences of nematodes and soil-borne pathogens were severe. The samples were taken at the end of the cropping season. The soil from Tongzhou was composed of 73.24% sand, 5.83% silt, and 20.93% clay, with an organic matter content of 9.12 g·kg$^{-1}$ soil and a pH of 7.1; the soil from Daxing was composed of 81.44% sand, 10.34% silt, 8.22% clay, with an organic matter content of 14.50 g·kg$^{-1}$ soil and a pH of 8.2. The soil was sieved through a 2 mm mesh and then mixed thoroughly before use. The soil moisture was adjusted to 15% (w/w).

The two types of biochar (BC-1 and BC-2) were made at 500 °C from Crofton weed and wood pellets, respectively. BC-1 was composed of 86.48% C, 11.70% O, 1.10% H, and 0.72% N, with an SSA of 382.21 m$^{-2}$·g$^{-1}$ and a pH of 10.5. BC-2 was composed of 82.94% C, 4.13% O, 1.70% H, and 0.42% N, with an SSA of 36.14 m$^{-2}$·g$^{-1}$ and a pH of 10.2. Particularly, BC-1 has a very low H/C (hydrogen atom/carbon atom ratio) value (0.01), while BC-2 with a high H/C value (0.25). The specific surface area (SSA) was determined by a V-Sorb 2800P surface area (Gold APP Instruments Corporation, China), and elemental composition was measured using a CHN element analyzer (vario PYRO cube, Elementar Analysensysteme Gmbh, Germany) [30]. The biochar was sieved through a 2 mm screen before use. An analytical standard of MITC (98.0% purity,) was provided by Damas-beta (ShenZhen, China). DZ 98 MG was obtained from Nantong Shizhuang Chemical Co., Ltd., JiangShu, China. Sodium sulfate anhydrous and ethyl acetate (both analytical grade) were obtained from Beijing Chemical Works, Beijing, China.

### 3.2. Emission Determination

The columns were constructed of two open stainless steel columns (each 6.0 cm high with 15 cm internal diameter) [31]; one was used as a soil column, while the other was used as a chamber to collect the emitted gas. MITC emissions were measured after MITC was injected into soil columns at 6.0 cm depth and the application of BC-1 or BC-2 to the surface soil (1.5 cm depth). Soil (water content 14% w/w) was packed into each soil column to a bulk density of 1.3 g·cm$^{-3}$. Pre-weighed (5.4 g, corresponded to biochar rates of 1%) BC-1 or BC-2 were uniformly mixed into the top 1.5 cm layer of the soil column before packing. Soil without biochar was used as a blank control (CK). After MITC (100 mg·kg$^{-1}$ soil) was injected, the soil column and the chamber were sealed together using aluminum foil tape. Each treatment was repeated three times. Then, the soil columns were placed in incubators at 28 °C.

After 0.25, 0.5, 1, 2, 3, 4, 8, 12, 16 and 24 h, 0.5 mL of gas was taken from the chamber and from the soil (at 6.0 cm depth) by a gastight syringe to determine the MITC concentrations. Previous gas-sealed tests have indicated that gas contentions in chambers remain unchanged more than 5 days, longer than our testing time of 24 h. Each gas sample was put into a 20 mL clear headspace vial containing 0.2 g of sodium sulfate that was immediately crimp-sealed with an aluminum cap and Teflon-faced butyl-rubber septum. The headspaces of the vials containing soil gas samples were analyzed using an Agilent 7890A gas chromatograph (GC) coupled with an Agilent 7694E headspace sampler and a micro electron capture detector (Agilent Technologies Inc., Palo Alto, CA, USA). The gas chromatography conditions were the same as those in the degradation experiments (Section 3.4), and autosampler headspace conditions were as follows: 1.0 mL sample loop; 125, 130, and 135 °C for sample equilibration, loop, and transfer line temperatures, respectively.

*3.3. Dose-Response Experiment*

The soil samples were amended with BC-1 and BC-2 (Tongzhou soil amended with BC-1, Daxing soil amended with BC-2, respectively) at rates of 0.1%, 0.25%, 0.5%, 1%, 2%, 5%, and 10% (w/w), in order to determine the effects of various amendment rates of biochar on the efficacy of DZ fumigation. Six hundred grams of soil amended with or without biochar was weighed into 2.5-L desiccators. DZ was uniformly mixed into the soil of each desiccator at the rate of 100 mg·kg$^{-1}$ soil (corresponding to a field rate of 360 kg·ha$^{-1}$). Ten *Abutilon theophrasti* seeds and fifteen *Digitaria sanguinalis* seeds were buried at 2 cm below the soil surface in each desiccator [32] after the application of DZ, and the desiccator was immediately sealed with a cover. The desiccators were placed in incubators at 28 °C. The desiccators were opened after 7 days of incubation, and the residual fumigant was released for a day. The weed height was measured with a caliper. *Fusarium* spp. and *Phytophthora* spp. were isolated from the soil and assessed quantitatively using the methods described by Komada [33] and Masago et al. [34], respectively. Root-knot nematodes (*Meloidogyne* spp.) were extracted from 100 g subsamples using the methods described by Liu [35].

To determine the effects of MTTC in controlling soil-borne pests in soil amended with biochar, DZ was applied at rates of 25, 50, 75, 100, 125, 150, and 175 mg·kg$^{-1}$ soil (corresponding to rates of 90, 180, 270, 360, 450, 540, 630 kg·ha$^{-1}$ in the field; its recommended dosage is 294–450 kg·ha$^{-1}$) amended with or without 1% BC-1 or 0.5% BC-2 (Tongzhou soil amended with BC-1, Daxing soil amended with BC-2). Other operations were similar to the experiment described above.

The nematode mortality was calculated according to Equation (1):

$$X = \frac{N_1}{N_1 + N_2},\tag{1}$$

where X is nematode mortality (%), $N_1$ is the number of dead nematodes, and $N_2$ is the number of live nematodes. The numbers of dead and live nematodes were counted under a dissecting microscope.

The corrected nematode mortality was calculated according to Equation (2):

$$Y = \frac{X_1 - X_2}{100 - X_2} \times 100,\tag{2}$$

where Y is the corrected nematode mortality (%), $X_1$ is the nematode mortality in treatments (%), and $X_2$ is the nematode mortality in the control (%).

The efficacy of controlling fungi or weeds was calculated according to Equation (3):

$$Y = \frac{X_1 - X_2}{X_1} \times 100, \tag{3}$$

where Y is the efficacy in controlling fungi or weeds, $X_1$ is the fungal population or weed height in the control, and $X_2$ is the fungal population or weed height in treated plots.

### 3.4. Degradation Experiment

A laboratory incubation experiment was conducted to determine the effect of biochar amendments on the degradation of MITC, using a method similar to that reported in Qin et al. [36]. Soil samples (Tongzhou soil) at 10% water content (w/w) were amended with biochar BC-1 or BC-2 at the rate of 1% (w/w). Eight grams of amended or unamended soil was placed in a 20-mL clear headspace vial, and 5 μL of ethyl acetate containing 48 mg·mL$^{-1}$ of MITC was added to each vial. The vials were sealed with an aluminum cover and Teflon-faced septa immediately after fumigant application. The treated vials were then inverted and placed in incubators at 30 °C. After incubation for 0.5, 1, 3, 4, 7, 10, 15, 30, 40, 60 and 90 days, three replicate samples from each treatment were removed from the incubator and immediately stored at −80 °C until the analysis of the MITC concentration. In the same experiment, we also determined the degradation rate of MITC in pure biochar using the following method. A total of 0.2 g biochar was weighed into a 20 mL clear headspace vial, and the vials were crimp-sealed with an aluminum cap and Teflon-faced butyl-rubber septum immediately after adding 5 μL standard solution of MITC-ethyl acetate. Other operations were similar to those described in the experiment above.

The extraction procedure for soil samples was similar to the methodology described by Hadiri [37]. Eight grams of anhydrous sodium sulfate and 8 mL ethyl acetate were added to each frozen vial, and the vials were recapped immediately. The vials were vortexed for 30 s, placed in a table concentrator and shaken for 60 min, and then subsequently allowed to settle for 1 h. After settling, the supernatants were filtered (using a 0.22-μm nylon syringe filter) into 2 mL vials for the fumigant analysis by Agilent 7890 A gas chromatography. A gas chromatograph with a micro-ECD (GC-μECD) and an HP-5 capillary column (30 m length × 320 μm × 0.25 μm film thickness) (Agilent Technologies, Inc.) was used to analyze the MITC. The detector and inlet temperatures were 300 °C and 250 °C, respectively. The oven temperature was held at 60 °C for 10 min. Under these conditions, the retention time of MITC was 5.4 min. Preliminary experiments indicated that the MITC recovery efficiency in the soil ranged from 89% to 103% using the above procedures. The LOD (lowest detectable limit) and LOQ (limit of quantitation) for MITC in the soil samples were 0.004 and 0.0170 mg·kg$^{-1}$, respectively. First-order kinetics were used to fit the MITC degradation behavior, using the following equation:

$$C = C_0 \times \exp(-kt), \tag{4}$$

where C (mg·kg$^{-1}$) and $C_0$ (mg·kg$^{-1}$) are the concentrations in soil at time $t$ (day) and time $t_0$ (day), respectively; $k$ (day$^{-1}$) is the first-order rate constant; and $t$ is the incubation time (day):

$$t_{1/2} = 0.693/k. \tag{5}$$

When using the first-order kinetics model, the half-life of degradation was calculated using Equation (5). The first-order kinetics model was fitted using Origin Pro 8.0 software (version 8.0, OriginLab Corporation, Wellesley Hills, MA, USA).

*3.5. Adsorption Experiment*

For the adsorption experiment, 0.1 g biochar was weighed into a number of 20 mL clear headspace vials. The two types of biochar (BC-1 and BC-2,) and one activated charcoal (AC) were tested at three temperature settings (15, 30 and 45 °C). All of the vials for each sampling time were placed in a container. The container was constructed of two open stainless steel cylindrical boxes of 12 cm ID, each 4 cm high, similar to the permeability cell described by Papiernik et al. [31]. Before assembling the container, approximately 0.2 g of pure solid MITC was added to a 2 mL vial, which was placed in the bottom of the container. This amount of fumigant was sufficient to generate a saturated vapor phase inside the container. The container was sealed with aluminum tape and carefully placed in incubators with the required temperature setting. After incubation for 3, 6, 12, 24, 48, 72 and 96 h, a container was removed from the incubator. The aluminum tape was removed and the container was opened and ventilated for half an hour in a fume hood. The biochar was then extracted and analyzed for adsorbed MITC using the same method described in the degradation experiments.

The Bangham model [24] was tested against the adsorption kinetic data, and the rate constant $k$ of adsorption was calculated using the Bangham Equation (6), as follows:

$$\ln\left(\ln\frac{q_t}{q_t - q_e}\right) = \ln k - z \times \ln t \tag{6}$$

where $q_t$ (mg·g$^{-1}$) is the amount of MITC adsorption at time $t$ (h); $q_e$ (mg·g$^{-1}$) is the calculated amount of adsorption at equilibrium; $t$ (h) is the time in hours; $k$ (h$^{-z}$) is the rate constant of adsorption; and $z$ is a constant in relation to the adsorbent.

## 4. Conclusions

Both of the biochars used in this study (BC-1 and BC-2) can significantly affect the degradation rate of MITC (reduce or accelerate). The dose-response experiments indicated that there were no negative effects on the control of pathogens, nematodes and weeds when the biochar amendment was less than 1% or 0.5% for BC-1 and BC-2, respectively, in soil at conventional dosages of the fumigant dazomet. The information obtained in this study will be useful for evaluating the effects of biochar on the bioavailability and efficacy of MITC.

**Acknowledgments:** This study was supported by the National Natural Science Foundation Project of China (31572035). The authors would like to express their thanks to Melanie Miller for help with the English language.

**Author Contributions:** Wensheng Fang and Qiuxia Wang designed the study and wrote the protocol; Dawei Han and Dongdong Yan carried out determination of the biochar characteristics; Wensheng Fang, Bin Huang, Dawei Han and Xiaoman Liu performed most of the experiments; Wensheng Fang, Jun Li and Meixia Guo managed the literature search and analyses; Wensheng Fang, Qiuxia Wang and Dawei Han analyzed the data; Wensheng Fang, Qiuxia Wang and Aocheng Cao were responsible for the overall design and wrote the article.

**Conflicts of Interest:** The authors declare no conflict of interest.

## References

1.  Ruzo, L.O. Physical, chemical and environmental properties of selected chemical alternatives for the pre-plant use of methyl bromide as soil fumigant. *Pest Manag. Sci.* **2006**, *62*, 99–113. [CrossRef] [PubMed]
2.  Martin, F.N. Development of alternative strategies for management of soilborne pathogens currently controlled with methyl bromide. *Annu. Rev. Phytopathol.* **2003**, *41*, 325–350. [CrossRef] [PubMed]
3.  Ajwa, H.A.; Klose, S.; Nelson, S.D.; Minuto, A.; Gullino, M.L.; Lamberti, F.; Lopez-Aranda, J.M. Alternatives to methyl bromide in strawberry production in the United States of America and the Mediterranean region. *Phytopathol. Mediterr.* **2003**, *42*, 220–244.
4.  Saeed, I.A.M.; Rouse, D.I.; Harkin, J.M.; Smith, K.P. Effects of soil water content and soil temperature on efficacy of metham-sodium against verticillium dahliae. *Plant Dis.* **1997**, *81*, 773–776. [CrossRef]

5.  Frick, A.; Zebarth, B.; Szeto, S. Behavior of the soil fumigant methyl isothiocyanate in repacked soil columns. *J. Environ. Qual.* **1998**, *27*, 1158–1169. [CrossRef]
6.  Kiely, T.; Donaldson, D.; Grube, A. *Pesticides Industry Sales and Usage*; Office of Prevention, Pesticides and Toxic Substances, United States Environment Protection Agency: Washington, DC, USA, 2004; p. 16.
7.  Wang, Q.; Yan, D.; Liu, P.; Mao, L.; Wang, D.; Fang, W.; Li, Y.; Ouyang, C.; Guo, M.; Cao, A. Chloropicrin emission reduction by soil amendment with biochar. *PLoS ONE* **2015**, *10*, e0129448. [CrossRef] [PubMed]
8.  Wang, Q.; Mao, L.; Wang, D.; Yan, D.; Ma, T.; Liu, P.; Zhang, C.; Wang, R.; Guo, M.; Cao, A. Emission reduction of 1,3-dichloropropene by soil amendment with biochar. *J. Environ. Qual.* **2014**, *43*, 1656–1662. [CrossRef] [PubMed]
9.  Shen, G.; Ashworth, D.J.; Gan, J.; Yates, S.R. Biochar amendment to the soil surface reduces fumigant emissions and enhances soil microorganism recovery. *Environ. Sci. Technol.* **2016**, *50*, 1182–1189. [CrossRef] [PubMed]
10. Graber, E.R.; Tsechansky, L.; Khanukov, J.; Oka, Y. Sorption, volatilization, and efficacy of the fumigant 1,3-dichloropropene in a biochar-amended soil. *Soil Sci. Soc. Am. J.* **2011**, *75*, 1365. [CrossRef]
11. Jablonowski, N.D.; Krutz, J.L.; Martinazzo, R.; Zajkoska, P.; Hamacher, G.; Borchard, N.; Burauel, P. Transfer of atrazine degradation capability to mineralize aged 14C-labeled atrazine residues in soils. *J. Agric. Food Chem.* **2013**, *61*, 6161–6166. [CrossRef] [PubMed]
12. Spokas, K.A.; Koskinen, W.C.; Baker, J.M.; Reicosky, D.C. Impacts of woodchip biochar additions on greenhouse gas production and sorption/degradation of two herbicides in a Minnesota soil. *Chemosphere* **2009**, *77*, 574–581. [CrossRef] [PubMed]
13. Yu, X.; Pan, L.; Ying, G.; Kookana, R.S. Enhanced and irreversible sorption of pesticide pyrimethanil by soil amended with biochars. *J. Environ. Sci.* **2010**, *22*, 615–620. [CrossRef]
14. Kookana, R.S. The role of biochar in modifying the environmental fate, bioavailability, and efficacy of pesticides in soils: A review. *Aust. J. Soil Res.* **2010**, *48*, 627–637. [CrossRef]
15. Nag, S.K.; Kookana, R.; Smith, L.; Krull, E.; Macdonald, L.M.; Gill, G. Poor efficacy of herbicides in biochar-amended soils as affected by their chemistry and mode of action. *Chemosphere* **2011**, *84*, 1572–1577. [PubMed]
16. Fang, W.; Wang, Q.; Han, D.; Liu, P.; Huang, B.; Yan, D.; Ouyang, C.; Li, Y.; Cao, A. The effects and mode of action of biochar on the degradation of methyl isothiocyanate in soil. *Sci. Total Environ.* **2016**, *565*, 339–345. [PubMed]
17. Primo, P.D.; Gamliel, A.; Austerweil, M.; Steiner, B.; Beniches, M.; Peretz-Alon, I.; Katan, J. Accelerated degradation of metam-sodium and dazomet in soil: Characterization and consequences for pathogen control. *Crop Prot.* **2003**, *22*, 635–646. [CrossRef]
18. Warton, B.; Matthiessen, J.N. The crucial role of calcium interacting with soil pH in enhanced biodegradation of metam-sodium. *Pest Manag. Sci.* **2005**, *61*, 856–862. [CrossRef] [PubMed]
19. Zhou, L.; Hebert, V.R.; Miller, G.C. Gas-phase reaction of methyl isothiocyanate and methyl isocyanate with hydroxyl radicals under static relative rate conditions. *J. Agric. Food Chem.* **2014**, *62*, 1792–1795.
20. Bourke, J.; Manleyharris, M.; Fushimi, C.; Dowaki, K. Do all carbonized charcoals have the same chemical structure? 2. A model of the chemical structure of carbonized charcoal. *Ind. Eng. Chem. Res.* **2007**, *46*, 5954–5967. [CrossRef]
21. Lehmann, J.D.; Joseph, S. *Biochar for Environmental Management: Science and Technology*; Earthscan: London, UK, 2009; Volume 25, pp. 15801–15811.
22. Fang, G.; Zhu, C.; Dionysiou, D.D.; Gao, J.; Zhou, D. Mechanism of hydroxyl radical generation from biochar suspensions: Implications to diethyl phthalate degradation. *Bioresour. Technol.* **2015**, *176*, 210–217. [CrossRef] [PubMed]
23. Chen, B.; Johnson, E.J.; Chefetz, B.; Zhu, L.; Xing, B. Sorption of polar and nonpolar aromatic organic contaminants by plant cuticular materials: Role of polarity and accessibility. *Environ. Sci. Technol.* **2005**, *39*, 6138–6146. [CrossRef] [PubMed]
24. Wang, L.; Cao, B.; Wang, S.; Yuan, Q. $H_2S$ catalytic oxidation on impregnated activated carbon: Experiment and modelling. *Chem. Eng. J.* **2006**, *118*, 133–139. [CrossRef]
25. Kołodyńska, D.; Wnętrzak, R.; Leahy, J.J.; Hayes, M.H.B.; Kwapiński, W.; Hubicki, Z. Kinetic and adsorptive characterization of biochar in metal ions removal. *Chem. Eng. J.* **2012**, *197*, 295–305. [CrossRef]

26. Dungan, R.S.; Gan, J.; Yates, S.R. Accelerated degradation of methyl isothiocyanate in soil. *Water Air Soil Pollut.* **2003**, *142*, 299–310. [CrossRef]

27. Gan, J.; Papiernik, S.K.; Yates, S.R.; Jury, W.A. Temperature and moisture effects on fumigant degradation in soil. *J. Environ. Qual.* **1999**, *28*, 1436–1441. [CrossRef]

28. Wang, Q.; Gao, S.; Wang, D.; Spokas, K.; Cao, A.; Yan, D. Mechanisms for 1,3-dichloropropene dissipation in biochar-amended soils. *J. Agric. Food Chem.* **2016**, *64*, 2531–2540. [CrossRef] [PubMed]

29. Wang, Q.; Fang, W.; Yan, D.; Han, D.; Li, Y.; Ouyang, C.; Guo, M.; Cao, A. The effects of biochar amendment on dimethyl disulfide emission and efficacy against soil-borne pests. *Water Air Soil Pollut.* **2016**, *227*, 1–9. [CrossRef]

30. Silber, A.; Levkovitch, I.; Graber, E.R. pH-dependent mineral release and surface properties of cornstraw biochar: Agronomic implications. *Environ. Sci. Technol.* **2010**, *44*, 9318–9323. [CrossRef] [PubMed]

31. Papiernik, S.K.; Yates, S.; Gan, J. An approach for estimating the permeability of agricultural films. *Environ. Sci. Technol.* **2001**, *35*, 1240–1246. [CrossRef] [PubMed]

32. Klose, S.; Acosta, V. Microbial community composition and enzyme activities in a sandy loam soil after fumigation with methyl bromide or alternative biocides. *Soil Biol. Biochem.* **2006**, *38*, 1243–1254. [CrossRef]

33. Komada, H. Development of a selective medium for quantitative isolation of Fusarium oxysporum from natural soil. *Rev. Plant Prot. Res.* **1975**, *8*, 114–124.

34. Masago, H.; Yoshikawa, M.; Fukada, M. Selective inhibition of *Pythium* spp. on a medium for direct isolation of *Phytophthora* spp. from soils and plants. *Phytopathology* **1977**, *67*, 425–428. [CrossRef]

35. Liu, W. *Plant Pathogenic Nematodes*; China Agriculture Press: Beijing, China, 2004; Volume 2000, p. 373.

36. Qin, R.; Gao, S.; Ajwa, H.; Hanson, B.D.; Trout, T.J.; Wang, D.; Guo, M. Interactive effect of organic amendment and environmental factors on degradation of 1,3-dichloropropene and chloropicrin in soil. *J. Agric. Food Chem.* **2009**, *57*, 9063–9070. [CrossRef] [PubMed]

37. Hadiri, N.E.; Ammati, M.; Chgoura, M.; Mounir, K. Behavior of 1,3-dichloropropene and methyl isothiocyanate in undisturbed soil columns. *Chemosphere* **2003**, *52*, 893–899. [CrossRef]

*Article*

# Effects of Biochar Amendment on Chloropicrin Adsorption and Degradation in Soil

Pengfei Liu [1], Qiuxia Wang [2,*], Dongdong Yan [2], Wensheng Fang [2], Liangang Mao [2], Dong Wang [3], Yuan Li [2], Canbin Ouyang [2], Meixia Guo [2] and Aocheng Cao [2,*]

[1]  Southwest Research and Design Institute of Chemical Industry Co., Ltd., Chengdu 610225, China; liupengfei0912@163.com

[2]  Plant Protection Institute of Chinese Academy of Agricultural Sciences, State Key Laboratory for Biology of Plant Disease and Insect Pest, Beijing 100193, China; 13260176634@163.com (D.Y.); fws0128@163.com (W.F.); maoliangang@126.com (L.M.); liyuancaas@126.com (Y.L.); oycb@ippcaas.cn (C.O.); guomeixia@126.com (M.G.)

[3]  U.S. Department of Agriculture—Agricultural Research Service (USDA-ARS), Water Management Research Unit, Parlier 93648, CA, USA; dong.wang@ars.usda.gov

*  Correspondences: qxwang@ippcaas.cn (Q.W.); caoac@vip.sina.com (A.C.); Tel.: +86-10-6289-4863 (Q.W.); +86-10-6281-5940 (A.C.)

Academic Editor: Mejdi Jeguirim

Received: 10 August 2016; Accepted: 18 October 2016; Published: 26 October 2016

**Abstract:** The characteristics of biochar vary with pyrolysis temperature. Chloropicrin (CP) is an effective fumigant for controlling soil-borne pests. This study investigated the characteristics of biochars prepared at 300, 500, and 700 °C by *michelia alba* (*Magnolia denudata*) wood and evaluated their capacity to adsorb CP. The study also determined the potential influence of biochar, which was added to sterilized and unsterilized soils at rates of 0%, 1%, 5%, and 100%, on CP degradation. The specific surface area, pore volume, and micropores increased considerably with an increase in the pyrolytic temperature. The adsorption rate of biochar for CP increased with increasing pyrolytic temperature. The maximum adsorption amounts of CP were similar for the three biochars. Next, the study examined the degradation ability of the biochar for CP. The degradation rate constant (k) of CP increased when biochar was added to the soil, and k increased with increased amendment rate and pyrolysis temperature. The results indicate that biochar can accelerate CP degradation in soil. The findings will be instructive in using biochar as a new fertilizer in fumigating soil with CP.

**Keywords:** chloropicrin; biochar; pyrolysis temperature; adsorption; degradation

---

## 1. Introduction

Biochar is produced during pyrolysis of organic material under oxygen-limited and relatively low temperature conditions [1]. The application of biochar to agricultural soil can improve soil fertility and the nitrogen-use ratio; selectively immobilize nutrient elements [2]; adsorb heavy metals and organic contaminants; transform $Cu^{2+}$ and $Zn^{2+}$ for plant uptake [3,4]; and change the soil physicochemical properties, including pH, organic carbon, and cation exchange capacity [5,6]. Biochar has received increasing recognition as a soil amendment for sequestering carbon to mitigate climate change [7,8]. In addition, biochar can affect the fate of pesticides in soil. Yang et al. and Zhang et al. suggested that wheat straw-derived biochar affects the degradation of organic chemicals in soil, presumably through adsorption, and indicated that biochar can decrease the microbial degradation of Diuron [9,10]. Jones et al. found that biochar suppresses simazine biodegradation and reduces simazine leaching [11]. However, Jablonowski et al. observed that increased degradation of atrazine residues likely occurred through the transfer of atrazine-adapted soil microflora from different soils and regions to non-adapted soil [12]. Therefore, the influence of biochar on pesticides produces variable results.

Chloropicrin (CP) is applied as a fumigant to many crops to control soil-borne disease and nematodes [13,14]. The influence of biochar on CP degradation in soil remains poorly understood. Biochar, which is a type of charcoal, possesses a high specific surface area (SSA) that may decrease CP degradation because of its strong capacity to adsorb this fumigant. However, chloropicrin is a polyhalide that can be degraded by free radical-induced reactions [15], and biochar typically contains high quantities of free radicals [16] that could potentially accelerate CP degradation. Thus, the application of biochar as a new fertilizer into the soil could accelerate the degradation of CP, and, accordingly, affect the efficacy of CP to soil borne pests. Biochar produced under different temperatures has different properties. For example, biochar produced under high pyrolysis temperatures typically has a relatively high surface area and aromaticity and can, therefore, resist decomposition. This type of biochar is known as a green adsorbent [2,17,18]. In contrast, low-temperature pyrolysis favors a greater recovery of C and several other nutrients (e.g., N, K, and S) that are increasingly lost at higher temperatures [19].

The objectives of this study were to determine the structure and surface chemical properties of biochar prepared under different pyrolysis temperatures (300, 500, and 700 °C, marked as B300, B500, and B700, respectively) and to evaluate the effects of amending soil with these biochars on CP degradation and adsorption. Additionally, the study evaluated the interplay between CP degradation and adsorption. The results will be instructive in how to use biochar as a new fertilizer in fumigating field with CP.

## 2. Results and Discussion

### 2.1. The Composition and Structure of Biochar

The Boehm titration results (Table 1) indicates that, in general, the carboxyl concentration decreased and the phenol hydroxyl concentration increased as the pyrolytic temperature increased. The phenol hydroxyl and carboxyl concentrations changed dramatically from 300 °C to 500 °C. The lactone concentration did not change significantly with changes in temperature, but these molecules were present in the highest concentration of all the oxygen-containing functional groups. In addition, the changes in carboxyl and lactone concentrations were not significant between B500 and B700. The elemental composition is shown in Table 1. An increase in the pyrolytic temperature resulted in a decrease in the hydrogen and oxygen content, and an increase in the carbon content. The calculated H/C values were lower under high temperatures, indicating that the biochar was highly carbonized and showed a highly aromatic structure. The higher H/C values for B300 and B500 indicates that most of the carbon-containing compounds had not decomposed under low pyrolytic temperatures, including lignin and cellulose components [20,21]. Additionally, the (O + N)/C and O/C ratios decreased sharply with increasing pyrolytic temperature, indicating that the biochar had high polarity and was hydrophilic under high temperatures [22,23].

The SSA, pore size, and ash content of the different biochars are shown in Table 1. The SSA and pore volume increased and the pore size decreased as the pyrolytic temperature increased. In addition, the scanning electron microscopy (SEM) images showed that the pore volume (Figure 1A,C,E) and the number of micro-pores on the biochar surfaces (Figure 1B,D,F) increased. The pore canal gradually collapsed as the pyrolytic temperature increased which caused the pore volume to increase. At high pyrolytic temperatures, an immense heat flow was released from the inner feedstock and burst through the micro-pores. The changes in the properties of the biochars produced under different temperatures were due to the decomposition of hemicellulose, cellulose, and lignin. The corresponding temperatures for maximum decomposition of hemicellulose and cellulose were 300 °C and 355 °C, respectively, while the temperature for lignin decomposition ranged between 150 to 900 °C [20,21]. Therefore, as the temperature increased from 300 °C to 700 °C, the surface characteristics of biochar changed dramatically, which caused obvious differences in surface properties. Intense heat was released from the inner parts of the feedstock and burst through the micropores at high pyrolytic temperatures, resulting in an increase in the number of micropores [1].

**Table 1.** Characteristics of biochar produced at different temperatures.

| Types | Oxygen-Containing Functional Groups | | |
|---|---|---|---|
| | Carboxyl (mmoL/g) | Phenol Hydroxyl (mmoL/g) | Lactones (mmoL/g) |
| B300 | 0.37 ± 0.03 [1] a | 0.01 ± 0.001 a [2] | 0.56 ± 0.01 a |
| B500 | 0.29 ± 0.01 b | 0.32 ± 0.01 b | 0.44 ± 0.01 b |
| B700 | 0.26 ± 0.01 b | 0.39 ± 0.01 c | 0.47 ± 0.03 b |

| | Elemental composition | | | | | | |
|---|---|---|---|---|---|---|---|
| | Carbon (mol %) | Hydrogen (mol %) | Nitrogen (mol %) | Oxygen (mol %) | (O + N)/C [3] | O/C | H/C |
| B300 | 67.97 a | 2.68 a (B500) | 0.94 a | 24.28 a | 0.37 | 0.36 | 0.04 |
| B500 | 76.17 b | 2.36 b (B300) | 0.99 b | 14.35 b | 0.20 | 0.19 | 0.04 |
| B700 | 83.55 c | 1.62 c | 0.97 c | 6.58 c | 0.09 | 0.08 | 0.01 |

| | SSA, pore size and ash content | | | | |
|---|---|---|---|---|---|
| | SSA (m$^2$/g) [4] | BJH pore size (nm) | Pore volume (cm$^3$/g) | Ash content (%) | pH (1:2.5) |
| B300 | 1.81 | 15.59 | 0.00034 | 4.62 ± 0.01 a | 7.42 ± 0.05 a |
| B500 | 68.36 | 5.34 | 0.02 | 5.86 ± 0.14 b | 9.39 ± 0.08 b |
| B700 | 364.63 | 4.35 | 0.13 | 7.74 ± 0.01 c | 10.14 ± 0.11 c |

[1] Data were expressed as means ± S.E. ($n$ = 3); [2] The difference was analyzed by ANOVA based on Duncan multiple comparison ($p < 0.05$); [3] H/C: atomic ratio of hydrogen to carbon. O/C: atomic ratio of oxygen to carbon. (O + N)/C: atomic ratio of the sum of nitrogen and oxygen to carbon; [4] The specific surface area (SSA) of biochar was measured using the BET-N$_2$ method. The pore volume and pore diameter were calculated under an absorption capacity with a relative pressure of 0.99 and the Barrett-Joyner-Halendar (BJH) method, respectively.

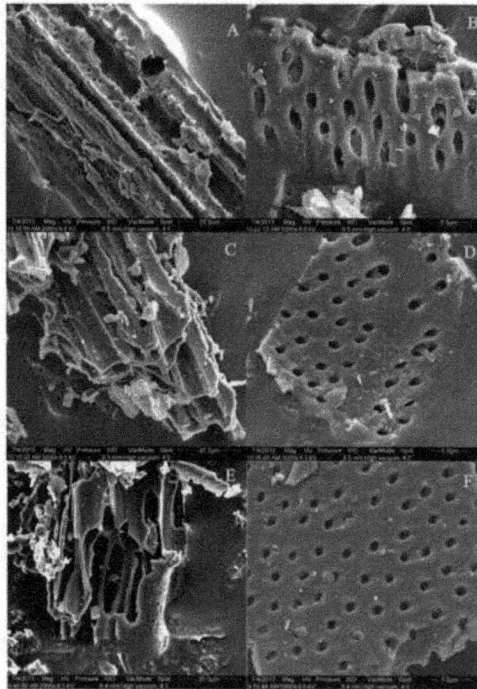

**Figure 1.** SEM images of biochars under different pyrolytic temperatures. B300 is shown in (**A,B**); B500 is shown in (**C,D**); and B700 is shown in (**E,F**). (**A,C,E**) are magnified 2000× and (**B,D,F**) are magnified 5000×.

The ash content of the biochars increased gradually with an increase in temperature (Table 1), which indicates that at higher temperatures there was a greater decomposition of organic matter and a reduction in the biochar yield rate. However, compared with other feedstocks (e.g., chicken manure, rice straw), wood has a lower ash content [24] and, therefore, has good potential as a feedstock. Additionally, the pH of the biochars increased with an increase in the pyrolysis temperature, which indicates that the biochars were alkaline, as indicated by Lehmann [25].

*2.2. Adsorption Kinetics*

The Lagergren pseudo-first-order model fit the adsorption kinetics of biochar to CP well (Table 2). The adsorption equilibrium time was approximately 36 h (Figure 2) and 24 h in biochar and active carbon, respectively, at 30 °C. The adsorption rate constants were 0.01, 0.07, and 0.24 min$^{-1}$ for B300, B500, and B700, respectively, which shows that the adsorption rate gradually increased as the pyrolytic temperature increased. This result may be due to changes in the biochar characteristics, such as the number of micropores and adsorption sites on the surface of the biochar. An increase in the pyrolytic temperature from 300 °C to 700 °C did not result in significant differences in the quantity of CP on the three biochars at equilibrium. The adsorption capacity of the biochars was considerably lower than that of active carbon. The large SSA of biochar appears to be in accordance with its high adsorption capacity for CP; however, this result conflicts with the adsorption results of B700, whose SSA is 364.63 m$^2 \cdot$g$^{-1}$. Therefore, other factors, such as the number of adsorption sites on the adsorbent, may account for the adsorption performance of biochar. The following section discusses how the properties of CP adsorption to biochar were studied by determining the adsorption isotherm.

**Table 2.** Fitting results of adsorption kinetics of biochar for chloropicrin (CP).

| Biochar Type | q$_e$ (mg/g) | k (min$^{-1}$) | R$^2$ |
|:---:|:---:|:---:|:---:|
| B300 | 23.81 | 0.01 | 0.98 [1] |
| B500 | 25.72 | 0.07 | 0.87 |
| B700 | 26.03 | 0.24 | 0.74 |
| AC | 176.86 | 0.27 | 0.88 |

[1] The adsorption kinetics was fitted by Lagergren pseudo-first-order kinetics equation.

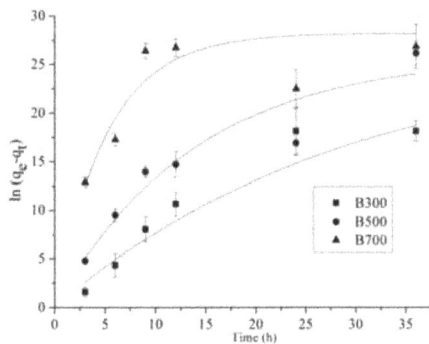

**Figure 2.** The fitting curve of adsorption kinetics using the Lagergren pseudo-first-order model.

## 2.3. Adsorption Isotherm

The isotherm equation was fitted using three models, and the Freundlich equation provided the best degree of fit (Table 3, Figures 3 and 4). Therefore, the discussion below focuses mainly on the parameters of this equation. According to the principles of the Freundlich model, the process of CP adsorption to biochar is associated with heterogeneous adsorption and biochar surfaces that contain sites with different adsorption energies. The parameter n increased with an increase in the pyrolytic temperature. The larger the value of n, the more nonlinear the adsorption isotherm became, and the more its behavior deviated from linearity. The $k_f$ value increased with an increase in pyrolytic temperature, which indicated the following adsorption capacity trend: B700 > B500 > B300. This result was not in agreement with the adsorption kinetics data. However, the sum of the amount of CP in the gas and biochars was considerably lower than the applied amounts (Table 4), which indicated that the CP was degraded by the biochar. The degradation ratio increased as the pyrolytic temperature increased, and 8%–25%, 14%–36%, and 44%–80% of applied CP was degraded in B300, B500, and B700, respectively. The adsorption capacity of CP to B700 was strong, and the degradation ability was high; therefore, the maximum adsorption amounts of CP in B300, B500, and B700 were similar.

**Table 3.** Fitting results of different adsorption isotherm models [1].

| Biochar Type | Langmuir | | | Freundlich | | | D-R | | |
|---|---|---|---|---|---|---|---|---|---|
| | $q_{max}$ | $k_L$ | $R^2$ | $k_f$ | n | $R^2$ | $q_{max}$ | E | $R^2$ |
| B300 | 0.51 | 0.004 | 0.82 | 0.26 | 1.1 | 0.82 | 4.79 | 0.29 | 0.67 |
| B500 | 2.24 | 0.1 | 0.86 | 0.53 | 3 | 0.93 | 1.73 | 0.49 | 0.58 |
| B700 | 1.97 | 1.64 | 0.92 | 1.1 | 5.04 | 0.94 | 1.81 | 3.78 | 0.82 |

[1] The adsorption isotherms employed the Langmuir, Freundlich, and Dubinin-Radushkevitch (D-R) models to describe the adsorption capacity of biochar. Compared with other equations, the Freundlich equation provided the best degree of fit.

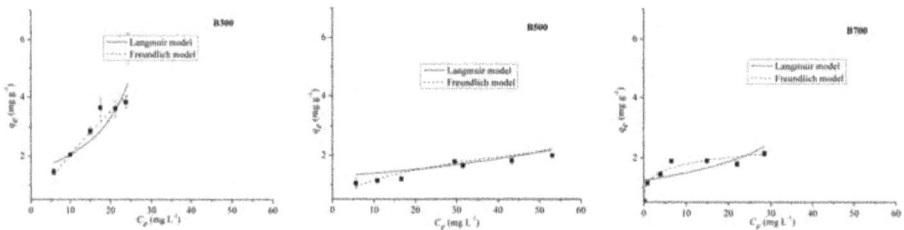

**Figure 3.** The fitting curve of the adsorption isotherm using Freundlich and Langmuir equations.

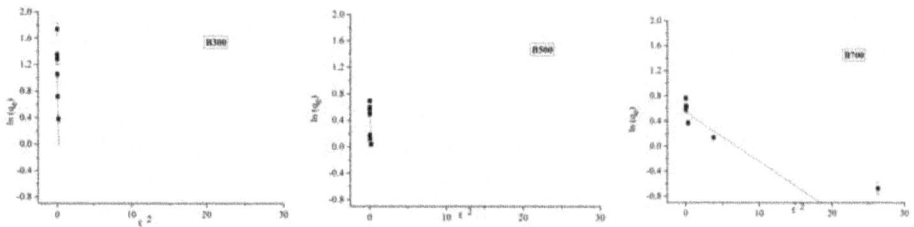

**Figure 4.** The fitting curve of the adsorption isotherm using Dubinin-Radushkevitch (D-R) models.

**Table 4.** The fate of CP after application to the biochars.

| Biochar Types | Applied Amount (mg) | Amount in Gas Phase (mg) | Amount in Biochar (mg) | Degradation Amount (mg) | Degradation Ratio (%) | Average Degradation Ratio (%) |
|---|---|---|---|---|---|---|
| B300 | 0.4 | 0.11 ± 0.003 [1] | 0.22 ± 0.01 | 0.07 ± 0.01 | 16.92 | |
| | 0.6 | 0.2 ± 0.01 | 0.31 ± 0.002 | 0.09 ± 0.01 | 15.36 | |
| | 0.8 | 0.3 ± 0.05 | 0.43 ± 0.02 | 0.07 ± 0.01 | 8.72 | |
| | 1 | 0.25 ± 0.18 | 0.55 ± 0.05 | 0.2 ± 0.02 | 20.46 | 17.46 a [2] |
| | 1.2 | 0.43 ± 0.03 | 0.54 ± 0.04 | 0.23 ± 0.05 | 19.52 | |
| | 1.4 | 0.48 ± 0.02 | 0.57 ± 0.03 | 0.35 ± 0.01 | 25.01 | |
| | 1.6 | 0.49 ± 0.04 | 0.85 ± 0.08 | 0.26 ± 0.04 | 16.26 | |
| B500 | 0.4 | 0.11 ± 0.01 | 0.16 ± 0.03 | 0.13 ± 0.04 | 33.24 | |
| | 0.6 | 0.22 ± 0.005 | 0.17 ± 0.003 | 0.21 ± 0.005 | 35.58 | |
| | 0.8 | 0.33 ± 0.05 | 0.18 ± 0.006 | 0.29 ± 0.05 | 36.13 | |
| | 1 | 0.59 ± 0.06 | 0.27 ± 0.002 | 0.14 ± 0.06 | 14.23 | 25.65 b |
| | 1.2 | 0.63 ± 0.15 | 0.25 ± 0.01 | 0.32 ± 0.14 | 26.84 | |
| | 1.4 | 0.87 ± 0.06 | 0.27 ± 0.02 | 0.26 ± 0.07 | 18.66 | |
| | 1.6 | 1.06 ± 0.09 | 0.3 ± 0.005 | 0.24 ± 0.08 | 14.86 | |
| B700 | 0.4 | 0.003 ± 0.0004 | 0.08 ± 0.02 | 0.32 ± 0.02 | 80.13 | |
| | 0.6 | 0.02 ± 0.004 | 0.17 ± 0.002 | 0.41 ± 0.006 | 68.53 | |
| | 0.8 | 0.08 ± 0.005 | 0.22 ± 0.01 | 0.5 ± 0.01 | 63.10 | |
| | 1 | 0.13 ± 0.01 | 0.28 ± 0.01 | 0.59 ± 0.02 | 58.68 | 59.35 c |
| | 1.2 | 0.30 ± 0.10 | 0.28 ± 0.006 | 0.62 ± 0.10 | 51.38 | |
| | 1.4 | 0.44 ± 0.06 | 0.27 ± 0.01 | 0.69 ± 0.07 | 49.36 | |
| | 1.6 | 0.57 ± 0.08 | 0.32 ± 0.01 | 0.71 ± 0.08 | 44.29 | |

[1] Data were expressed as mean ± S.E. ($n$ = 3); [2] The data was analyzed using univariate analyses based on the Duncan multiple comparison and differences were considered significant in different biochars at $p < 0.05$.

## 2.4. The Effects of Different Biochar Types and Rates on CP Degradation Kinetics

The degradation of CP (Figure 5) followed first-order kinetics in unamended soils and those that were amended with biochar, and the correlation coefficients were greater than 0.99 (Table 5). The degradation rate constant ($k$) and half-life ($t_{1/2}$) was 1.22 d$^{-1}$ and 0.57 d, respectively, in unamended soil, which was comparable to the findings of Gan et al., who reported $t_{1/2}$ values of 1.5, 4.3, and 0.2 d in Arlington sandy loam, Carsitas loamy sand, and Waukegen silt loam, respectively [26]. The degradation rate constants in soils amended with 1% B300, B500, and B700 were 2.30, 2.44, and 4.05 d$^{-1}$, respectively, and the associated half-lives were 0.30, 0.28, and 0.17 d. The soil amended with biochar showed a drastic increase in the degradation rate constant in comparison to the soil without biochar, in other words, 89% to 232%. These results indicate that biochar added to soil can accelerate the degradation of CP, and this acceleration increases with the biochar production temperature.

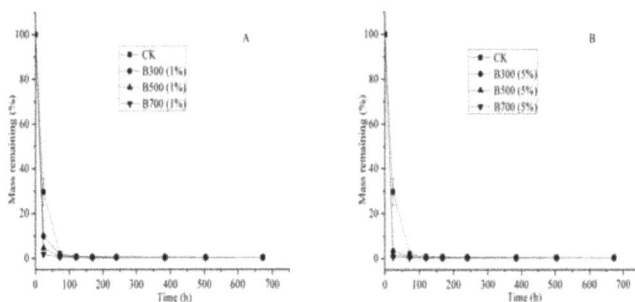

**Figure 5.** Degradation trend of CP in soil amended with different biochar types, (**A**) was 1% biochar content and (**B**) was 5%.

**Table 5.** Fitting results of CP degradation in different biochar-amended soils using a first-order model.

| Treatment | k (d$^{-1}$) | t$_{1/2}$ (d) | R$^2$ |
|---|---|---|---|
| Soil (CK) | 1.22 ± 0.02 [1] | 0.57 | >0.99 |
| Soil + 1% B300 | 2.30 ± 0.07 | 0.30 | >0.99 |
| Soil + 1% B500 | 2.44 ± 0.05 | 0.28 | >0.99 |
| Soil + 1% B700 | 4.05 ± 0.21 | 0.17 | >0.99 |
| Soil + 5% B300 | 3.42 ± 0.15 | 0.20 | >0.99 |
| Soil + 5% B500 | 4.17 ± 0.32 | 0.17 | >0.99 |
| Soil + 5% B700 | 5.10 ± 0.49 | 0.14 | >0.99 |
| Sterilized soil | 0.19 ± 0.01 | 3.65 | 0.98 |
| Sterilized soil + 1% B300 | 1.23 ± 0.02 | 0.56 | >0.99 |
| Sterilized soil + 1% B500 | 0.91 ± 0.03 | 0.76 | >0.99 |
| Sterilized soil + 1% B700 | 0.93 ± 0.03 | 0.75 | >0.99 |

[1] Data were expressed as mean ± S.E. ($n$ = 3). The significant difference of values of k for each group was determined using a 95% confidence interval around the estimates ($p < 0.05$).

The degradation rate constants for soils amended with 5% B300, B500, and B700 were 3.42, 4.17, and 5.10 d$^{-1}$ and the half-lives were 0.20, 0.17, and 0.14 d, respectively. The degradation rate constants for the above treatments were 1.49, 1.71, and 1.26 times higher than those for soils amended with 1% B300, B500, and B700, respectively. These results indicate that the degradation rate increased as the biochar amendment rate increased.

In unamended sterilized soil, the degradation rate constant was 0.19 d$^{-1}$ and the half-life was 3.65 d. The degradation rate constant was 84% lower than that of unsterilized soil. In sterilized soils amended with 1% B300, B500, and B700, the degradation rate constants were 1.23, 0.91, and 0.93 d$^{-1}$ and the half-lives were 0.56, 0.76, and 0.75 d, respectively. The degradation rate constants for sterilized soils were 86%–448% lower than those for the unsterilized soil treatments. These results indicate that biodegradation played an important role in CP degradation, and the sterilization effect on CP degradation in soil amended with biochar was greater than that in unamended soil. Gan et al. reported that biodegradation accounts for 68%–92% of CP degradation in soil [26]. The edaphon that degrade CP are composed of pseudomonades that rely on dehalogenation to generate nitromethane. The processes are as follows: $Cl_3CNO_2 \rightarrow Cl_2CHNO_2 \rightarrow ClCH_2NO_2 \rightarrow CH_3NO_2$ [27], and the end products are $CO_2$, $NO_3$, and $Cl^{-1}$. Biochar produced under a high pyrolytic temperature has numerous surface micropores and larger interior channels, which are suitable for microorganism colonization and biodegradation [28,29]. The sterilization effect on CP degradation was therefore significant in soil amended with B700. Additionally, biochars react with microbes, soil organic matter, and minerals within days of addition to soil [1]. There is a formation of organomineral phases that contain nanoparticles especially of redox active Fe and compounds with high contents of quinones/phenols that can catalyse redox reactions [30–33]. Microbes that can be involved in these redox reactions and the mineral oxide nanoparticles can break down CP through biotic and abiotic oxidation.

In addition, the degradation rate constants for sterilized soil with biochar amendment were significantly higher than those for sterilized soil without biochar and were similar to those for unamended and unsterilized soil. After sterilization at high temperatures, most of the soil microorganisms are nearly killed, and the effects of biodegradation may be ignored. The above results suggest that chemical degradation of CP occurred when soils were amended with biochar. The chemical degradation capacity of biochar was similar to the microorganism biodegradation capacity.

The fitting results of CP degradation in 0.2 g of pure biochar are shown in Figure 6. The degradation rate constants for B300, B500, and B700 were 0.04, 0.30, and 1.53 d$^{-1}$ and the half-lives were 17.33, 2.31, and 0.45 d, respectively. The results indicate that biochar can degrade CP and the degradation rate constants increased as the biochar production temperature increased. This finding confirms the occurrence of a chemical reaction between CP and the biochar.

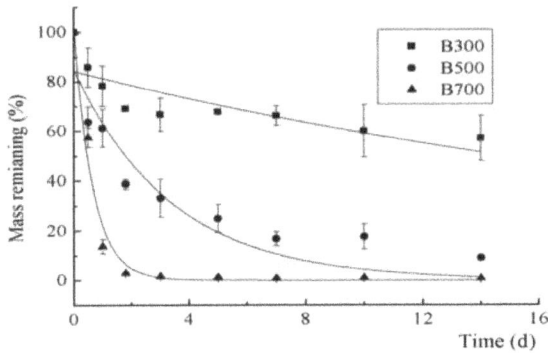

**Figure 6.** Fitting curves of the CP degradation in biochar produced under different temperatures. The first-order kinetics model was used to fit degradation data.

## 2.5. The Degradation Products of CP in Soil and Biochar

Chloropicrin and its main dechlorinated product (dichloronitromethane, DICP) were detected in soil and biochar (Figure 7). The peak areas of CP and DICP were used to express the changing concentrations. Both CP and DICP degraded much more rapidly in B700 than in the other materials. The CP degraded more rapidly in B500 than in B300 and pure soil. The difference between the CP concentration in B500 and in the soil was not significant. The DICP concentration in B500 increased up to 10 h after incubation and then decreased from 25 h after incubation; these values were considerably higher than those measured in pure soil. The DICP concentration in B300 increased over time and was higher than that measured in soil 12 h after incubation. The results showed that biochar has the ability to accelerate the dechlorination process of CP.

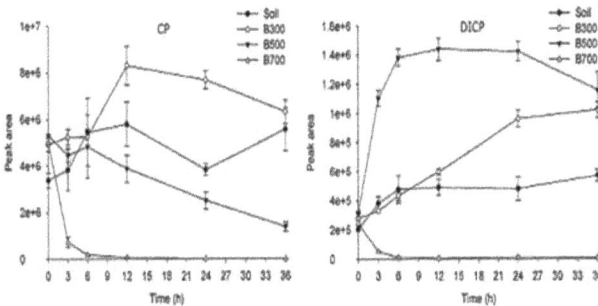

**Figure 7.** CP and dichloronitromethane (DICP) concentrations in soil and biochar. The peak areas of CP and DICP were used to express the changing concentrations. The significant difference of values in each biochar was determined using a 95% confidence interval around the estimates ($p < 0.05$). The CP in B500 and soil was not significantly different, but there was a significant difference in B300 and B700. The differences of DICP were considered significant in different biochar and soil.

The dechlorination process is generally induced by radical reactions. Considerable research has been conducted on the hydroxyl-radical-induced degradation reactions of fumigants [15,34,35]. Mezyk and Cole researched the reaction process and mechanism between CP and free radicals. These authors reported that free radicals react with CP in the following chain reactions: $CP + H' \rightarrow {}^{\cdot}CCl_2NO_2 + H^+ + Cl^-$; $CP + H' \rightarrow {}^{\cdot}CCl_3 + H^+ + NO^{2-}$; $CP + OH' \rightarrow$ intermediate products $\rightarrow {}^{\cdot}CCl_3 + ONOOH$.

In the process of pyrolysis, the biochar surface forms many free radicals [16] that are generated by oxygen atoms and inorganic impurities of the feedstock [36]. The free radical content varies with pyrolytic temperature. At pyrolytic temperatures between 300 °C and 600 °C, the free radical content gradually increases and tends to maximize at 500 °C–600 °C [16,37,38]. The CP degradation rate in soil amended with B700 was, therefore, the fastest.

## 3. Materials and Methods

### 3.1. Biochar Characteristics

Clean and dry michelia alba (*Magnolia denudata*) wood was collected as a feedstock for biochar production. As one of many potential feedstocks, the properties of michelia alba wood are more stable, which results in consistency under different pyrolytic temperatures. The wood was sawn into small pieces (<2 cm) and charred in a muffle furnace at different temperatures (300, 500, and 700 °C) under oxygen-limited conditions for 4 h. The charred residuals were marked as B300, B500, and B700 and ground into powder using a high-speed smashing machine. The biochar powder was then passed through a 2-mm sieve.

The surface oxygen-containing functional groups were titrated using Boehm titration [39]. The surface structure was surveyed using an environment scanning electron microscope (QUANTA 200 FEG, FEI Co., Hillsboro, OR, USA). The specific surface area (SSA) and pore volume were measured using a V-Sorb 2800P surface area analyzer (Gold APP Instrument Co., Beijing, China). The contents (%) of carbon, oxygen, and hydrogen of the biochar samples were measured by elemental analyzer (Vario EL III, Elementar Analysensysteme GmbH, Hanau, Germany). The samples were placed in a muffle furnace at 900 °C for 5 h to determine the ash content.

### 3.2. Adsorption Kinetics

Activated carbon and 0.15 g of B300, B500, and B700 were weighed into 20-mL headspace vials. Three replicates of each type of carbon, in other words, a total of twelve vials, were placed in a container that was constructed in two stainless steel cylindrical boxes. A clear vial containing 5 mL of pure CP liquid was placed in the cylindrical boxes to form a saturated gas concentration. The cylindrical boxes were then transferred to an incubator at 30 °C. After incubating the samples for 3, 6, 12, 24, or 36 h, the cylindrical boxes were removed and placed in an airing chamber for 30 min. The CP adsorbed by the biochar was extracted using the extraction method described in the degradation experiment section.

The Lagergren pseudo-first-order kinetics equation was used to fit the data as follows:

$$\ln (q_t - q_e) = \ln q_e - kt, \tag{1}$$

where $q_t$ is the amount of CP adsorbed at time t; $q_e$ is the calculated adsorption amount at equilibrium (mg/g), and k is the adsorption rate constant ($\text{min}^{-1}$).

### 3.3. Adsorption Isotherm Experiment

Adsorption isotherms were determined using the headspace-gas chromatography technique [40]. Activated carbon and 0.15 g of B300, B500, and B700 were weighted into 20-mL headspace vials. Various quantities of CP were injected into small 2-ml vials to generate different gas concentrations; these vials were placed inside the 20-mL headspace vials after the addition of the biochar. The 20-mL vials were capped with an aluminum cover and a Teflon-faced rubber septum. Each treatment was repeated three times. When the CP liquid was completely volatilized, the initial gas concentrations of CP in the headspace vials were 20, 30, 40, 50, 60, 70, and 80 mg/L. The preliminary experiment showed that adsorption equilibrium was reached after 48 h in a dark incubator at 30 °C. After the samples were incubated for 48 h, the gas concentration of the vials was analyzed using gas chromatography (GC) with an electron capture detector (ECD) ECD detector and automated headspace sampler. The GC procedure was the same as the degradation experiment described, and the auto-sampler headspace conditions were as follows: 100 °C oven temperature; 105 and 110 °C loop/valve and transfer line temperatures, respectively; 1.0 mL sample loop; 5 min vial equilibration; 0.5 min loop filling; 0.05 min loop equilibration; 0.1 min pressurization; 0.5 min injection; and low shake mode for 1 min. Then, the vials were withdrawn from the headspace sampler, the aluminum covers were removed, and the vials were placed in an airing chamber for 30 min to extract the CP adsorbed by the biochar. The extraction method was similar to the degradation experiment.

The Langmuir, Freundlich, and Dubinin-Radushkevitch (D-R) models were used to describe the adsorption isotherms. The Langmuir isotherm model is described using the following equation:

$$q_e = \frac{q_{max} K_L C_e}{1 + K_L C_e} \tag{2}$$

where $q_e$ is the amount adsorbed per unit mass; $q_{max}$ represents the maximum adsorbed concentration; the parameter $k_L$ represents the affinity constant or Langmuir constant, and $C_e$ is the concentration in the gas phase. The equation is a measure of how strongly an adsorbate molecule is attracted to a surface.

The equation for the Freundlich isotherm is:

$$q_e = k_f C_e^{(1/n)}, \tag{3}$$

This model assumes a heterogeneous adsorption surface with sites of different adsorption energies. $k_f$ represents the adsorption equilibrium constant, which indicates the adsorption capacity and represents the strength of the adsorptive bond. n is the heterogeneity factor, which represents the bond distribution. Generally, $n > 1$ represents favorable adsorption.

The D-R equation is expressed as:

$$q_e = q_{max} \exp(-k\varepsilon^2), \tag{4}$$

$$\varepsilon = RT\ln\left(1 + \frac{1}{C_e}\right) \tag{5}$$

where k is a constant related to the adsorption energy $(mol^2/kJ^2)$; $q_{max}$ is the maximum adsorption capacity (mg/g); and $\varepsilon$ is the Polanyi potential (J/mol). The mean free energy of adsorption (E) was calculated from the k values:

$$E = (2k)^{-(1/2)}, \tag{6}$$

*3.4. Soil and Incubation Experiments*

The sandy loam soil (2.7% clay, 30.5% silt, and 69.5% sand) was collected from the top 20 cm of soil in a cucumber greenhouse in Fangshan District, Beijing and passed through a 2-mm sieve, and the soil was stored at −80 °C until it was added to the vial. The prepared soil had a 12.5% moisture content (*w/w*) and the following physicochemical properties: 7.30 pH, 3.03% organic matter, and 0.20 ds/m electrical conductivity (EC).

Laboratory incubation experiments were conducted to determine the effects of different biochar types (B300, B500, and B700), rates, and soil sterilization on the degradation of CP. The preparation of soil or biochar prior to fumigant injection was as follows for each experiment:

(1) The sandy loam soil was amended with B300, B500, and B700 at rates of 1% and 5% (*w/w*, dry weight basis).

(2) The sandy loam soil with or without the 1% (*w/w*, dry weight basis) B300, B500, or B700 amendment was autoclaved at 121 °C for 1 h.

(3) Biochar (0.2 g of B300, B500, and B700) was weighed into a 20-mL glass vial. The vials with the biochar were disinfected under ultraviolet light for 30 min.

Soil (8 g) with or without biochar was placed into a 20-mL glass vial. The fumigant solution was prepared by dissolving 150 g/L of CP (99.5% purity, Dalian Dyestuffs & Chemicals Co., Ltd., Dalian, China) in ethyl acetate (AR, Sinopharm Chemical Reagent Co., Ltd., Beijing, China). Fumigant solution (5 mL) was injected into each of the above vials containing 8 g of soil or vials containing 0.2 g biochar and immediately capped with an aluminum cover and a Teflon-faced rubber septum; this led to limited oxygen conditions. The fumigant concentration in the soil was 93.75 μg/g. After treatment, the vials were inverted and placed in 30 °C incubators. Each treatment was repeated three times.

After incubation for 1, 3, 5, 7, 10, 14, 21, and 28 d, the vials were withdrawn and then immediately stored at −80 °C until the extraction of the CP residue. The residue extraction method was as follows: Anhydrous sodium sulfate (8 g; AR, Sinopharm Chemical Reagent Co., Ltd.) and 8 mL of ethyl acetate were added to each frozen vial, after which the vials were recapped immediately. The vials were vortexed for 30 s and then placed on a table concentrator for 30 min of shaking and subsequent settling for 2 h. After settling, 1–1.5 mL of supernatant was transferred into a 2-mL amber glass vial. Preliminary experiments showed that this extraction method has an extraction efficiency of 86%–95%. A gas chromatograph with a micro-ECD (GC-μECD) and HP-5 capillary column (30 m length × 320 μm × 0.25 μm film thickness) (Agilent Technologies, Santa Clara, CA, USA) was used to analyze CP. The detector and inlet temperatures were higher than 250 °C. The oven temperature was held at 60 °C for 6 min [41].

Typically, a first-order kinetics model is used in studies of pesticide degradation [42]. Therefore, we selected a first-order kinetics model to fit the degradation of CP in biochar-amended soil:

$$C = C_0 \, e^{-kt}, \tag{7}$$

where C and $C_0$ are the concentrations in the soil at times t (d) and time $t_0$, respectively; k is the first-order rate constant (d$^{-1}$), and t is the incubation time (d).

When using first-order kinetics models, the half-life of degradation is calculated using the equation:

$$t_{1/2} = \frac{0.693}{k} \tag{8}$$

The first-order kinetics model was fit using Origin Pro 8.0 software.

*3.5. The Degradation Products of CP in Biochar and Soil*

The biochars and soil (1 g of each) were placed in 20-mL headspace vials. Fumigant solution (5 μL, i.e., 150 g/L of CP in ethyl acetate) was injected into each vial, which was immediately capped with an aluminum cover and a Teflon-faced rubber septum. Each treatment was repeated three times. The vials were then transferred to 30 °C incubators. After incubation for 3, 6, 12, 24, and 36 h, the samples were withdrawn from the incubators and immediately stored at −80 °C until the extraction of the CP residue and its degradation products. The extraction method was similar to that used in the degradation experiment except that 5 mL of ethyl acetate was used instead of 8 mL. A GC-MS (Agilent 7890A, Agilent Technologies, Santa Clara, CA, USA) equipped with a split/splitless injection port and a 5975C mass selective detector was used to determine CP and its degradation products. A DB-5MS (30 m × 0.25 mm ID × 0.25 μm film thickness) column (Agilent Technologies, Santa Clara, CA, USA) was used to separate the chemicals, and the flow rate of the carrier gas with helium (99.999%) was 0.8 mL/min. The temperature of the injection port was held constant at 250 °C, and the splitless mode started 1 min after injection. The primary oven was held constant at 40 °C for 4 min and then increased to 150 °C by 15 °C/min, followed by a rate of 30 °C/min until a temperature of 250 °C was reached. The quadrupole, ion source, and transfer line temperatures were 150 °C, 230 °C, and 280 °C, respectively. Electron impact ionization (EI, Agilent Technologies, Santa Clara, CA, USA) with an energy of 70 eV and electron multiplier voltage of 1494 V was used. The full scan mode, with a scan range of 50–250 $m/z$ and a solvent delay of 3.6 min, was used to obtain the MS spectrum. The MSD

*Energies* **2016**, *9*, 869

Productivity ChemStation software (Version E.02.00.493, Agilent Technologies, Santa Clara, CA, USA) was used to control the instrument and collect data.

## 4. Conclusions

In summary, biochar produced under different pyrolytic temperatures exhibited variable characteristics. Under high pyrolytic temperatures, biochar possesses larger SSA and pore volume and more micropores. These characteristics resulted in different adsorption rates of CP to biochar. The adsorption isotherm fitted the Freundlich model very well, and the $k_f$ value indicated the following adsorption capacity trend: B700 > B500 > B300. However, the actual measurements indicated that there was no significant difference between the three biochars because biochar produced under higher temperature has a high adsorption capacity and possesses a strong CP degradation ability. When biochar was added to soil, it greatly shortened the degradation half-life of chloropicrin. Degradation of CP further accelerated as the rate of addition and pyrolytic temperature of the biochar increased. The ability of biochar to adsorb CP could, therefore, not reduce the degradation rate of CP in biochar-amended soil. The mechanisms by which biochar accelerated the degradation of CP were shown through the detection of degradation products and could likely be attributed to free radical chemical reactions and microbial processes. The findings in the present study show that biochar amendments to soil are beneficial but can decrease the effect of CP on soil-born pests as a fumigant by accelerating degradation. The atmospheric emission of CP is an air pollutant that can cause severe irritation of the eyes, mucous membranes, skin, and lungs, and plays a role in forming tropospheric ozone [43]. Therefore, reducing CP emissions is crucial for protection against the harmful nature of CP. The present study showed that soil amendment with biochar is a promising method to reduce CP emissions. Further testing on the CP efficacy and emission reduction by biochar needs to be conducted in field experiments.

**Author Contributions:** Pengfei Liu and Qiuxia Wang designed the study and wrote the protocol. Wensheng Fang and Liangang Mao carried out to determinate the biochar characteristics. Pengfei Liu, Wensheng Fang, Dongdong Yan, and Liangang Mao performed most experiments. Pengfei Liu, Canbin Ouyang, Yuan Li, and Meixia Guo managed the literature searches and analyses. Pengfei Liu, Qiuxia Wang, and Dong Wang analyzed the data. Pengfei Liu, Qiuxia Wang, and Aocheng Cao were responsible for the overall design and wrote the article.

**Conflicts of Interest:** The authors declare no competing financial interests.

## References

1. Downie, A.; Crosky, A.; Munroe, P. Physical properties of biochar. In *Biochar for Environmental Management: Science & Technology*; Lehmann, J., Joseph, S., Eds.; Earthscan: London, UK, 2009; Chapter 2; pp. 13–32.
2. Novak, J.M.; Busscher, W.J.; Laird, D.L.; Ahmedna, M.; Watts, D.W.; Niandou, M.A.S. Impact of biochar amendment on fertility of a southeastern Coastal Plain soil. *Soil Sci.* **2009**, *174*, 105–112. [CrossRef]
3. Uchimiya, M.; Lima, I.; Klasson, K.; Wartelle, L. Contaminant immobilization and nutrient release by biochar soil amendment: Roles of natural organic matter. *Chemosphere* **2010**, *80*, 935–940. [CrossRef] [PubMed]
4. Karami, N.; Clemente, R.; Moreno-Jiménez, E.; Lepp, N.W.; Beesley, L. Efficiency of green waste compost and biochar soil amendments for reducing lead and copper mobility and uptake to ryegrass. *J. Hazard. Mater.* **2011**, *191*, 41–48. [CrossRef] [PubMed]
5. Chan, K.Y.; Van Zwieten, L.; Meszaros, I.; Downie, A.; Joseph, S. Using poultry litter biochars as soil amendments. *Soil Res.* **2008**, *46*, 437–444. [CrossRef]
6. Liang, B.; Lehmann, J.; Solomon, D.; Kinyangi, J.; Grossman, J.; O'Neill, B.; Skjemstad, J.O.; Thies, J.; Luizão, F.J.; Petersen, J.; et al. Black carbon increases cation exchange capacity in soils. *Soil Sci. Soc. Am. J.* **2006**, *70*, 1719–1730. [CrossRef]
7. Laird, D.A. The charcoal vision: A win–win–win scenario for simultaneously producing bioenergy, permanently sequestering carbon, while improving soil and water quality. *Agron. J.* **2008**, *100*, 178–181. [CrossRef]
8. Case, S.D.; McNamara, N.P.; Reay, D.S.; Whitaker, J. The effect of biochar addition on $N_2O$ and $CO_2$ emissions from a sandy loam soil—The role of soil aeration. *Soil Biol. Biochem.* **2012**, *51*, 125–134. [CrossRef]

9. Yang, Y.; Sheng, G.; Huang, M. Bioavailability of diuron in soil containing wheat-straw-derived char. *Sci. Total Environ.* **2006**, *354*, 170–178. [CrossRef] [PubMed]

10. Zhang, P.; Sheng, G.; Wolf, D.C.; Feng, Y. Reduced biodegradation of benzonitrile in soil containing wheat-residue-derived ash. *J. Environ. Qual.* **2004**, *33*, 868–872. [CrossRef] [PubMed]

11. Jones, D.L.; Edwards-Jones, G.; Murphy, D.V. Biochar mediated alterations in herbicide breakdown and leaching in soil. *Soil Biol. Biochem.* **2011**, *43*, 804–813. [CrossRef]

12. Jablonowski, N.D.; Krutz, J.L.; Martinazzo, R.; Zajkoska, P.; Hamacher, G.; Borchard, N.; Burauel, P. Transfer of atrazine degradation capability to mineralize aged [14]C-labeled atrazine residues in soils. *J. Agric. Food Chem.* **2013**, *61*, 6161–6166. [CrossRef] [PubMed]

13. Harris, D.C. Control of Verticillium wilt and other soil-borne diseases of strawberry in Britain by chemical soil disinfestation. *J. Hortic. Sci. Biotechnol.* **1990**, *65*, 401–408. [CrossRef]

14. Johnson, M.O.; Godfrey, G.H. Chloropicrin for nematode control. *Ind. Eng. Chem.* **1932**, *24*, 311–313. [CrossRef]

15. Cole, S.K.; Cooper, W.J.; Fox, R.V.; Gardinali, P.R.; Mezyk, S.P.; Mincher, B.J.; O'Shea, K.E. Free radical chemistry of disinfection byproducts. 2. Rate constants and degradation mechanisms of trichloronitromethane (chloropicrin). *Environ. Sci. Technol.* **2007**, *41*, 863–869. [CrossRef] [PubMed]

16. Bourke, J.; Manley-Harris, M.; Fushim, C.; Dowaki, K.; Nunoura, T.; Antal, M.J. Do all carbonized charcoals have the same chemical structure? 2. A model of the chemical structure of carbonized charcoal. *Ind. Eng. Chem. Res.* **2007**, *46*, 5954–5967. [CrossRef]

17. Gaskin, J.W.; Steiner, C.; Harris, K.; Das, K.C.; Bibens, B. Effect of low-temperature pyrolysis conditions on biochar for agricultural use. *Trans. ASABE* **2008**, *51*, 2061–2069. [CrossRef]

18. Mohan, D.; Rajput, S.; Singh, V.K.; Steele, P.H.; Pittman, C.U., Jr. Modeling and evaluation of chromium remediation from water using low cost bio-char, a green adsorbent. *J. Hazard. Mater.* **2011**, *188*, 319–333. [CrossRef] [PubMed]

19. Keiluweit, M.; Nico, P.S.; Johnson, M.G.; Kleber, M. Dynamic molecular structure of plant biomass-derived black carbon (biochar). *Environ. Sci. Technol.* **2010**, *44*, 1247–1253. [CrossRef] [PubMed]

20. Raveendran, K.; Ganesh, A.; Khilar, K.C. Pyrolytic characteristics of biomass and biomass components. *Fuel* **1996**, *75*, 987–998. [CrossRef]

21. Yang, H.; Yan, R.; Chen, H.; Lee, D.H.; Zheng, C. Characteristics of hemicellulose, cellulose and lignin pyrolytic. *Fuel* **2007**, *86*, 1781–1788. [CrossRef]

22. Chun, Y.; Sheng, G.; Chiou, C.; Xing, B. Compositions and sorptive properties of crop residue-derived chars. *Environ. Sci. Technol.* **2004**, *38*, 4649–4655. [CrossRef] [PubMed]

23. Samsuri, A.W.; Sadegh-Zadeh, F.; Seh-Bardan, B.J. Characterization of biochars produced from oil palm and rice husks and their adsorption capacities for heavy metals. *Int. J. Environ. Sci. Technol.* **2014**, *11*, 967–976. [CrossRef]

24. Chen, B.; Zhou, D.; Zhu, L. Transitional adsorption and partition of nonpolar and polar aromatic contaminants by biochars of pine needles with different pyrolytic temperatures. *Environ. Sci. Technol.* **2008**, *42*, 5137–5143. [CrossRef] [PubMed]

25. Lehmann, J. Bio-energy in the black. *Front. Ecol. Environ.* **2007**, *5*, 381–387. [CrossRef]

26. Gan, J.; Yates, S.R.; Ernst, F.F.; Jury, W.A. Degradation and volatilization of the fumigant chloropicrin after soil treatment. *J. Environ. Qual.* **2000**, *29*, 1391–1397. [CrossRef]

27. Castro, C.E.; Wade, R.S.; Belser, N.O. Biodehalogenation. The metabolism of chloropicrin by Pseudomonas sp. *J. Agric. Food Chem.* **1983**, *31*, 1184–1187. [CrossRef]

28. Steinbeiss, S.; Gleixner, G.; Antonietti, M. Effect of biochar amendment on soil carbon balance and soil microbial activity. *Soil Biol. Biochem.* **2009**, *41*, 1301–1310. [CrossRef]

29. Lehmann, J.; Rillig, M.C.; Thies, J.; Masiello, C.A.; Hockaday, W.C.; Crowley, D. Biochar effects on soil biota—A review. *Soil Biol. Biochem.* **2011**, *43*, 1812–1836. [CrossRef]

30. Joseph, S.D.; Camps-Arbestain, M.; Lin, Y.; Munroe, P.; Chia, C.H.; Hook, J.; van Zwieten, L.; Kimber, S.; Cowie, A.; Singh, B.P.; et al. An investigation into the reactions of biochar in soil. *Soil Res.* **2010**, *48*, 501–515. [CrossRef]

31. Joseph, S.; Husson, O.; Graber, E.R.; Donne, S.W. The electrochemical properties of biochars and how they affect soil redox properties and processes. *Agronomy* **2015**, *5*, 322–340. [CrossRef]

32. Graber, E.R.; Tsechansky, L.; Lew, B.; Cohen, E. Reducing capacity of water extracts of biochars and their solubilization of soil Mn and Fe. *Eur. J. Soil Sci.* **2014**, *65*, 162–172. [CrossRef]

33. Mukome, F.N.D.; Kilcoyne, A.L.D.; Parikh, S.J. Alteration of biochar carbon chemistry during soil incubations: SR-FTIR and NEXAFS investigation. *Soil Sci. Soc. Am. J.* **2014**, *78*, 1632–1640. [CrossRef]

34. Swancutt, K.L.; Dail, M.K.; Mezyk, S.P.; Ishida, K.P. Absolute kinetics and reaction efficiencies of hydroxyl-radical-induced degradation of methyl isothiocyanate (MITC) in different quality waters. *Chemosphere* **2010**, *81*, 339–344. [CrossRef] [PubMed]

35. Mezyk, S.P.; Helgeson, T.; Cole, S.K.; Cooper, W.J.; Fox, R.V.; Gardinali, P.R.; Mincher, B.J. Free radical chemistry of disinfection-byproducts. 1. Kinetics of hydrated electron and hydroxyl radical reactions with halonitromethanes in water. *J. Phys. Chem. A* **2006**, *110*, 2176–2180. [CrossRef] [PubMed]

36. Brennan, J.K.; Bandosz, T.J.; Thomson, K.T.; Gubbins, K.E. Water in porous carbons. *Colloid Surf. A* **2001**, *187*, 539–568. [CrossRef]

37. Feng, J.W.; Zheng, S.; Maciel, G.E. EPR investigations of the effects of inorganic additives on the charring and char/air interactions of cellulose. *Energy Fuels* **2004**, *18*, 1049–1065. [CrossRef]

38. Emmerich, F.G.; Rettori, C.; Luengo, C.A. ESR in heat treated carbons from the endocarp of babassu coconut. *Carbon* **1991**, *29*, 305–311. [CrossRef]

39. Boehm, H.P. Some aspects of the surface chemistry of carbon blacks and other carbons. *Carbon* **1994**, *32*, 759–769. [CrossRef]

40. Wu, J.; Strömqvist, M.E.; Claesson, O.; Fängmark, I.E.; Hammarström, L.G. A systematic approach for modelling the affinity coefficient in the Dubinin–Radushkevich equation. *Carbon* **2002**, *40*, 2587–2596. [CrossRef]

41. Wang, Q.X.; Wang, D.; Tang, J.T.; Yan, D.D.; Zhang, H.J.; Wang, F.Y.; Guo, M.X.; Cao, A.C. Gas-phase distribution and emission of chloropicrin applied in gelatin capsules to soil columns. *J. Environ. Qual.* **2010**, *39*, 917–922. [CrossRef] [PubMed]

42. Ma, Q.L.; Gan, J.; Papiernik, S.K.; Becker, J.O.; Yates, S.R. Degradation of soil fumigants as affected by initial concentration and temperature. *J. Environ. Qual.* **2001**, *30*, 1278–1286. [CrossRef] [PubMed]

43. Goldman, L.R.; Mengle, D.; Epstein, D.M.; Fredson, D.; Kelly, K.; Jackson, R.J. Acute symptoms in persons residing near a field treated with the soil fumigants methyl bromide and chloropicrin. *West. J. Med.* **1987**, *147*, 95–98. [PubMed]

*energies*

MDPI

*Article*

# Reduction of Furfural to Furfuryl Alcohol in Liquid Phase over a Biochar-Supported Platinum Catalyst

Ariadna Fuente-Hernández, Roland Lee, Nicolas Béland, Ingrid Zamboni
and Jean-Michel Lavoie *

Industrial Research Chair on Cellulosic Ethanol and Biocommodities (CRIEC-B),
Department of Chemical & Biotechnological Engineering, Université de Sherbrooke,
Sherbrooke, QC J1K2R1, Canada; ariad22p@yahoo.com (A.F.-H.); roland.lee2@hotmail.com (R.L.);
nicolas.beland@gmail.com (N.B.); ingrid.rocio.zamboni.corredor@USherbrooke.ca (I.Z.)
*  Correspondence: jean-michel.lavoie2@usherbrooke.ca; Tel.: +1-819-821-8000 (ext. 65505)

Academic Editor: Mejdi Jeguirim
Received: 7 December 2016; Accepted: 23 February 2017; Published: 28 February 2017

**Abstract:** In this work, the liquid phase hydrogenation of furfural has been studied using a biochar-supported platinum catalyst in a batch reactor. Reactions were performed between 170 °C and 320 °C, using 3 wt % and 5 wt % of Pt supported on a maple-based biochar under hydrogen pressure varying from 500 psi to 1500 psi for reaction times between 1 h and 6 h in various solvents. Under all reactive conditions, furfural conversion was significant, whilst under specific conditions furfuryl alcohol (FA) was obtained in most cases as the main product showing a selectivity around 80%. Other products as methylfuran (MF), furan, and trace of tetrahydrofuran (THF) were detected. Results showed that the most efficient reaction conditions involved a 3% Pt load on biochar and operations for 2 h at 210 °C and 1500 psi using toluene as solvent. When used repetitively, the catalyst showed deactivation although only a slight variation in selectivity toward FA at the optimal experimental conditions was observed.

**Keywords:** biochar; furfural; furfuryl alcohol (FA); hydrogenation; maple; platinum catalyst

## 1. Introduction

Production of second-generation biofuels as cellulosic ethanol should involve the valorization of every macromolecular fraction of the biomass in order to be economical [1]. Contrarily to the C6 carbohydrates, which are ideal candidates for classical fermentation, valorization of C5 sugars remains a challenge, either through biological or chemical pathways. Many biological alternatives have been investigated including fermentation to ethanol using, as an example, *E. coli* [2], *Z. mobilis* [3], and *P. stipitis* [4]. However, most of these approaches are limited by the kinetics of the fermentation, which usually involves longer fermentation periods as compared to their classical C6 counterparts.

As an alternative to the biological conversion of C5 sugars, another approach leading to their conversion could imply chemical processes. In a recent review, Fuente-Hernández et al. investigated the possible option of the conversion of xylose including reduction, oxidation, acid and base treatments [5]. Amongst the latter, acid treatments, leading to furfural have been thoroughly investigated and reported in literature [6,7]. Furfural with an annual global demand ranging between 20 kton/year and 30 kton/year can be used as chemical but could as well be used as a platform chemical for other compounds including but not limited to levulinic acid [8]. However, conversion of furfural to levulinic acid is not possible and must go through a reduction of the aldehyde function of furfural into an alcohol, thus producing furfuryl alcohol (FA).

Previous work from open literature reported that catalytic hydrogenation of furfural could be carried out either in the liquid [9] or vapor phase [10,11]. As for the industrial processes, they are generally conducted at high temperatures and pressures at which operations both in liquid and gas phase were reported. In terms of catalyst, the most popular for the reduction of furfural at industrial levels are Ni and Cu/Cr-based catalysts, although these catalysts exhibit a moderate activity towards FA [12,13]. At bench level, numerous reports have been made on the hydrogenation of furfural in liquid-phase using noble metal-based catalysts, such as palladium [14], platinum [9,13,15,16], iridium and ruthenium [17], rhodium [18], and zirconium oxide [19]. Reactions were performed either with or without a solvent (using different solvent types) sometimes even using a second metal (or a promoter) to improve the activity and/or the selectivity [13,20,21]. Amongst the non-noble metals that have been reported to selectively hydrogenate $\alpha,\beta$-unsaturated aldehydes, iron [22] and nickel [13] were the most commonly cited. The best example of the duality of two metals was reported when a combination of nickel and copper showed interesting characteristics of chemo-regio-, and, stereoselectivity for hydrogenation reactions [13,20,21].

The choice of the support is also a key aspect that may lead to significant changes in catalytic activity. The most conventional supports are either acidic or basic oxides such as silica [22], alumina [14,23], and porous metal (Raney type) [24]. However, other types of supports (such as carbon) were also used for the production of reducing catalyst for chemoselective hydrogenations. Although the most common carbon support would certainly be activated carbon [25], biochar could as well be a cheap, stable, carbon-rich compound that could be considered to this purpose. The latter can be produced by "thermo" processes for biomass conversion such as torrefaction, pyrolysis, or gasification. It is made from renewable material and the already available inorganics that could be found in trace amounts in the biochar could influence the output of a catalytic reaction when used as support. Although activated carbon has been used as a catalytic support for hydrogenation reactions most probably as a mean to standardize catalyst synthesis, there is, to the best of our knowledge no report on the utilization of biochar as a support for hydrogenation reactions.

In this work, we report on the liquid phase hydrogenation of furfural to FA, which was used as a model compound of biomass-derived feedstock, by using a biochar-supported platinum catalyst. In addition, various reaction parameters such as metal loading in the catalyst, operating pressure of hydrogen, reaction time, solvent choice, and reaction temperature were studied to optimize furfural conversion, FA selectivity, and to determine kinetic parameters for catalysis reaction. Specific attention was given to the regeneration of the catalyst in order to link with further downstream industrial applications.

## 2. Results and Discussion

### 2.1. Support Synthesis and Functionalization

Biochar (BC) produced from torrefaction (slow pyrolysis) may contain bio-oils and traces of metal both from the pyrolysis process as well as from the original biomass. Bio-oils were removed using an Acid treatment (Section 3.1) prior to catalytic metal impregnation [26].

### 2.2. Catalyst Characterization

The scanning electron microscopy (SEM) morphology of both biochar support and platinum catalyst Pt/BC revealed that the metal particles were not homogeneously distributed in the support, as shown in Figure 1. This lack of homogeneity in the distribution of the Pt can be seen from its tendency to agglomerate (Figure 1b).

Energy-dispersive X-ray (EDX) spectroscopy microanalysis was performed in order to determine the elemental analysis of the surface samples. The biochar (support without impregnation) spectrum revealed the presence of C, O, Al, K and Ca elements in trace amounts (Figure 1a). As for comparison, the metal loading deposited during the impregnation is shown in Figure 1b. The EDX results indicate the presence of platinum as well as sulfur coming from the biochar treatment with the sulfuric acid solution. Copper most likely coming from the original biomass is not present in Figure 1a, which can be explained by the low amounts contained in the support, the heterogeneity of the sample, and by the size of the scan frame. The chlorine present in the impregnated catalyst probably comes from the platinum precursor.

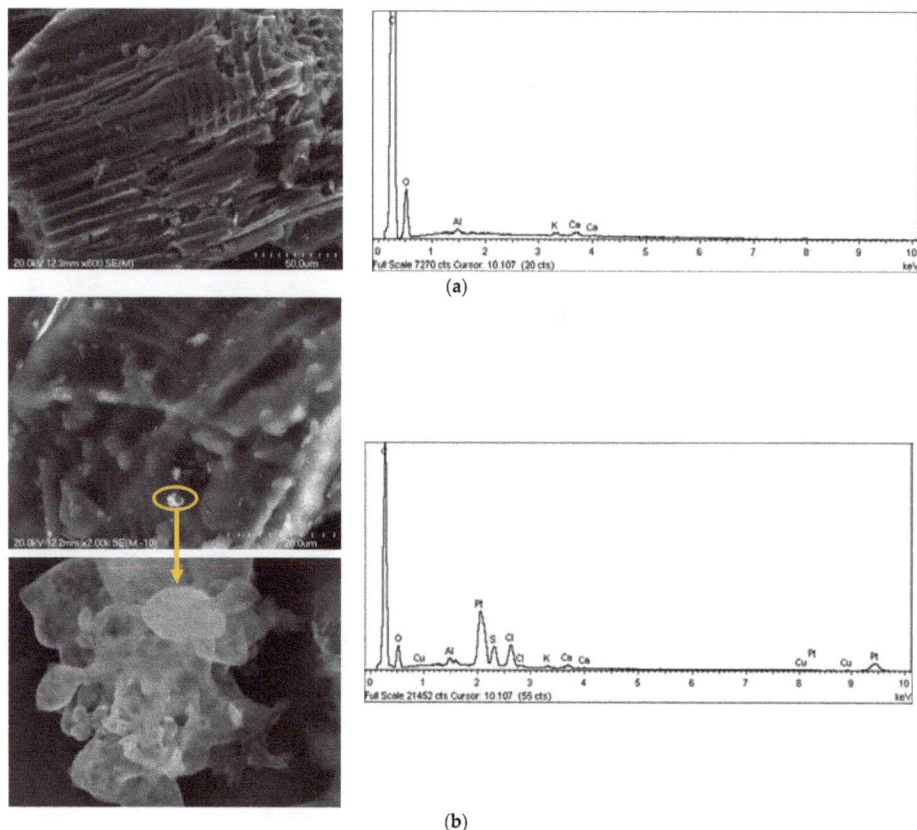

(a)

(b)

**Figure 1.** Scanning electron microscopy (SEM) images and energy-dispersive X-ray (EDX) spectra of biochar for the: (**a**) unimpregnated support; and (**b**) impregnated with Pt 3 wt % prior to the reaction.

### 2.3. Catalyst Test with Pt/BC

The pathway leading to the products observed during furfural hydrogenation can be simplified as suggested in Figure 2. The main products were FA, tetrahydrofurfuryl alcohol (THFA), furan (F), methyltetrahydrofuran (MTHF), and methylfuran (MF). Table 1 summarizes the obtained results from the reaction of furfural over Pt/BC 3 wt % catalyst in correlation to the operation parameters.

Hydrogenation rates have been previously reported to increase proportionally with temperatures [24] however, no clear tendency of this nature was observed for the actual experiments. For the experiments involving the catalyst with a 3 wt % platinum loading, the optimal temperature was reached at 210 °C, showing a maximum conversion of 60.8% with 79.2% selectivity to FA after 2 h (Entry 9, Table 1). The yield of FA drops from 48.2% to 16.1% (Entry 8) when temperature decreases from 210 °C to 170 °C. Even if a higher conversion value is observed at 320 °C (Entry 13), the selectivity toward FA is 16.8% and the then FA yield decreases from 48.2% to 11.8%. Moreover, results from Entries 10 and 11 also showed that increasing the reaction time only leaded to a slight modification of the conversion rate and an increased portion of furan, resulting in a decrease in selectivity to FA. This behavior was observed as well after addition of a higher amount of catalyst (Entry 5, Table 1).

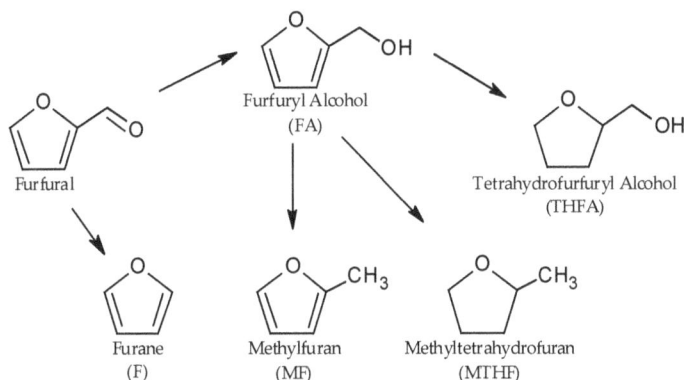

**Figure 2.** Possible reaction pathways for furfural hydrogenation.

**Table 1.** Experimental conditions, conversion, and selectivity for liquid phase furfural hydrogenation in toluene using the Pt/BC 3 wt % catalyst. FA: furfuryl alcohol; TOF: turn-over frequency; THFA: tetrahydrofurfuryl alcohol; MF: methylfuran; and F: furan.

| Entry | Catalyst (mmol) | P (psi) | T (°C) | t (h) | Conversion (%) | FA Yield (%) | TOF (s⁻¹) | Selectivity (%) | | | |
|---|---|---|---|---|---|---|---|---|---|---|---|
| | | | | | | | | FA | THFA | MF | F |
| 1 | 0.025 | 500 | 210 | 2 | 20.1 | 6.9 | $4.0 \times 10^{-2}$ | 34.2 | - | 6.4 | 40.1 |
| 2 | 0.025 | 500 | 250 | 2 | 18.4 | 3.9 | $3.7 \times 10^{-2}$ | 21.2 | - | 12.6 | 63.4 |
| 3 | 0.025 | 1000 | 210 | 2 | 35.6 | 25.7 | $7.1 \times 10^{-2}$ | 72.2 | - | 6.6 | 20.1 |
| 4 | 0.025 | 1000 | 250 | 2 | 36.4 | 17.0 | $7.3 \times 10^{-2}$ | 46.7 | 0.4 | 14.7 | 33.8 |
| 5 | 0.058 | 1000 | 250 | 2 | 52.0 | 13.8 | $4.5 \times 10^{-2}$ | 26.6 | 0.1 | 13.0 | 58.8 |
| 6 | 0.058 | 1000 | 250 | 1 | 30.1 | 11.4 | $5.2 \times 10^{-2}$ | 38.0 | - | 14.5 | 46.2 |
| 7 | 0.025 | 1000 | 250 | 3 | 40.8 | 20.2 | $5.4 \times 10^{-2}$ | 49.6 | 0.2 | 19.1 | 29.8 |
| 8 | 0.025 | 1500 | 170 | 2 | 19.4 | 16.1 | $3.9 \times 10^{-2}$ | 83.2 | 0.5 | 5.1 | 9.0 |
| 9 | 0.025 | 1500 | 210 | 2 | 60.8 | 48.2 | $1.2 \times 10^{-2}$ | 79.2 | 0.4 | 8.5 | 11.0 |
| 10 | 0.025 | 1500 | 210 | 4 | 66.9 | 55.4 | $6.7 \times 10^{-2}$ | 82.8 | 1.1 | 0.9 | 13.4 |
| 11 | 0.025 | 1500 | 210 | 6 | 69.4 | 49.3 | $4.6 \times 10^{-2}$ | 71.1 | 0.7 | 12.8 | 14.7 |
| 12 | 0.025 | 1500 | 250 | 2 | 33.4 | 27.1 | $6.7 \times 10^{-2}$ | 81.1 | 1.0 | 8.8 | 7.7 |
| 13 | 0.025 | 1500 | 320 | 2 | 70.0 | 11.8 | $1.4 \times 10^{-2}$ | 16.8 | 0.6 | 35.2 | 43.2 |
| 14 | 0.025 | 1500 | 300 | 3 | 59.9 | 10.1 | $8.0 \times 10^{-2}$ | 16.9 | 1.2 | 41.6 | 37.3 |

Variation of the temperature in either direction from 210 °C impacts conversion and selectivity to FA (Entries 8, 9 and 12–14, Table 1) thus increasing the formation of byproducts as MF, furan, as well as other unknown compounds. One hypothesis explaining this phenomenon is probably related to catalyst selectivity that can be affected by adsorption of furfural and/or FA and by products on the surface of the catalyst's active sites.

Increasing $H_2$ pressure from 500 psi to 1500 psi was shown to increase the conversion of furfural from 20.1% to 60.8% (Entries 1, 3 and 9, Table 1). The formation of byproducts at higher pressures of hydrogen does not appear to significantly affect the selectivity to FA, the latter increasing to 79.2% at 1500 psi conditions that were considered optimal for this catalyst and support. Results also showed that almost no THFA was formed for most of the tests performed with the Pt/BC catalyst, indicating that the latter was not intrinsically selective to C–C double bonds thus prioritizing reduction of C=O or C–OH bonds. The tests using different amounts of catalyst (Pt/BC 1 and 5 wt %) were performed in the conditions that were shown optimal for a 3 wt % loading of catalyst (toluene as solvent, 210 °C, 1500 psi, for 2 h and 4 h), as presented in Table 2.

**Table 2.** Furfural hydrogenation results using Pt/BC 3 and 5 wt % as catalyst.

| Entry | Catalyst (mmol) | t (h) | Conversion (%) | FA Yield (%) | TOF (s$^{-1}$) | Selectivity (%) | | | |
|---|---|---|---|---|---|---|---|---|---|
| | | | | | | FA | THFA | MF | F |
| 9 | 0.025 (3%) | 2 | 60.8 | 48.2 | $1.2 \times 10^{-2}$ | 79.2 | 0.4 | 8.5 | 11.0 |
| 10 | | 4 | 66.9 | 55.4 | $6.7 \times 10^{-2}$ | 82.8 | 1.1 | 0.9 | 13.4 |
| 15 | 0.044 (5%) | 2 | 40.5 | 31.3 | $4.6 \times 10^{-2}$ | 77.3 | 0.5 | 4.3 | 12.0 |
| 16 | | 4 | 45.3 | 32.4 | $2.5 \times 10^{-2}$ | 71.6 | 0.4 | 6.2 | 16.6 |

Using a higher percentage of Pt in the catalyst but overall using the same amount of platinum based catalyst did not favor an increased conversion of furfural to FA compared with 3 wt % catalyst (Entries 9 and 10, Table 2). Furthermore, the use of higher quantity of impregnated metal (5 wt %) was shown to reduce conversion of approximately 20% in both cases (Entries 15 and 16, Table 2). The reduced homogeneity of Pt distribution, being a consequence of the greater impregnation concentration, results in non-favorable conditions for furfural hydrogenation. Selectivity to FA was also shown to be significantly reduced and chromatograms showed the presence of new unknown byproducts, which did not occur at lower (3 wt %) Pt content.

It should be noted that biochar per se was not inert in furfural hydrogenation, which was elucidated by testing only support at the same experimental conditions as above. Hydrogenation of furfural with the support alone results in conversion of 3.3% and selectivity to FA of 21.3%, using the optimal conditions (identified by the tests on a 3 wt % loading of platinum). The biochar carbon surfaces and pores possess complex structures containing metals and oxygen groups, as well, the biochar has a tendency for fixation of metallic ion thus leading to a concentration of the plant inorganic content. As well, this functionality also allows the biochar to be a very efficient support for other types of metals, either for catalytic purposes or for soil remediation [27]. Overall, both the textural properties and the trace metal content in the biochar could be generating some activity. Furthermore, it is important to note that the support is stable under the reaction conditions and it does not interact with the solvent. Duplicate experiments were performed showing the rates and selectivity to the various products to be within an error margin of around 5%, thus confirming reproducibility of the results.

The most important solvent effects in the hydrogenation of $\alpha,\beta$-unsaturated aldehydes are usually related to solvent polarity, solubility of hydrogen, and interactions between the catalyst and the solvent as well as solvation of reactants in the bulk liquid phase [28]. Further and as noted previously, both the support and the type of solvent can affect selectivity. As for this work, the solvent effect on the furfural hydrogenation was investigated with the Pt/BC 3 wt % at 210 °C and 250 °C with a H$_2$ pressure of 1500 psi for 2 h using toluene, isopropanol, isobutanol, and hexane. Results obtained for the hydrogenation of furfural using other solvents are shown in Table 3.

**Table 3.** Selectivity and conversion for furfural hydrogenation with Pt/B 3 wt % in toluene, isopropanol, isobutanol, and hexane at 1500 psi H$_2$ pressure for 2 h.

| Entry | Solvent | T (°C) | Conversion (%) | FA Yield (%) | TOF (s$^{-1}$) | Selectivity (%) | |
|---|---|---|---|---|---|---|---|
| | | | | | | FA | MF |
| 12 | Toluene | 250 | 33.4 | 27.1 | $6.7 \times 10^{-2}$ | 81.1 | 8.8 |
| 9 | | 210 | 60.8 | 48.1 | $1.2 \times 10^{-1}$ | 79.2 | 8.5 |
| 17 | Isopropanol | 250 | 37.5 | 21.1 | $7.5 \times 10^{-2}$ | 56.4 | 23.4 |
| 18 | | 210 | 42.9 | 34.9 | $8.6 \times 10^{-2}$ | 81.3 | 9.0 |
| 19 | Isobutanol | 250 | 57.1 | 50.6 | $1.1 \times 10^{-1}$ | 88.6 | - |
| 20 | | 210 | 82.8 | 49.6 | $1.7 \times 10^{-1}$ | 59.9 | - |
| 21 | Hexane | 250 | 66.4 | 16.1 | $1.3 \times 10^{-1}$ | 24.2 | - |
| 22 | | 210 | 52.3 | 41.9 | $1.0 \times 10^{-1}$ | 80.2 | - |

Utilization of isopropanol has led to the formation of dimerization products (Figure 3) that did not occur in toluene. The selectivity toward these products was close to 18% and 10% at 250 °C and 210 °C, respectively. However, in both cases, the main product remained FA. Although conversion did increase, the difference was not significantly higher than what was noted for toluene at 250 °C (Entry 17 Table 3, and Entry 12 Table 3), and the selectivity to FA was shown to decrease, as a result of significant quantities of MF as well as dimerization products formed. At 210 °C (Entries 9 and 18 Table 3), conversion decreased to 23% in isopropanol although selectivity remained close to 80% in both cases. Isopropanol has a significant effect on the product's selectivity and formation of by-products is more pronounced. A hypothesis that could explain such observation would be that the greater availability of hydrogen in the media could lead to furfural and/or FA dimerization via a higher availability of radicals that could generate the end products depicted in Figure 3 below (identified by gas chromatograph coupled to mass spectrometer (GC-MS)).

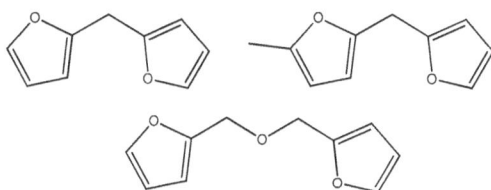

**Figure 3.** Observed dimers structures.

Utilization of bulky alcohols (as isobutanol) are supposed to lead to greater hydrogen solubility although it was not reported to correlate with an increased hydrogenation rate. The % conversion in isobutanol seems to corroborate such assumptions. At 250 °C, an increase in both conversion and selectivity toward alcohol was observed (Entry 19 Table 3). However, at 210 °C, the selectivity was reduced by about 20%.

Hexane (Entry 22, Table 3) gave results that were comparable to toluene at 210 °C (Entry 9, Table 3) however, two liquid phases were obtained following the reaction and the possibility of finding products in both phases thus hinders quantification. The increased conversion noted at 250 °C (Entry 21, Table 3) over 210 °C in hexane as compared to the other solvents is probably a result of the increased solubility of furfural in hexane as well as the formation of other unidentified condensation products.

All of the solvents that were used in this research lead to the production of MF and furan. However, dimerization and/or condensation products formation was affected by the solvent type and occurred essentially with a polar solvent.

*2.4. Catalyst Reactivation Test*

From Figure 4a it can be seen that the activity of reactivated Pt/BC catalyst is retained in the first recycling (column 2) with regard to furfural hydrogenation in toluene at 210 °C for 2 h (column 1), indicating no significant deactivation of the catalyst. On recycling a second time (catalyst undergoes three consecutive reactions with intermediary reactivation step, column 3, Figure 4a), there is a decrease in the conversion of furfural by almost 20%. However, the selectivity to FA is not affected by the recycling of the catalyst (columns 1–3, Figure 4b).

At increased temperature (320 °C), conversion increased (column 4) compared to 210 °C (column 1), but the reactivation of the catalyst is not as effective as with the lower temperatures (there is a greater loss in the conversion of almost 12%) and selectivity to FA is also seen to be reduced (less than 20%). That indicates that temperature plays an important role in both catalyst activity and selectivity, further that deactivation of the catalyst occurs more readily at higher temperatures. Deactivation is assumed to be the result from a poisoning caused by strong chemisorption of compounds at the metal surface that cannot be solved during the reactivation step. The experiments shown in Figure 4 indicate that this catalyst has potential for recycling in furfural hydrogenation following process optimization.

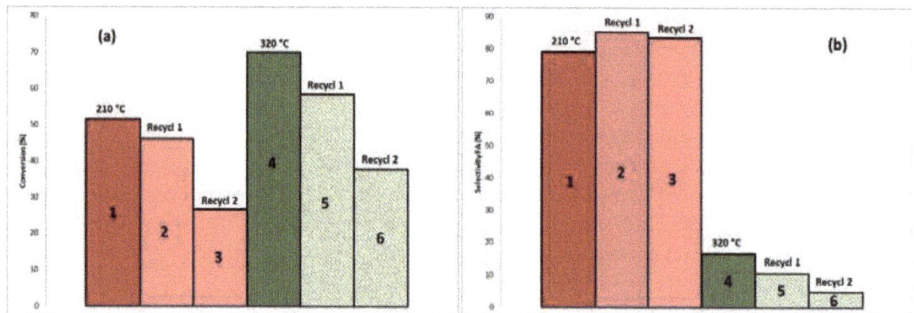

Figure 4. (a) Conversion; and (b) selectivity to FA at 210 °C (columns 1–3) and at 320 °C (columns 4–6).

## 3. Experimental

### 3.1. Support Preparation

The biochar support was produced by torrefaction (slow pyrolysis) of sugar maple (*Acer saccharum*) sawdust in a continuous pilot scale Auger-type reactor under $CO_2$ stream at atmospheric pressure. Biomass was obtained from Scierie Joseph Audet Inc. (Sainte-Rose-de-Watford, QC, Canada). Heating was performed countercurrent with a temperature gradient ranging inside the reactor from 230 °C to 400 °C with a residency time of about 5 min. Prior to impregnation, biochar was treated twice by sonication in a 50% aqueous sulfuric acid solution. Finally, samples were washed with water and toluene, and then dried in an oven at 120 °C overnight.

### 3.2. Catalyst Preparation

The platinum metal loading of the catalyst was achieved by wet impregnation using an aqueous solution of tetraammineplatinum (II) chloride hydrate (98%, Sigma Aldrich, Saint Louis, MO, USA) in appropriate concentration so as to obtain 3 wt % and 5 wt % Pt on the resulting catalyst. The impregnated sample was dried at 120 °C overnight to remove water. Prior to reaction, catalysts were activated in situ in flowing $H_2$ (50 mL/min) at 250 °C, for 1 h.

### 3.3. Catalyst Characterization

The morphology and the elemental analysis of the catalyst were analyzed using a SEM model Hitachi S-4700 (Hitachi, Toyo, Japan) microscope equipped with an EDX (X-max 500 mm², Oxford Instruments, Buckinghamshire, UK) system and operating with a voltage of 20 kV. The EDX used in this experiment was an Oxford model X-Max 50 mm².

### 3.4. Catalytic Test

Liquid-phase hydrogenation of furfural was conducted in a 100 mL continuous stirred-tank reactor (CSTR) equipped with a reagent injection port, a gas inlet, and a vent (Figure 5). Hydrogenations were performed between 200 °C and 300 °C under a 500–1500 psi atmosphere of $H_2$ (Praxair, purity 4.5), stirring at 600 rpm, for 1–6 h using toluene (99.99%, bought from Anachemia, Lachine, QC, Canada), isobutanol (99%, bought from Alfa Aesar, Ward Hill, MA, USA), *n*-hexane (95%, bought from Anachemia, Lachine, QC, Canada), or 2-propanol (99.99%, bought from Anachemia) as solvents. The reactor was loaded with 30 mL of solvent and 3 mL of furfural and the latter was reacted with different amounts of catalyst (0.025–0.058 mmol of Pt equivalent). The reactor was preheated to the desired temperature for the catalyst activation with a low flow of hydrogen, sample addition to the hot reactor and hydrogen pressurization was performed simultaneously.

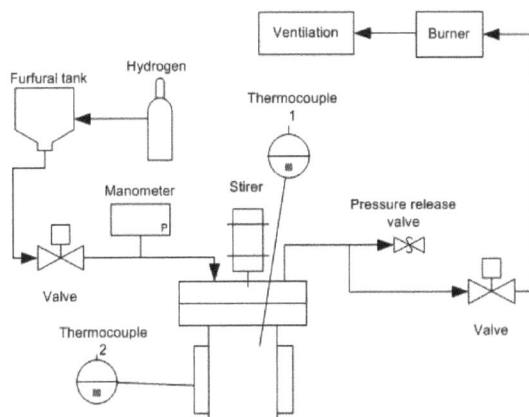

**Figure 5.** Process and instrumentation diagram for the liquid-phase hydrogenation of furfural.

The evolution of the reaction was followed using a GC-2014 gas chromatograph equipped (Shimadzu, Guelph, ON, Canada) with a flame ionization detector (FID), temperature programmer, and a capillary column Zebron ZB-5MS (L = 30 m × I.D. = 0.25 mm × df = 0.25 μm) (Phenomenex, Torrance, CA, USA) with helium as carrier gas. The reaction products were FA, THFA, furan, MTHF, and MF.

Identification of the compounds present in the reaction mixture was performed using a Bruker Scion SQ GC-MS (Bruker Daltonics Inc., Milton, ON, Canada) equipped with Bruker CombiPAL autosampler, zebron capillary column (ZB-5MS, 60 m length, 0.25 mm I.D., 0.25 μm film thickness), and a mass spectrum detector (Bruker SQ) using helium as carrier gas (Praxair, purity 5.0).

To compare the activity exhibited by the catalysts, the turn-over frequency (TOF) was calculated based on furfural conversion (Equation (1)):

$$TOF = \frac{mmol\, furfural\, converted}{time * mmol\, catalyst\, initial} \tag{1}$$

*3.5. Catalyst Regeneration Tests*

To investigate whether the catalyst could be reactivated, three consecutive experiments were performed at 210 °C and 320 °C. Between the runs, the catalyst was washed repeatedly with toluene and dried. Reactivation of the catalyst was done each time as described in Section 3.2.

**4. Conclusions**

Results from this work show that biochar can be efficiently used as catalyst support in addition to being an economically and environmentally sound approach for furfural hydrogenation. The platinum-based catalyst produced from wet impregnation of a platinum salt on maple biochar proved to work well during liquid phase hydrogenation of furfural to FA and can be improved to favor the FA selectivity. Variation of temperature had an impact, both on conversion and selectivity to FA. Longer reaction times decrease selectivity to FA, with little or no effect on conversion showing that FA is an intermediate toward other reduction products as MF. Results also showed that solvent polarity tends to increase the hydrogenation rates but also the formation of dimerization and/or condensation products. Reaction conditions of 210 °C for 2 h and 1500 psi, with toluene as the solvent and 3 wt % Pt

*Energies* **2017**, *10*, 286

content, were found to be optimal for Pt/BC catalyst to favor furfural conversion at high selectivity to FA. Finally, the Pt/BC catalyst showed potential to be reactivated for further furfural hydrogenation reaction with a high selectivity to FA.

**Acknowledgments:** We would like to acknowledge the Industrial Chair in Cellulosic Ethanol and Biocommodities for financial support. The authors are also thankful to the Scierie Joseph Audet Inc. for providing the sugar maple samples. Finally, the authors are grateful to Charles Bertrand, research professional at the Materials Characterization Center, University of Sherbrooke, for his help with the SEM-EDX experiments.

**Author Contributions:** Ariadna Fuente-Hernández performed the catalyst loading, characterization and testing in reactor. She is also responsible for writing the part of the article that concerns these aspects. Roland Lee conceived and designed the experiments and wrote the first version of the article. Nicolas Béland co-designed, assembled and operated the pilot scale torrefaction unit. He produced the catalyst support and contributed to the corresponding part of the manuscript. Ingrid Zamboni contributed to the overall manuscript and catalyst interpretation. Jean-Michel Lavoie is the principal investigator of this work. He contributed to the overall concept, designed the experiments in collaboration with Ariadna Fuente-Hernández, and provided useful tips along the way, contributed to the design of the torrefaction unit. He reviewed the manuscript and composed both introduction and conclusion.

**Conflicts of Interest:** The authors declare no conflict of interest.

## References

1. Lee, R.; Lavoie, J.M. From first- to third-generation biofuels: Challenges of producing a commodity from a biomass of increasing complexity. *Anim. Front.* **2013**, *3*, 6–11. [CrossRef]
2. Sánchez, O.J.; Montoya, S. Production of Bioethanol from Biomass: An Overview. In *Biofuel Technologies*; Springer: Berlin/Heidelberg, Germany, 2013; pp. 397–441.
3. Su, R.; Ma, Y.; Qi, W.; Zhang, M.; Wang, F.; Du, R.; Yang, J.; Zhang, M.; He, Z. Ethanol Production from High-Solid SSCF of Alkaline-Pretreated Corncob Using Recombinant Zymomonas mobilis CP4. *BioEnergy Res.* **2013**, *6*, 292–299. [CrossRef]
4. Tran, A.; Chambers, R.P. Red oak wood derived inhibitors in the ethanol fermentation of xylose by Pichia stipitis CBS 5776. *Biotechnol. Lett.* **1985**, *7*, 841–845. [CrossRef]
5. Fuente-Hernandez, A.; Corcos, P.O.; Beauchet, R.; Lavoie, J.M. Biofuels and co-products out of hemicelluloses. In *Liquid, Gaseous and Solid Biofuels-Conversion Techniques*; Fang, Z., Ed.; InTech: Rijeka, Croatia, 2013; p. 346.
6. Yang, W.; Li, P.; Bo, D.; Chang, H.; Wang, X.; Zhu, T. Optimization of furfural production from D-xylose with formic acid as catalyst in a reactive extraction system. *Bioresour. Technol.* **2013**, *133*, 361–369. [CrossRef] [PubMed]
7. Gürbüz, E.; Gallo, J.; Alonso, D.; Wettstein, S.; Lim, W.; Dumesic, J.A. Conversion of Hemicellulose into Furfural Using Solid Acid Catalysts in γ-Valerolactone. *Angew. Chem. Int. Ed.* **2013**, *52*, 1270–1274. [CrossRef] [PubMed]
8. Cai, C.M.; Zhang, T.; Kumar, R.; Wyman, C.E. Integrated furfural production as a renewable fuel and chemical platform from lignocellulosic biomass. *J. Chem. Technol. Biotechnol.* **2014**, *89*, 2–10. [CrossRef]
9. O'Driscoll, A.; Leahy, J.J.; Curtin, T. The influence of metal selection on catalyst activity for the liquid phase hydrogenation of furfural to furfuryl alcohol. *Catal. Today* **2015**, *279*, 194–201. [CrossRef]
10. Jiménez-Gómez, C.P.; Cecilia, J.A.; Durán-Martín, D.; Moreno-Tost, R.; Santamaría-González, J.; Mérida-Robles, J.; Mariscal, R.; Maireles-Torres, P. Gas-phase hydrogenation of furfural to furfuryl alcohol over Cu/ZnO catalysts. *J. Catal.* **2016**, *336*, 107–115. [CrossRef]
11. Li, M.; Hao, Y.; Cárdenas-Lizana, F.; Keane, M.A. Selective production of furfuryl alcohol via gas phase hydrogenation of furfural over Au/Al2O3. *Catal. Commun.* **2015**, *69*, 119–122. [CrossRef]
12. Wu, J.; Shen, Y.; Liu, C.; Wang, H.; Geng, C.; Zhang, Z. Vapor phase hydrogenation of furfural to furfuryl alcohol over environmentally friendly Cu–Ca/SiO2 catalyst. *Catal. Commun.* **2005**, *6*, 633–637. [CrossRef]
13. Vaidya, P.; MahajaniInd, V. Kinetics of Liquid-Phase Hydrogenation of Furfuraldehyde to Furfuryl Alcohol over a Pt/C Catalyst. *Ind. Eng. Chem. Res.* **2003**, *42*, 3881–3885. [CrossRef]
14. Sitthisa, S.; Pham, T.; Prasomsri, T.; Sooknoi, T.; Mallinson, R.; Resasco, D. Conversion of furfural and 2-methylpentanal on Pd/SiO2 and Pd–Cu/SiO2 catalysts. *J. Catal.* **2011**, *280*, 17–27. [CrossRef]
15. Kijenski, J.; Winiarek, P. Selective hydrogenation of α,β-unsaturated aldehydes over Pt catalysts deposited on monolayer supports. *Appl. Catal. A Gen.* **2000**, *193*, L1–L4. [CrossRef]

16. Nagaraja, B.; Siva Kumar, V.; Shasikala, V.; Padmasri, A.; Sreedhar, B.; Raju, B.; Rama Rao, K. A highly efficient Cu/MgO catalyst for vapour phase hydrogenation of furfural to furfuryl alcohol. *Catal. Commun.* **2003**, *4*, 287–293. [CrossRef]

17. Zhou, Z.; Ma, Q.; Zhang, A.; Wu, M. Synthesis of water-soluble monotosylated ethylenediamines and their application in ruthenium and iridium-catalyzed transfer hydrogenation of aldehydes. *Appl. Organomet. Chem.* **2011**, *25*, 856–861. [CrossRef]

18. Nakagawa, Y.; Tomishige, K. Production of 1,5-pentanediol from biomass via furfural and tetrahydrofurfuryl alcohol. *Catal. Today* **2012**, *195*, 136–143. [CrossRef]

19. Inada, K.; Shibagaki, M.; Nakanishi, Y.; Matsushita, H. The catalytic reduction of aldehydes and ketones with 2-propanol over silica-supported zirconium catalyst. *ChemInform* **1994**, *25*. [CrossRef]

20. Kijenski, J.; Winiarek, P.; Paryjczak, T.; Lewicki, A.; Mikołajska, A. Platinum deposited on monolayer supports in selective hydrogenation of furfural to furfuryl alcohol. *Appl. Catal. A Gen.* **2002**, *233*, 171–182. [CrossRef]

21. Nagaraja, B.; Padmasri, A.; David Raju, B.; Rama Rao, K. Vapor phase selective hydrogenation of furfural to furfuryl alcohol over Cu–MgO coprecipitated catalysts. *J. Mol. Catal. A Chem.* **2007**, *265*, 90–97. [CrossRef]

22. Li, H.; Luo, H.; Zhuang, L.; Dai, W.; Qiao, M. Liquid phase hydrogenation of furfural to furfuryl alcohol over the Fe-promoted Ni-B amorphous alloy catalysts. *J. Mol. Catal. A Chem.* **2003**, *203*, 267–275. [CrossRef]

23. Sitthisa, S.; An, W.; Resasco, D. Selective conversion of furfural to methylfuran over silica-supported Ni-Fe bimetallic catalysts. *J. Catal.* **2011**, *284*, 90–101. [CrossRef]

24. Baijuna, L.; Lianhaia, L.; Bingchuna, W.; Tianxia, C.; Iwatani, K. Liquid phase selective hydrogenation of furfural on Raney nickel modified by impregnation of salts of heteropolyacids. *Appl. Catal. A Gen.* **1998**, *171*, 117–122. [CrossRef]

25. Mäki-Arvela, P.; Hájek, P.; Salmi, J.; Murzin, T.; Yu, D. Chemoselective hydrogenation of carbonyl compounds over heterogeneous catalysts. *Appl. Catal. A Gen.* **2005**, *292*, 1–49. [CrossRef]

26. Dehkhoda, A.; West, A.; Ellis, N. Biochar based solid acid catalyst for biodiesel production. *Appl. Catal. A Gen.* **2010**, *382*, 197–204. [CrossRef]

27. Beesley, L.; Moreno-Jiménez, E.; Gomez-Eyles, J. Effects of biochar and greenwaste compost amendments on mobility, bioavailability and toxicity of inorganic and organic contaminants in a multi-element polluted soil. *Environ. Pollut.* **2010**, *158*, 2282–2287. [CrossRef] [PubMed]

28. Von Arx, M.; Mallat, T.; Baiker, A. Unprecedented selectivity behaviour in the hydrogenation of an α,β-unsaturated ketone: Hydrogenation of ketoisophorone over alumina-supported Pt and Pd. *J. Mol. Catal. A Chem.* **1999**, *148*, 275–283. [CrossRef]

*energies*

MDPI

*Article*

# The Potential of Activated Carbon Made of Agro-Industrial Residues in NOx Immissions Abatement

Imen Ghouma [1,2], Mejdi Jeguirim [1,*] (ID), Uta Sager [3] (ID), Lionel Limousy [1], Simona Bennici [1], Eckhard Däuber [3], Christof Asbach [3], Roman Ligotski [4], Frank Schmidt [4] and Abdelmottaleb Ouederni [2]

[1]  Institut de Sciences des Matériaux de Mulhouse, Université de Haute-Alsace, 15 Rue Jean Starcky, F-68057 Mulhouse, France; imenghouma83@gmail.com (I.G.); lionel.limousy@uha.fr (L.L.); simona.bennici@uha.fr (S.B.)
[2]  Department of Chemical Engineering, National School of Engineers (ENIG), University of Gabes, Avenue Omar Ibn El Khattab, Gabes 6029, Tunisia; mottaleb.ouederni@enig.rnu.tn
[3]  Institut für Energie-und Umwelttechnik e.V., Bliersheimer Str. 58-60, D-47229 Duisburg, Germany; sager@iuta.de (U.S.); daeuber@iuta.de (E.D.); asbach@iuta.de (C.A.)
[4]  Nanopartikel Prozesstechnik, Universität Duisburg-Essen, Lotharstr. 1, D-47057 Duisburg, Germany; roman.ligotski@uni-due.de (R.L.); frank.schmidt@uni-duisburg.de (F.S.)
*  Correspondence: mejdi.jeguirim@uha.fr; Tel.: +33-389-608-661

Received: 22 June 2017; Accepted: 25 September 2017; Published: 28 September 2017

**Abstract:** The treatment of NOx from automotive gas exhaust has been widely studied, however the presence of low concentrations of NOx in confined areas is still under investigation. As an example, the concentration of $NO_2$ can approximate 0.15 ppmv inside vehicles when people are driving on highways. This interior pollution becomes an environmental problem and a health problem. In the present work, the abatement of $NO_2$ immission is studied at room temperature. Three activated carbons (ACs) prepared by physical ($CO_2$ or $H_2O$) or chemical activation ($H_3PO_4$) are tested as adsorbents. The novelty of this work consists in studying the adsorption of $NO_2$ at low concentrations that approach real life immission concentrations and is experimentally realizable. The ACs present different structural and textural properties as well as functional surface groups, which induce different affinities with $NO_2$. The AC prepared using water vapor activation presents the best adsorption capacity, which may originate from a more basic surface. The presence of a mesoporosity may also influence the diffusion of $NO_2$ inside the carbon matrix. The high reduction activity of the AC prepared from $H_3PO_4$ activation is explained by the important concentration of acidic groups on its surface.

**Keywords:** activated carbon; $NO_2$ adsorption; ambient temperature; low $NO_2$ concentrations; textural properties-surface chemistry characterization

## 1. Introduction

The increase of the threatening substances emission to the atmosphere has become a major environmental problem. Among the various harmful gases, nitrogen oxides ($NO_x$) have a negative impact through the smog and acid rain formations as well as the decrease of the superior ozone layer [1,2]. Several methods have been applied for nitrogen oxides elimination including the reduction of $NO_x$ by selective non-catalytic reduction (SNCR) and selective catalytic reduction (SCR) [3]. However, these techniques have several drawbacks such as costs and technical complexity. Therefore, the separation by adsorption at reduced temperatures is receiving increasing attention. Several adsorbents including zeolites, perovskites and carbonaceous materials were tested for $NO_x$

treatment [4–6]. However, studies on activated carbons (AC) performance for $NO_x$ removal showed promising results in term of adsorption efficiencies [7–12].

Several lignocellulosic precursors were used for the preparation of activated carbons dedicated to the $NO_2$ removal at ambient temperature [7–12]. These carbonaceous adsorbents showed interesting adsorption capacities ranging between 40 mg/g and 120 mg/g. However, the performance of these activated carbons was generally evaluated at 500–1000 ppm $NO_2$ concentrations in presence of air under dry and wet conditions. These experimental conditions are far from the ones encountered at ambient air especially near urban traffics and industrial plants.

Despite the emission reduction measures implemented for automotive and industrial exhaust gases [13–16], vehicle occupants and urban population are particularly affected by high or even increasing nitrogen oxides immissions. As a result, the interest in adsorptive cabin air filters and HVAC (Heating, Ventilation and Air Conditioning) filters is rising. This interest leads to more research on the adsorptive separation of low concentrations of nitrogen oxides at ambient temperatures. In this context, Sager et al. have examined the performance of different activated carbons in cabin air filter for the elimination of several pollutants [17–21]. In particular, authors examined the adsorption of low concentrations of $NO_2$ on modified activated carbon, prepared of polymer as base material, at ambient temperatures. They have showed that the elimination of $NO_2$ present in the air, on activated carbons at ambient temperature, and after repeated adsorption cycles, can be increased by the infiltration of metal oxide nanoparticles into the sorbent. In this case the regeneration of the filter was assumed to be connected to the redox properties of the sorbent that can act also as reduction catalyst towards $NO_x$ [18–21]. However, the role of surface chemistry of the activated carbon on the $NO_2$ adsorption/reduction was not examined in detail.

Due to industrial activities and the urban traffic, $NO_x$ emissions are also a problem in developing countries such as Tunisia that has few resources and is rather agricultural and agro-industrial outside the urban agglomerations. One approach to counteract both problems is the development of an efficient sorbent for the separation of $NO_x$ from a locally available raw precursor; in this case, the agro-industrial residue olive stones have interesting potential for activated carbons preparation [22]. Ghouma et al. [10] have prepared activated carbon based on olive stones by water vapor activation. The adsorption tests with 500 ppm $NO_2$ have shown that the prepared activated carbon has a capacity as high as the ones available in literature. Furthermore, it was demonstrated that, besides microporosity, the surface functional groups are strongly related to the $NO_2$ adsorption capacity. However, the investigation of the AC performance at low concentrations as well as the determination of the precise contribution of the different activated carbon characteristics to $NO_2$ adsorption performance are still missing.

Based on the described considerations and experience, a study was started with regard to which extent the potential of activated carbon from olive stones can be enhanced by the different activation methods resulting in modifications of the functional groups on the inner surface and surface characteristics.

For that purpose, three differently activated carbons prepared from olive stones were evaluated with breakthrough tests using $NO_2$ (5 ppmv as inlet concentration) as the adsorptive at ambient temperature and 50% of relative humidity (RH). Two activated carbons were prepared by gas phase activation. A first sample was prepared by water vapor treatment, while a second was activated with carbon dioxide ($CO_2$). A third activated carbon was prepared by chemical activation with phosphoric acid ($H_3PO_4$). The differently activated sorbents were characterized with regard to their pore structure, morphology and carbon surface chemistry using nitrogen adsorption, scanning electron microscopy, Fourier transform infrared spectroscopy, and temperature-programmed desorption coupled with mass-spectrometry. The results of the breakthrough tests were correlated to the results of AC characterization with the aim to identify the key-parameters influencing the adsorption capacity, and then to adjust the carbon activation and modification methods to obtain better performing sorbents.

## 2. Materials and Methods

### 2.1. Samples Preparation

Carbon materials were prepared from olive pomace provided by a Tunisian olive oil factory located in Zarzis. The raw precursor was washed abundantly with hot distilled water and dried at ambient temperature for 24 h. Then, the dried olive stones were crushed and sieved to 1–3 mm particle size. Different activated carbons were prepared through chemical and physical activations of olive stones according to the following protocols.

#### 2.1.1. Chemical Activation

A granular $H_3PO_4$-activated sample was prepared according to the optimized protocol reported by Limousy et al. [22]. Briefly, olive stones were soaked in an aqueous solution of orthophosphoric acid (50%, $w/w$) at the weight ratio (1:3). The suspension was stirred at 110 °C for 9 h. Then, the filtered material was dried and carbonized in nitrogen flow at 170 °C for 30 min and finally at 410 °C for 2 h 30 min. The resulting carbon, denoted as AC-$H_3PO_4$, was then washed abundantly with distilled water until the elimination of all acid traces, and was dried overnight at 110 °C. The yield of chemical activation method was 33 wt %. Referring to the recent work published on chemically activated carbon, it is shown that this yield is more important that KOH activation of biomass. Travis et al. [23] have synthesized a series of KOH activated carbons from spent coffee grounds. Sample yields ranged from 11 to 16 wt % with yields lower for higher KOH.

#### 2.1.2. Physical Activation

The carbonization and the activation steps were carried out in a fixed bed reactor in a stainless steel reactor placed in a vertical automated furnace equipped with a temperature controller, with an initial mass of biomass equal to 2 g. Firstly, the precursor was carbonized under nitrogen flow at 600 °C for 2 h. Subsequently, the activation of the resultant char was performed by switching the gas either to pure $CO_2$ or to water vapor. The targeted temperature was maintained to 750 °C for 6 h. Further details about physical activation procedure could be found in a previous work [10].

### 2.2. Characterization of Activated Carbons

#### 2.2.1. Pore Structure and Morphology Characterization

Characterization of the pore structure of the activated carbon samples was made by measurement of $N_2$ adsorption isotherms using an automatic gas sorption analyzer (ASAP 2010, Micrometrics). Specific surface area was calculated from the $N_2$ adsorption isotherms applying the Brunauer–Emmett–Teller (BET) equation and yield important information about structural features. The total pore volume was determined from the amount of nitrogen adsorbed at $P/P° = 0.99$. The t-plot method was applied to measure the total micropore volume.

Scanning electron microscopy (Philips model FEI model Quanta 400 SEM) was used to analyze morphology of the different activated carbons.

#### 2.2.2. Characterization of Carbon Surface Chemistry and Composition

Temperature Programmed Desorption Coupled with Mass Spectrometry (TPD-MS)

The surface chemistry of samples was analyzed by TPD-MS. The sample weighting 10 mg was placed in quartz tube in an oven and heat-treated with a linear heating rate of 5 °C/min under vacuum. The material surface chemistry was evaluated in the temperature range 25–900 °C. The gases evolved during the heating process were continuously analyzed quantitatively by a mass spectrometer (INFICON Transpector). The desorption rate of each gas as a function of temperature was determined from the TPD analysis. The total amount of each gas released was computed by time integration of the TPD curves.

Fourier Transform Infrared Spectroscopy (FTIR)

FTIR was used to characterize the main functional groups of the activated carbon surface using a spectrometer FTIR (Jasco FT-IR 4100 series spectrophotometer with a diffuse reflectance accessory manufactured by PIKE Technologies, Madison, WI, USA). During this characterization, samples of activated carbon were mixed with finely divided spectroscopic grade KBr. All the spectra were taken at a spectral resolution of 16 cm$^{-1}$ using minimum 30 scans.

*2.3. NO$_2$ Adsorption Experiments*

The different sorbents were tested by breakthrough experiments in a test set-up with a fixed-bed flow reactor with an inner diameter of 0.05 m (see Figure 1). In each test, 2 g of activated carbon was packed in the breakthrough column with a resulting bed length of about 2.5 mm. The very low bed length should represent the ultrathin sorbent layers of commercially available cabin air filters and filters for heating, ventilation and air conditioning purposes used for the reduction of ambient pollution. Then, an air stream of 23 °C and 50% relative humidity with 5 ppm$_V$ NO$_2$ was forced to pass through the sorbent layer with a velocity of 0.2 m/s. The NO$_2$ was directly supplied to the airflow from a reservoir using a mass flow controller. To avoid condensation and to achieve the required vapor pressure, the NO$_2$ reservoir, the tubing, and the mass flow controller were heated and kept at about 35 °C. At the inlet and outlet of the fixed-bed reactor, the concentrations of NO$_2$ and NO were measured with two nitrogen oxide analyzers (type AC 31M from Ansyco, Karlsruhe, Germany) using chemiluminescence. The measuring range of the analyzers is up to 10 ppmv and the lower detection limit is <1 ppb. As the precision of the measurement is ±1% of the upper range value at input concentrations of 5 ppm$_V$, the measurement uncertainty causes a lack of significance of breakthrough values below 2%. In the test rig the total pressure corresponds approximately to the atmospheric pressure. Before the tests, the activated carbon samples were conditioned in air stream at 23 °C and 50% relative humidity, for 15 min.

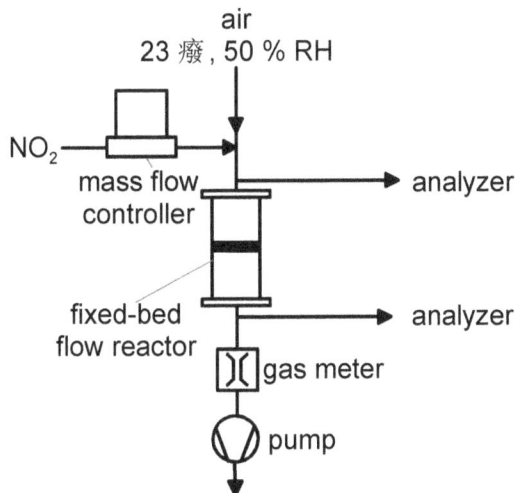

**Figure 1.** Sketch of the test device for breakthrough experiments.

### 3. Results and Discussion

#### 3.1. Activated Carbon Characterization

3.1.1. Morphology and Pore Structure Characterizations

Scanning electron microscopy (SEM) micrographs of AC-$H_3PO_4$, AC-$CO_2$ and AC-$H_2O$ are gathered in Figure 2. The three samples have shown remarkable differences in the surface morphology, thus revealing different activation mechanisms. The sample AC-$H_3PO_4$ was arranged in tightly compacted sheets (see Figure 2a). Actually, the phosphoric acid acts as an acid catalyst to promote bond cleavage and formation of cross-links via cyclization/condensation reactions. The deposited $H_3PO_4$ can react with organic species to form phosphate and polyphosphate bridges that connect and cross-link biopolymer fragments. The addition (or insertion) of phosphate groups drives a process of dilation that, after the removal of the acid, leaves the matrix in an expanded state characterized by an accessible pore structure [24]. Similar observations were observed during chemical activation. As an example, KOH reacts with carbon to yield carbonates/bicarbonates at intermediate temperatures and further increase in the temperature leads to the decomposition and gasification to create porosities [25,26]. In addition, during ZnO chemical activation, different reduction reactions occur to generate high porosities.

The physically activated carbons AC-$CO_2$ and AC-$H_2O$ show quite similar surface morphologies. Figure 2b,c shows irregular shaped particles with large concoidal cavities and smooth surfaces for both activated carbons.

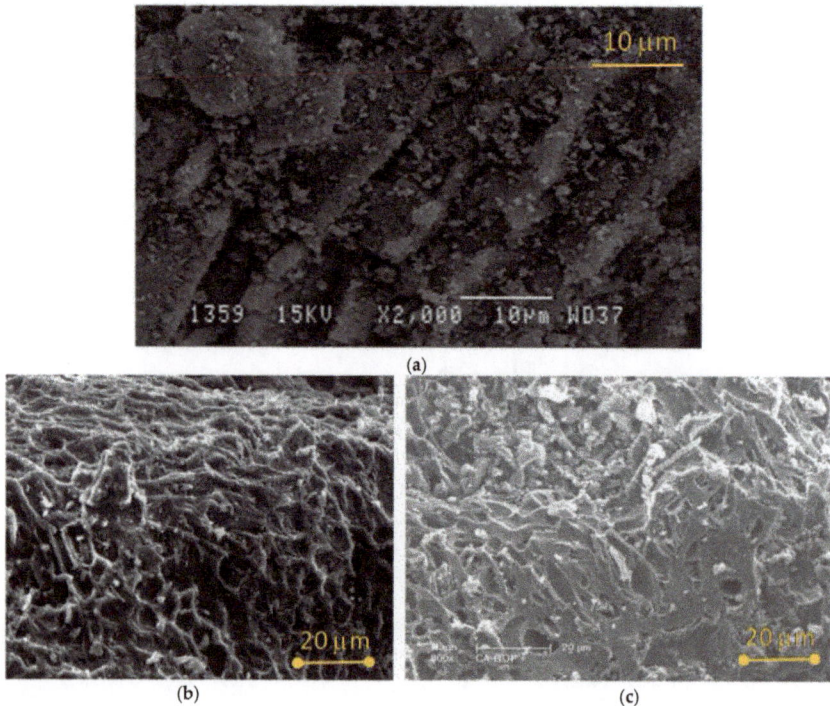

(a)

(b)

(c)

**Figure 2.** Scanning electron microscopy micrographs of sample: (a) AC-$H_3PO_4$; (b) AC-$CO_2$; and (c) AC-$H_2O$.

Figure 3a shows the nitrogen adsorption isotherms measured at $-196\,^\circ$C for the prepared carbon materials. All the ACs show Type-I isotherm according to the IUPAC classification except the isotherm related to the sample AC-H$_2$O which is typically of Type III. Therefore, all carbons are basically microporous. In addition, AC-H$_2$O exhibits mesopores in its internal structure. Such results were also observed by Roman et al. during the preparation of activated carbons from almond tree pruning through water vapor activation [27]. This results clearly indicates that the physical activation modes are different for H$_2$O and CO$_2$. Water vapor tends to react faster than carbon dioxide which leads to the formation of a highly mesoporous activated carbon. Such behavior for water vapor is due to the burning or also called as gasification of more carbon source at the given temperature, in the form of CO$_2$ (C + H$_2$O → CO/CO$_2$ + H$_2$). At the opposite, carbon dioxide generates only micropores. This may be due to both the difference of reactivity of these two molecules but also to the difference of diffusion coefficients [28]. Moreover, the AC-H$_3$PO$_4$ carbon shows the highest nitrogen uptake at −196 °C, thus displaying the highest surface area and the best-developed porosity. Such an observation is confirmed by the pore size distribution shown in Figure 3b. In fact, Figure 3b shows for AC-H$_3$PO$_4$ only the presence of micropores (size < 2 nm) with high micropores volume comparing to AC-CO$_2$. In contrast, AC-H$_2$O exhibits a large size range of mesoporosity created by water vapor activation. The textural properties of the carbon materials deduced from adsorption isotherms of N$_2$ at −196 °C are compiled in Table 1.

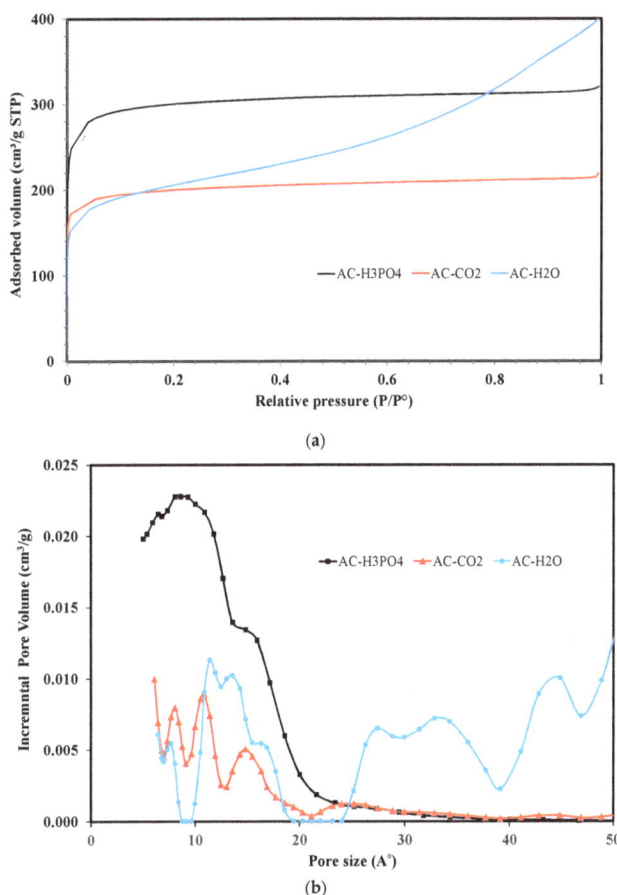

**Figure 3.** (**a**) Nitrogen adsorption/desorption isotherm of the different activated carbons; and (**b**) density functional theory (DFT) pore size distribution.

**Table 1.** Textural parameters of the activated carbons.

| Carbons | $S_{BET}$ (m$^2$/g) | $S_{ext}$ (m$^2$/g) | $S_\mu$ (m$^2$/g) | $V_{T\ pore}$ (cm$^3$/g) | $V_{micro}$ (cm$^3$/g) | $V_{meso}$ (cm$^3$/g) | % micro | $D_p$ (A) |
|---|---|---|---|---|---|---|---|---|
| AC-H3PO4 | 1178 | 11 | 1167 | 0.49 | 0.45 | 0.04 | 92 | 16.7 |
| AC-CO2 | 757 | 83 | 674 | 0.32 | 0.30 | 0.02 | 94 | 17.5 |
| AC-H2O | 754 | 291 | 463 | 0.58 | 0.28 | 0.30 | 48 | 32.2 |

% micro = ($V_{micro}$/$V_{Tpore}$) × 100, $D_p$: average pore diameter.

### 3.1.2. Characterization of Carbon Surface Chemistry

TPD-MS experiments give the evolution of $CO_2$ and CO emissions, as a result of the decomposition of the oxygen functionalities existing at the surface of the activated carbons. The determination of the amount of CO and $CO_2$ evolved gives an estimation of the amount of surface oxygen groups on the activated carbons. Moreover, the analysis of these emissions according to temperature indicates the presence of different oxygen surface groups.

The analysis of the desorption rates of $CO_2$ indicates the presence of different surface oxygen groups. In particular, the emission of $CO_2$ below 400 °C indicate the presence of carboxylic groups at the surface of the different activated carbons [29]. However, AC-$CO_2$ curve shows a $CO_2$ emission peak at 520 °C attributed to lactone group decomposition. In contrast, AC-$H_3PO_4$ and AC-$H_2O$ curves present $CO_2$ peaks at higher temperature (see Figure 4a) accompanied by CO emission (see Figure 4b). This decomposition indicates the presence of anhydride groups for the AC-$H_3PO_4$ and AC-$H_2O$ samples.

In a similar way, the analysis of the CO desorption rates shows the higher amount of surface oxygen groups for the AC-$H_3PO_4$. The CO decomposition could be attributed to carbonyl, quinone, ether and phenol groups. For the physically activated carbons, the CO peaks are obtained at different temperatures. Hence, the CO peak was obtained at 700 °C for the AC-$H_2O$ attributed to phenol group decomposition. In contrast, the CO peak for AC-$CO_2$ is obtained at 830 °C attributed to quinone groups.

The cumulated amounts of the emitted gases during TPD-MS are shown in Table 2. A significant amount of emitted hydrogen is observed for AC-$H_2O$. Such emission may be attributed to water vapor activations, which leads to a higher hydrogenation of the carbon surface due to the water gas shift reaction (production of $CO_2$ and $H_2$ at the surface of the AC).

The infrared spectra of the different activated carbons are shown in Figure 5. For the spectrum of AC-$H_2O$, the presence of a broad band located between 3020 and 3300 cm$^{-1}$ is observed. This band corresponds to the aromatic C-H groups. This band is related to the formation of the microcrystalline structure of this activated carbon. A second band between 1100 cm$^{-1}$ and 1300 cm$^{-1}$ assigned to the C-O groups present in the aromatic rings. The presence of a peak located at about 879 cm$^{-1}$ is also observed, which is attributed to the aliphatic C-H elongation. Another peak located at around 1462 cm$^{-1}$ is attributed to the elongation of the aromatic group. On the other hand, this spectrum shows the absence of significant peaks for the bands corresponding to the OH, C=O groups and the C-O-C groups. This behavior suggests that this activated carbon is primarily an aromatic polymer.

For the spectrum of AC-$CO_2$, the presence of an intense peak attributed to the hydroxyl groups located at 3140 cm$^{-1}$ is noted. Reddy et al. reported the same observations in their comparative study of the two activated carbons prepared from date palm cores activated by $CO_2$ and $H_3PO_4$ [30]. For the activated carbon AC-$H_3PO_4$, a broad and intense shoulder between 3000 and 3500 cm$^{-1}$ was observed. It was associated to the stretching vibrations of hydroxyl groups involved in hydrogen bonding [31]. The band at 1700–1710 cm$^{-1}$ is generally ascribed to the stretching vibrations of C=O bond in carboxylic acid and lactone groups. However, the peak at 1600 cm$^{-1}$ is attributed to a quinone structure. The band at 1250 cm$^{-1}$ has been assigned to C-O stretching and O-H bending modes of alcoholic, phenolic and carboxylic groups [32]. However, any peak at 1170 cm$^{-1}$ associated to the stretching vibration of the hydrogen bonding P=O contained in the group P-O-C (aromatic bond) [33] could not be identified, while a strong peak at 900–1000 cm$^{-1}$ can be assigned to P-OR ester species (for the AC-H3PO4 sample).

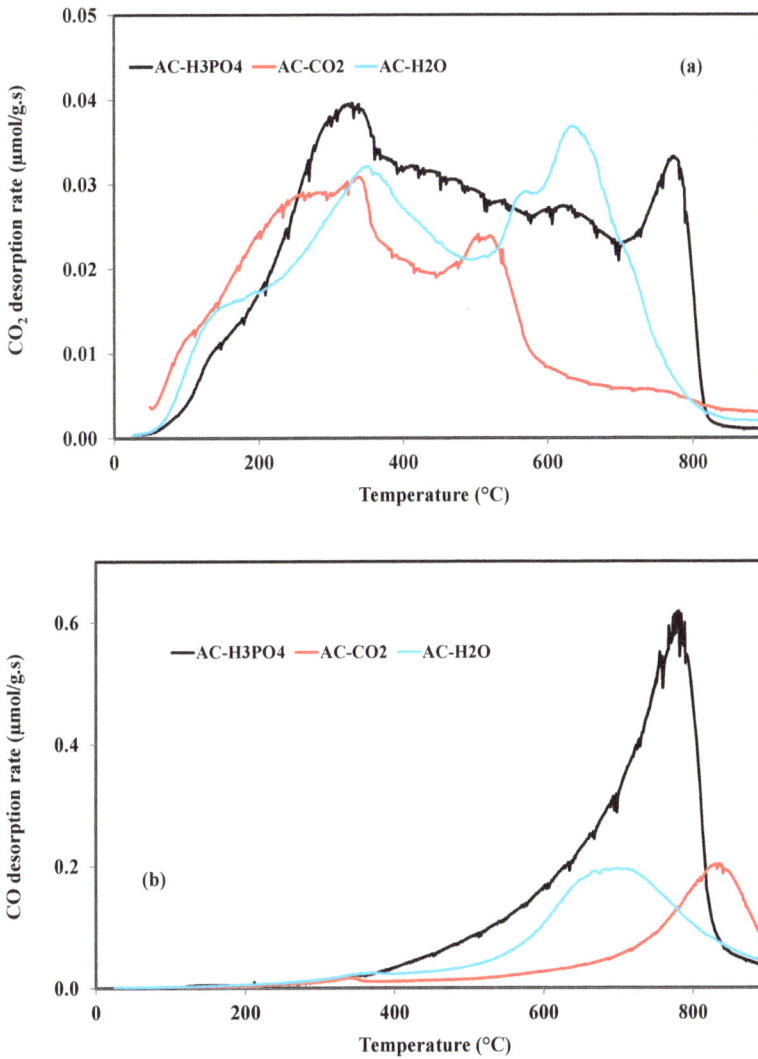

**Figure 4.** Desorption rates of: $CO_2$ (**a**); and CO (**b**) during temperature programmed desorption-mass spectroscopy (TPD-MS) of the different activated carbons.

**Table 2.** The cumulated amount of the emitted CO, $CO_2$, $H_2$ and $H_2O$ during the TPD-MS analysis for the different activated carbons.

| Sample | CO (mmol/g) | $CO_2$ (mmol/g) | $H_2O$ (mmol/g) | $H_2$ (mmol/g) |
|---|---|---|---|---|
| AC-$H_3PO_4$ | 3.43 | 0.72 | 3.46 | 1.80 |
| AC-$CO_2$ | 1.06 | 0.38 | 1.77 | 0.91 |
| AC-$H_2O$ | 1.25 | 0.39 | 1.06 | 6.76 |

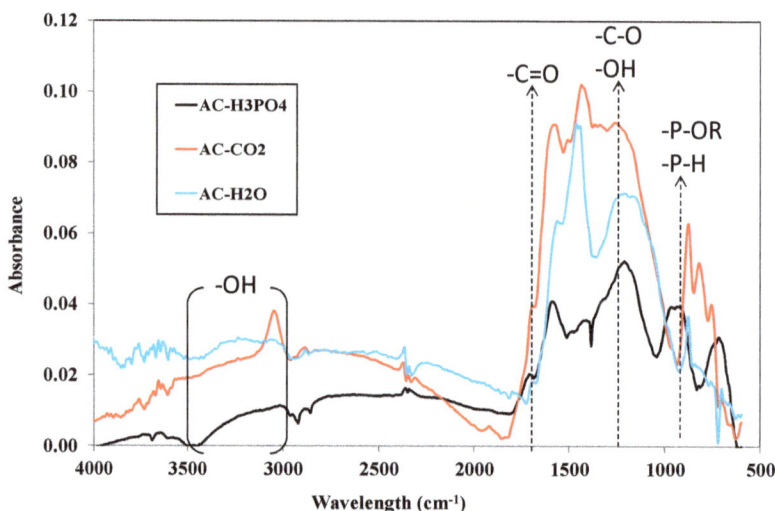

**Figure 5.** Fourier transform infrared spectroscopy (FTIR) spectra for the activated carbon samples.

### 3.2. $NO_x$ Adsorption on the Different Activated Carbons

Figure 6 shows the breakthrough curves of $NO_x$ through the thin sorbent layers of the three different activated carbons. The volumetric content of the sum parameter $NO_x = NO_2 + NO$ measured behind the sorbent layer is depicted in dependence of the experimental time. The inlet volumetric concentration of $NO_2$ was constantly kept at 5 ppm, whereas no NO was supplied. The breakthrough curves are averaged from repeated experiments with error bars indicating the standard deviation.

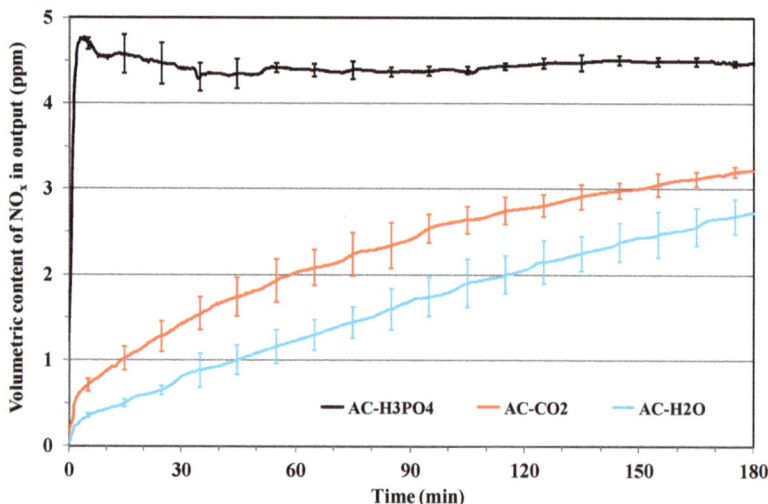

**Figure 6.** Breakthrough curves of $NO_x$ through three differently activated sorbent samples made from olive stones ($c_1$ $NO_2$ = 5 ppm, $c_1$ NO = 0 ppm, 23 °C, 50% RH, v = 0.2 m/s, $m_{Sorb}$ = 2 g).

As the following results show, for the assessment of the nitrogen oxide separation capacity of an activated carbon it is necessary to consider the volumetric fractions of the $NO_x$ present downstream the sorbent layer. Even though only $NO_2$ is supplied to the system, other nitrogen oxides-containing species (as NO) are formed due to reduction of $NO_2$ in the presence of activated carbon [11–17]. The reduction of $NO_2$ to NO is highly undesirable because the separation capacities of activated carbons for NO are negligible. Thus, a low activity in reducing $NO_2$ to NO is a desirable feature of activated carbons intended for $NO_2$ sorption.

For $AC-CO_2$ and $AC-H_2O$, the measured volumetric concentration of $NO_x$ regularly increases during the experiment. $AC-H_2O$ presents the lowest $NO_x$ breakthrough, reaching up to the 54% of the supplied $NO_2$. The characteristic of the $NO_x$ breakthrough curve of $AC-H_3PO_4$ is quite different. In the related experiments, in the first minutes, there is a deep increase of $NO_x$ measured at the outlet, then a slight decline and a subsequent stabilization at 90% of the supplied $NO_2$. The breakthrough curves of NO and $NO_2$ (see Figures 7 and 8) contribute to the explanation of the results.

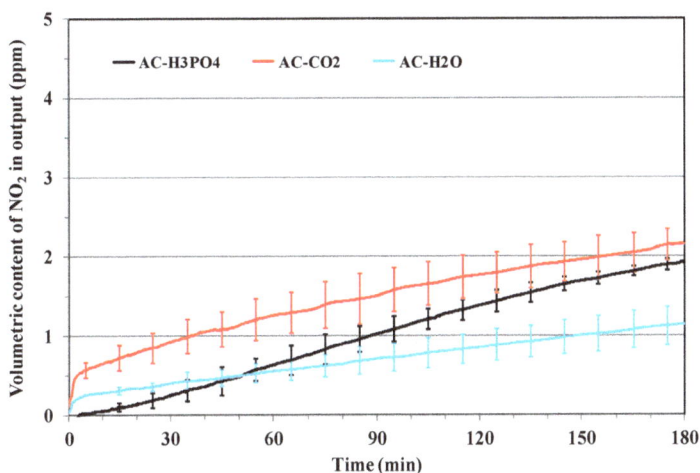

**Figure 7.** Breakthrough curves of $NO_2$ through three differently activated sorbent samples made from olive stones ($c_1$ $NO_2$ = 5 ppm, $c_1$ NO = 0 ppm, 23 °C, 50% RH, v = 0.2 m/s, $m_{Sorb}$ = 2 g).

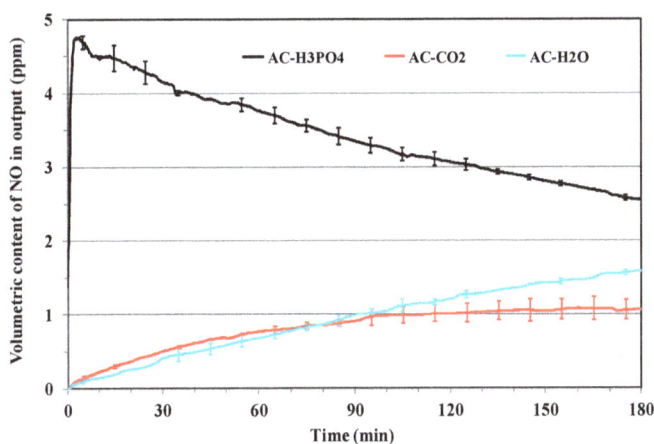

**Figure 8.** Breakthrough curves of NO through three differently activated sorbent samples made from olive stones ($c_1$ $NO_2$ = 5 ppm, $c_1$ NO = 0 ppm, 23 °C, 50% RH, v = 0.2 m/s, $m_{Sorb}$ = 2 g).

Figure 7 show the breakthrough curves of $NO_2$ in the same manner as in Figure 6 for $NO_x$ through the thin sorbent layers of the three different activated carbons. All curves increase regularly, for $NO_2$ the breakthrough of AC-$CO_2$ is the highest during the entire experimental duration of three hours. The breakthrough curves of the other two sorbents intersect after about one hour of testing. At the end of the experiment, AC-$H_2O$ presents the lowest $NO_2$ breakthrough.

Figure 8 shows the measured downstream volumetric contents of NO during the experiments on the three activated carbons. At this step, it is important to mention that, in a previous study performed in similar conditions, we have shown that $NO_2$ is reduced to NO but no $N_2$ is produced during the contact between $NO_2$ and the different activated carbons [10,11]. For AC-$CO_2$ and AC-$H_2O$, the measured volumetric concentration of NO regularly increases during the first half of the experiment, reaching up to the 20% of the supplied concentration of $NO_2$ for AC-$CO_2$ and AC-$H_2O$. In the second half time of the experiments, the NO content stabilized (the curve reach a plateau) for AC-$CO_2$, while it continues to increase on AC-$H_2O$. The NO breakthrough curve of AC-$H_3PO_4$ is quite different. In the related experiments, at the beginning, the supplied $NO_2$ is almost completely and easily reduced to NO passing through the sorbent layer. During the three hours of testing, the amount of formed NO slowly decreases to reach 50% of the measured $NO_2$ concentration. Thus, the $NO_x$ breakthrough characteristic of AC-$H_3PO_4$ shown in Figure 6 is determined by the reduction of $NO_2$ to NO by the sorbent.

To compare the total $NO_x$ sorption capacities of AC-$H_2O$, AC-$CO_2$ and AC-$H_3PO_4$, in Figure 9, the adsorbed amounts (in g) of $NO_x$ expressed by unit (in kg) of sorbent mass, are reported as a function of time. The mass of $NO_x$ adsorbed ($m_{ads}$) is calculated using the mass balance equation reported as follows:

$$m_{ads} = \sum_{0}^{i} \frac{(M_{NO_2} \cdot \Delta y_{NO_2 i}) + (M_{NO} \cdot \Delta y_{NO i})}{V_m} \cdot \dot{V} \cdot \Delta t_i \tag{1}$$

where $M_{NO2}$ and $M_{NO}$ are the molar masses of $NO_2$ and NO, $V_m$ the molar volume, $\Delta y_i$ the difference of the volumetric content in the inlet and the outlet of the fixed-bed reactor, $\dot{V}$ the flow through the reactor and $\Delta t_i$ a time interval.

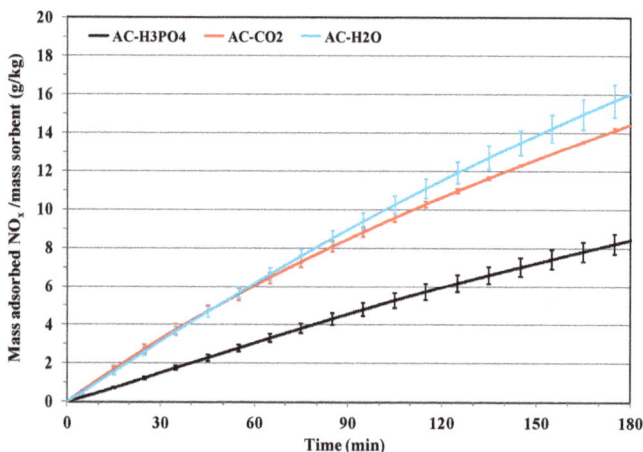

**Figure 9.** $NO_x$ capacities of three differently activated sorbent samples made from olive stones loaded for 3 h with $NO_2$ (5 ppm) at 23 °C, $c_1$ NO = 0 ppm, 50% RH with v = 0.2 m/s.

It is apparent that AC-$H_3PO_4$ has the lowest total capacity, while AC-$H_2O$ and AC-$CO_2$ have comparable capacities at the start of the experiment. However, at the end of the experiment, more $NO_x$ was clearly trapped on AC-$H_2O$ than on AC-$CO_2$.

The obtained capacities for $NO_2$ adsorption are in the same range as those available in the literature [34,35]. However, it is important to notice that the operating conditions in this present study are different from the ones available in the literature. In particular, the adsorption tests were performed at low inlet $NO_2$ concentrations (5 ppmv) while the ones in the literature were performed at high inlet $NO_2$ concentrations (500–1000 ppmv).

Effect of Porosity and Chemical Surface Groups

Porous texture and surface chemistry of activated carbons are crucial for adsorption process. However, the effect of texture parameters on $NO_2$ is not clearly shown in this present investigation. In fact, $AC-H_3PO_4$ has the highest microporous volume ($V\mu$ = 0.45 cm$^3$/g) and the lowest adsorption capacities ($C_{NO2}$ = 8.4 mg/g). In contrast, $AC-H_2O$ has the lowest microporous volume ($V\mu$ = 0.28 cm$^3$/g) and the highest adsorption capacities ($C_{NO2}$ = 16 mg/g). Such results indicate that the surface chemistry plays an important role in the interaction of $NO_2$ with activated carbon surface. In particular, the presence of high acidic surface groups in $AC-H_3PO_4$ seems to inhibit the $NO_2$ adsorption and favoring its reduction to NO. In a previous study, it was shown that physical activation leads to the formation of basic activated carbon surfaces while chemical using phosphoric acid activation favors acid activated carbon surfaces [35]. In fact, Boehm titration for $AC-H_3PO_4$ indicates the presence of 1.45 mmol/g of carboxylic groups and 0.70 mmol/g of phenol groups [22]. In contrast, the basic character of activated carbons favors the $NO_2$ adsorption since the total basicity of $AC-H_2O$ was equal to 1.86 mmol/g and the pHzc was equal to 10.8. Such role of basic groups could be confirmed through the comparison of the $AC-CO_2$ and $AC-H_2O$ performances. In fact, these activated carbons have quite similar textural properties but the adsorption capacity for $AC-H_2O$ ($C_{NO2}$ = 16 mg/g) is slightly higher than $AC-CO_2$ ($C_{NO2}$ = 14.4 mg/g). This difference is attributed to a higher amount of basic groups for $AC-H_2O$, as shown in the TPD-MS analysis. The role of surface chemistry on the interaction of $NO_2$ with activated carbons was already mentioned in previous investigations performed at higher $NO_2$ concentration (1000 ppm) [34–37]. However, no clear role of acidic and basic groups on the $NO_2$ adsorption capacity was identified.

To further clarify the role of the surface species towards $NO_2$ adsorption, FTIR analysis of the three samples before and after $NO_2$ adsorption were performed. Unfortunately, no evidence of N-containing species (generally detectable in the 1190–1600 cm$^{-1}$ range) has been found for any of the sample. In any case it is interesting to observe that for the $AC-H_2O$ and $AC-CO_2$ samples, that are those presenting the higher $NO_2$ adsorption capacity, the peaks related to C=O groups (ketones and quinones, at around 1500 cm$^{-1}$) and the -C-OH phenolic groups (at around 1000 cm$^{-1}$) disappear after adsorption of $NO_2$ (see FTIR spectra of AC-CO2 in Figure 10, reported as example). This behavior suggests the interaction of $NO_2$ with the surface oxygen groups as following:

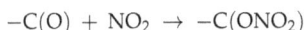

$$-C(O) + NO_2 \rightarrow -C(ONO_2)$$

The formation of -C(ONO$_2$) complexes is in agreement with the TPD performed in previous investigations after $NO_2$ adsorption on different AC. In fact, during TPD, the $NO_2$ and CO molecules always desorb at the same time, as well as the $CO_2$ and NO molecules, thus confirming the formation of a -C(ONO$_2$) surface complex [6]. This complex species can be formed on carbon presenting active sites (O-containing surface groups) with only one oxygen atom, such as those that indeed disappear in the FTIR spectra after adsorption test.

Another explanation can be the difference of textural properties between the different activated carbons. At the beginning of the adsorption process, the entire surface of the AC is accessible, thus the reduction process can be limited by external surface and the chemistry of the different AC (more important for $AC-H_3PO_4$ and similar for the others). The only difference between $AC-CO_2$ and $AC-H_2O$ corresponds to the presence of an important mesoporosity for the $AC-H_2O$ sample. Then, the accessibility of the adsorption sites (for $NO_2$) is higher for $AC-H_2O$ than for $AC-CO_2$, which may explain the difference of the breakthrough curves obtained for these two adsorbents. The reduction activity may be correlated to the external surface of the AC while the adsorption of $NO_2$ is more dependent on the diffusion process and the basic site concentration at the surface of the AC. We can also make the hypothesis that the kinetic of $NO_2$ reduction is faster than the adsorption process, or energetically favorable. Of course, this comment needs to be verified and validated by further experiments.

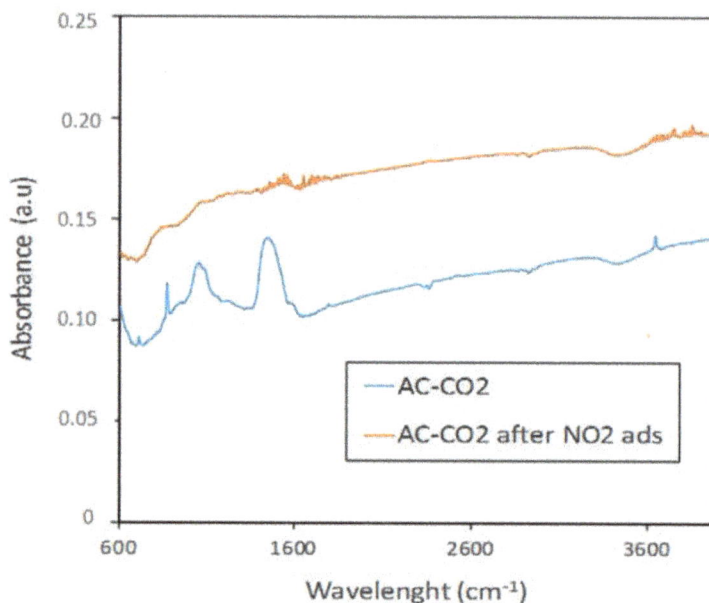

**Figure 10.** FTIR spectra of the AC-$CO_2$ sample before and after $NO_2$ adsorption.

## 4. Conclusions

In the present study, the importance of textural and surface properties of three different activated carbons is studied during the adsorption of $NO_2$ at room temperature and very low concentration. The results obtained during the different experiments indicate that both of these properties are responsible of $NO_2$ uptake and reduction to NO. The reduction rate of $NO_2$ is found to be very high when the activated carbon is prepared by $H_3PO_4$ activation. The presence of an important concentration of reducing groups at the surface of this AC (mainly anhydride groups) seems to be at the origin of this behavior. The adsorption of $NO_2$ is attributed to the presence of basic groups, which are more present when activation is carried out by a physical way. As AC-$H_2O$ presents a higher amount of basic groups in comparison to AC-$CO_2$, its adsorption capacity is higher. The difference observed for the breakthrough curve of $NO_2$ for AC-$CO_2$ and AC-$H_2O$ is explained by the mesoporous structure of the AC-$H_2O$ sample, which enables a better diffusion of $NO_2$ inside the activated carbon particles. This study is original because it is the only work that has been published on the adsorption of $NO_2$ on AC (produced from a biomass) at low concentration and room temperature. These results are very interesting from a mechanistic and reactivity point of view. From our knowledge, there is no study that relates the reactivity between $NO_2$ and the surface of an AC. Even if all the assumptions that have been done need to be verified and confirmed by further experimentations, this work enables a better understanding of $NO_2$ adsorption on biomass derived sorbents. The results can be used to choose more adapted biomasses as well as to tailor the surface properties of the sorbents as a function of their potential use (reduction catalyst, adsorbent, etc.). The next step of this work will be to investigate the form of the adsorbed species at the surface of the AC-H$_2$O by XPS and in situ DRIFTS experiments. Then, adsorption mechanisms could be approached.

**Acknowledgments:** The authors would like to thank the German Federal Ministry of Economics and Technology for financial support within the agenda for the promotion of industrial cooperative research and development (IGF) based on a decision of the German Bundestag. The access was opened by the IUTA e. V., Duisburg, and organized by the AiF (IGF-Project Nos. 14883 and 15751).

**Author Contributions:** All authors contributed equally to the work done.

**Conflicts of Interest:** The authors declare no conflict of interest.

# References

1. Wilkins, C.K.; Clausen, P.A.; Wolkoff, P.; Larsen, S.T.; Hammer, M.; Larsen, K.; Hansen, V.; Nielsen, G.D. Formation of strong airway irritants in mixtures of isoprene/ozone and isoprene/ozone/nitrogen dioxide. *Environ. Health Perspect.* **2001**, *109*, 937–941. [CrossRef] [PubMed]
2. Blondeau, P.; Iordache, V.; Poupard, O.; Genin, D.; Allard, F. Relationship between outdoor and indoor air quality in eight French schools. *Indoor Air* **2005**, *15*, 2–12. [CrossRef] [PubMed]
3. Gou, X.; Wu, C.; Zhang, K.; Xu, G.; Si, M.; Wang, Y.; Wang, E.; Liu, L.; Wu, J. Low Temperature Performance of Selective Catalytic Reduction of NO with NH3 under a Concentrated CO$_2$ Atmosphere. *Energies* **2015**, *8*, 12331–12341. [CrossRef]
4. Levasseur, B.; Ebrahim, A.M.; Burress, J.; Bandosz, T.J. Interactions of NO$_2$ at ambient temperature with cerium–zirconium mixed oxides supported on SBA-15. *J. Hazard. Mater.* **2011**, *197*, 294–303. [CrossRef] [PubMed]
5. Hodjati, S.; Vaezzadeh, K.; Petit, C.; Pitchon, V.; Kiennemann, A. Absorption/desorption of NO$_x$ process on perovskites: Performances to remove NO$_x$ from a lean exhaust gas. *Appl. Catal. B Environ.* **2000**, *26*, 5–16. [CrossRef]
6. Jeguirim, M.; Tschamber, V.; Brilhac, J.F.; Ehrburger, P. Interaction mechanism of NO$_2$ with carbon black: Effect of surface oxygen complexes. *J. Anal. Appl. Pyrolysis* **2004**, *72*, 171–181. [CrossRef]
7. Nowicki, P.; Pietrzak, R.; Wachowska, H. Sorption properties of active carbons obtained from walnut shells by chemical and physical activation. *Catal. Today* **2010**, *150*, 107–114. [CrossRef]
8. Pietrzak, R. Sawdust pellets from coniferous species as adsorbents for NO$_2$ removal. *Bioresour. Technol.* **2010**, *101*, 907–913. [CrossRef] [PubMed]
9. Belala, Z.; Belhachemi, M.; Jeguirim, M. Activated Carbon Prepared from Date Pits for the Retention of NO$_2$ at Low Temperature. *Int. J. Chem. React. Eng.* **2014**, *12*, 717–726. [CrossRef]
10. Ghouma, I.; Jeguirim, M.; Dorge, S.; Limousy, L.; Matei Ghimbeu, C.; Ouederni, A. Activated carbon prepared by physical activation of olive stones for the removal of NO$_2$ at ambient temperature. *C. R. Chim.* **2015**, *18*, 63–74. [CrossRef]
11. Belhachemi, M.; Jeguirim, M.; Limousy, L.; Addoun, F. Comparison of NO$_2$ removal using date pits activated carbon and modified commercialized activated carbon via different preparation methods: Effect of porosity and surface chemistry. *Chem. Eng. J.* **2014**, *253*, 121–129. [CrossRef]
12. Heschel, W.; Ahnert, F. Multicomponent adsorption of NO$_2$, SO$_2$ and butane from air on activated carbon. *Adsorpt. Sci. Technol.* **2002**, *20*, 353–370.
13. Labaki, M.; Issa, M.; Smeekens, S.; Heylen, S.; Kirschhock, C.E.A.; Villani, K.; Jeguirim, M.; Habermacher, D.; Brilhac, J.F.; Martens, J.A. Modeling of NO$_x$ adsorption–desorption–reduction cycles on a ruthenium loaded Na–Y zeolite. *Appl. Catal. B Environ.* **2010**, *97*, 13–20. [CrossRef]
14. Zouaoui, N.; Labaki, M.; Jeguirim, M. Diesel soot oxidation by nitrogen dioxide, oxygen and water under engine exhaust conditions: Kinetics data related to the reaction mechanism. *C. R. Chim.* **2014**, *17*, 672–680. [CrossRef]
15. Bennici, S.; Gervasini, A.; Ravasio, N.; Zaccheria, F. Optimization of Tailoring of CuO$_x$ Species of Silica Alumina Supported Catalysts for the Selective Catalytic Reduction of NO$_x$. *J. Phys. Chem. B* **2003**, *107*, 5168–5176. [CrossRef]
16. Limousy, L.; Mahzoul, H.; Brilhac, J.F.; Gilot, P.; Garin, F.; Maire, G. SO$_2$ sorption on fresh and aged SO$_x$ traps. *Appl. Catal. B Environ.* **2003**, *42*, 237–249. [CrossRef]

17. Sager, U.; Schmidt, F. Adsorption of Nitrogen Oxides, Water Vapour and Ozone onto Activated Carbon. *Adsorpt. Sci. Technol.* **2009**, *27*, 135–145. [CrossRef]
18. Sager, U.; Suhartiningsih; Schmidt, F. Einfluss der NO2-Dosierung auf Adsorptionsfiltertests. *Chem. Ing. Tech.* **2010**, *82*, 1737–1742. [CrossRef]
19. Sager, U.; Schmidt, W.; Schmidt, F. Suhartiningsih Catalytic reduction of nitrogen oxides via nanoscopic oxide catalysts within activated carbons at room temperature. *Adsorption* **2013**, *19*, 1027–1033. [CrossRef]
20. Sager, U.; Däuber, E.; Asbach, C.; Bathen, D.; Schmidt, F.; Weidenthaler, C.; Tseng, J.C.; Schmidt, W. Differences between the adsorption of NO2 and NO on modified activated carbon/Unterschiede bei der Adsorption von NO2 und NO an modifizierter Aktivkohle. *Air Qual. Control* **2014**, *74*, 181–184.
21. Sager, U.; Däuber, E.; Bathen, D.; Asbach, C.; Schmidt, F.; Tseng, J.-C.; Pommerin, A.; Weidenthaler, C.; Schmidt, W. Influence of the degree of infiltration of modified activated carbons with CuO/ZnO on the separation of NO2 at ambient temperatures. *Adsorpt. Sci. Technol.* **2016**, *34*, 307–319. [CrossRef]
22. Limousy, L.; Ghouma, I.; Ouederni, A.; Jeguirim, M. Amoxicillin removal from aqueous solution using activated carbon prepared by chemical activation of olive stone. *Environ. Sci. Pollut. Res.* **2017**, *24*, 9993–10004. [CrossRef] [PubMed]
23. Travis, W.; Gadipelli, S.; Guo, Z. Superior CO2 adsorption from waste coffee ground derived carbons. *RSC Adv.* **2015**, *5*, 29558–29562. [CrossRef]
24. Jagtoyen, M.; Derbyshire, F. Activated carbons from yellow poplar and white oak by H3PO4 activation. *Carbon* **1998**, *36*, 1085–1097. [CrossRef]
25. Srinivas, G.; Krungleviciute, V.; Guo, Z.-X.; Yildirim, T. Exceptional CO2 capture in a hierarchically porous carbon with simultaneous high surface area and pore volume. *Energy Environ. Sci.* **2013**, *7*, 335–342. [CrossRef]
26. Srinivas, G.; Burress, J.; Yildirim, T. Graphene oxide derived carbons (GODCs): Synthesis and gas adsorption properties. *Energy Environ. Sci.* **2012**, *5*, 6453–6459. [CrossRef]
27. Román, S.; Ledesma, B.; Álvarez-Murillo, A.; Al-Kassir, A.; Yusaf, T. Dependence of the Microporosity of Activated Carbons on the Lignocellulosic Composition of the Precursors. *Energies* **2017**, *10*, 542. [CrossRef]
28. Guizani, C.; Jeguirim, M.; Gadiou, R.; Escudero Sanz, F.J.; Salvador, S. Biomass char gasification by H2O, CO2 and their mixture: Evolution of chemical, textural and structural properties of the chars. *Energy* **2016**, *112*, 133–145. [CrossRef]
29. Figueiredo, J.L.; Pereira, M.F.R.; Freitas, M.M.A.; Órfão, J.J.M. Modification of the surface chemistry of activated carbons. *Carbon* **1999**, *37*, 1379–1389. [CrossRef]
30. Reddy, K.S.K.; Al Shoaibi, A.; Srinivasakannan, C. A comparison of microstructure and adsorption characteristics of activated carbons by CO2 and H3PO4 activation from date palm pits. *New Carbon Mater.* **2012**, *27*, 344–351. [CrossRef]
31. El-Hendawy, A.-N.A. Influence of HNO3 oxidation on the structure and adsorptive properties of corncob-based activated carbon. *Carbon* **2003**, *41*, 713–722. [CrossRef]
32. Shen, W.; Li, Z.; Liu, Y. Surface chemical functional groups modification of porous carbon. *Recent Pat. Chem. Eng.* **2008**, *1*, 27–40. [CrossRef]
33. Puziy, A.M.; Poddubnaya, O.I.; Martínez-Alonso, A.; Suárez-García, F.; Tascón, J.M.D. Synthetic carbons activated with phosphoric acid: I. Surface chemistry and ion binding properties. *Carbon* **2002**, *40*, 1493–1505. [CrossRef]
34. Nowicki, P.; Skibiszewska, P.; Pietrzak, R. NO2 removal on adsorbents prepared from coffee industry waste materials. *Adsorption* **2013**, *19*, 521–528. [CrossRef]
35. Nowicki, P.; Wachowska, H.; Pietrzak, R. Active carbons prepared by chemical activation of plum stones and their application in removal of NO2. *J. Hazard. Mater.* **2010**, *181*, 1088–1094. [CrossRef] [PubMed]
36. Nowicki, P.; Pietrzak, R. Carbonaceous adsorbents prepared by physical activation of pine sawdust and their application for removal of NO2 in dry and wet conditions. *Bioresour. Technol.* **2010**, *101*, 5802–5807. [CrossRef] [PubMed]
37. Gadipelli, S.; Guo, Z.X. Graphene-based materials: Synthesis and gas sorption, storage and separation. *Prog. Mater. Sci.* **2015**, *69*, 1–60. [CrossRef]

*energies*

MDPI

*Article*

# Green Carbon Composite-Derived Polymer Resin and Waste Cotton Fibers for the Removal of Alizarin Red S Dye

**Béchir Wanassi** [1,2], **Ichrak Ben Hariz** [3], **Camélia Matei Ghimbeu** [2,*] , **Cyril Vaulot** [2] **and Mejdi Jeguirim** [2,*]

[1]  Laboratoire du Génie Textile, LGTex, Université de Monastir, Ksar Hellal 5078, Tunisia; wanassi_b@yahoo.fr
[2]  Institut de Science des Matériaux de Mulhouse, UMR 7361 CNRS, UHA, 15, rue Jean-Starcky, 68057 Mulhouse, France; cyril.vaulot@uha.fr
[3]  Société Tunisienne des Industries de Raffinage, P8, 7021 Zarzouna, Tunisia; benhriz.ichrak@stir.com.tn
*   Correspondence: camelia.ghimbeu@uha.fr (C.M.G.); mejdi.jeguirim@uha.fr (M.J.);
    Tel.: +33-(0)3-89-60-87-43 (C.M.G.); +33-(0)3-89-60-86-61 (M.J.)

Academic Editor: Vijay Kumar Thakur
Received: 8 August 2017; Accepted: 25 August 2017; Published: 1 September 2017

**Abstract:** Phenolic resin and waste cotton fiber were investigated as green precursors for the successful synthesis using a soft template approach of a composite carbon with carbon nanofibers embedded in a porous carbon network with ordered and periodically pore structure. The optimal composite carbon (PhR/NC-1), exhibited a specific surface area of $394 \text{ m}^2 \cdot \text{g}^{-1}$ with the existence of both microporosity and mesoporosity. PhR/NC-1 carbon was evaluated as an adsorbent of Alizarin Red S (ARS) dye in batch solution. Various operating conditions were examined and the maximum adsorption capacity of $104 \text{ mg} \cdot \text{g}^{-1}$ was achieved under the following conditions, i.e., $T = 25 \,^{\circ}\text{C}$, $pH = 3$, contact time = 1440 min. The adsorption and desorption heat was assessed by flow micro-calorimetry (FMC), and the presence of both exothermic and endothermic peaks with different intensity was evidenced, meaning a partially reversible nature of ARS adsorption. A pseudo-second-order model proved to be the most suitable kinetic model to describe the ARS adsorption according to the linear regression factor. In addition, the best isotherm equilibrium has been achieved with a Freundlich model. The results show that the eco-friendly composite carbon derived from green phenolic resin mixed with waste cotton fibers improves the removal of ARS dye from textile effluents.

**Keywords:** green precursor; waste cotton; phenolic resin; anionic dye; adsorption

## 1. Introduction

Industrial activity development is unfortunately usually accompanied by pollution concerns [1,2]. This pollution is currently a major threat, negatively affecting human life and the environment [3]. Wastewater constitutes a major part of industrial waste [4,5] and textile effluents are considered some of the major polluting aqueous effluents due to their content of significant amounts of toxic dyes and auxiliary chemicals [6]. The water consumption for dyeing one kg of textile is around 70 L with 40% yield [7,8]. Therefore, as a consequence, the decolorization of textile wastewater has become required worldwide [9]. During the last decades, several studies were performed for the treatment of textile effluents using biological and chemical treatment methods. Biological methods are extensively applied in the textile industry [10] owing to their benefits such as low cost and ecofriendly concept, however, these treatments do not always meet the objectives due to the non-biodegradability of a wide range of textile dyes [11]. Chemical treatments are the most widely used in the decolorization of textile effluents owing to its ease of application [12,13]. These methods are usually applied at high pH values using ozone, peroxide or permanganate as oxidizing agents. Nevertheless, the high cost of these techniques and the instability of the oxidizing agents are significant drawbacks. Activated carbon (AC) was also widely used in the decolorization of textile effluents. Commercial or synthesized ACs have been used in the retention of dye from wastewater thanks to their textural properties (physical adsorption) [14] or chemical surface properties (chemical adsorption) [15].

Cotton fiber, a lignocellulosic biomass, has been investigated to produce bio-sourced AC. Classical techniques were used to improve the specific surface area and the porosity of carbon materials derived from cotton fibers. Duan et al. [16] used a mix of chemical and physical activation with $H_3PO_4$ and microwave treatment to develop the surface area of activated carbon fibers (ACF). Zheng et al. [17] employed an acidic pretreatment of waste cotton woven as precursor of ACF which was chemically activated. However, the use of chemical or physical activation generates a carbon material with heterogeneous and random pore size distributions which can limit its adsorption capacity for specific application. It has been demonstrated that the heterogeneous distributions of both micro- and mesopore size of conventional carbon material affects its adsorption ability [18]. To overcome this problem it is necessary to obtain carbon materials with controlled pore architectures in order to improve their adsorption capacity. There are two main types of template approaches—hard and soft-templates—which allow one to obtain carbon materials with controlled pore structures. The first attempt to prepare controlled porous carbon through the hard-template method was investigated by Kyotani et al. [19]. The channels of a Y zeolite were used as a hard template and the Brunauer-Emmett-Teller (BET) surface area of the obtained carbon exceeded 2000 $m^2 \cdot g^{-1}$. In other studies different forms of zeolitic frameworks like ammonium-form zeolite Y, mordenite and ZSM-5 were investigated, showing promising potential to produce suitable carbon material architectures for a wide range of application [20–22]. However, the multiple step synthesis reactions along with the harsh conditions used to remove the hard-template have driven researchers to look for more convenient routes for carbon preparation.

In this regard, the use of the soft-template route as an alternative has become more suitable to produce controlled porous carbons. The use of soft templates presents many advantages like the reduction of the number of synthesis steps and the facility of removal of the templates by simple thermal annealing. Ghimbeu Matei et al. recently proposed a green approach to produce carbon materials with ordered architecture pores and controlled pore size distributions using environmentally friendly phenolic-resin precursors [23]. He used melamine to produce honeycomb porous carbons for electrode material production [24]. Shu prepared a nanoporous carbon for vanadium-based active electrode materials using polyvinylpyrrolidone (PVP) as soft template [25]. Ma et al. elaborated a mesoporous carbon composite through a soft-template route for electrochemical methanol oxidation [26]. Balach et al. employed a cationic polyelectrolyte as soft template to synthetize a porous carbon with a hierarchical nanostructure [27]. In general, the soft-template approach consists in the use of specific molecules to produce carbon materials with desired porosity. During a calcination process these molecules are degraded, resulting in a regular pore size distribution [28–30].

Recently, many research studies were focused on the utilization of bio-sourced AC as eco-friendly and low cost adsorbents for dyes from textile effluent. Regdi et al. used *Persea americana* as a precursor of an AC to remove cationic dye from textile wastewater [31]. Macedo et al. synthesized a mesoporous activated carbon derived from coconut coir to remove Remazol Yellow dye from textile effluent [32]. Noorimothagh used an activated carbon prepared from Iranian milk vetch for the adsorption of Acid Orange 7 dye using a batch flow mode experimental reactor [33]. Georgin et al.'s [34] studies focused on the elaboration of activated carbon through a microwave irradiation-pyrolysis using peanut shell raw precursor which was further used as an adsorbent of Direct Black 38 organic dye.

The utilization of composite materials based on AC in the adsorption of dyes from textile effluents has received significant attention in recent years. Carbon composites were investigated as new potential materials that can increase the AC adsorption capacity. Singh et al. investigated the adsorption of Acid Red 131 dye using a TiO$_2$-activated carbon (TiO$_2$/AC) nanocomposite prepared through a sol-gel process [35]. The adsorption behavior of Alizarin Red S (ARS) in aqueous solution was investigated by Fayazi et al. by means of an activated carbon/$\gamma$-Fe$_2$O$_3$ nanocomposite [36]. Sandeman et al. examined the adsorption of anionic Methyl Orange, cationic Methylene Blue and Congo Red using a porous poly(vinyl alcohol)/AC composite [37]. All these works highlight that the adsorption behavior of carbon materials is influenced by several parameters such as their surface chemical groups and their reactivity [38,39], textural characteristics [40] and the composition of the lignocellulosic precursors [41,42].

The main purpose of this work was to optimize the production of an eco-friendly composite carbon using a green phenolic resin as matrix and waste cotton fiber as reinforcement. The composite carbon morphology and textural properties were analyzed. Then, the optimized composite carbon was used to effectively remove ARS from aqueous solution. The impact of the experimental conditions such as temperature, pH, contact time, and initial dye concentration on the adsorption capacity was studied. The interactions of carbon composite network with the ARS dye were assessed by flow micro-calorimetry (FMC).

## 2. Materials and Methods

### 2.1. Phenolic Resin/Cotton Composite Carbon Preparation

Activated carbon (NC) was synthetized from waste cotton non-woven using the procedure as reported in our previous study [43]. This material was used for comparison purposes. For the preparation of phenolic resin, phloroglucinol (1,3,5-benzenetriol, C$_6$H$_6$O$_3$), glyoxylic acid monohydrate (C$_2$H$_2$O$_3$·H$_2$O), Pluronic F127 triblock copolymer [poly(ethylene oxide)-block–poly (propylene oxide)-block–poly(ethylene oxide), PEO$_{106}$-PPO$_{70}$-PEO$_{106}$, M$_w$ = 12,600 Da], and absolute ethanol (C$_2$H$_5$OH) were obtained from Sigma-Aldrich (Lyon, France). Briefly, the carbon precursors, phloroglucinol (0.41 g) and glyoxylic acid monohydrate (0.30 g) and the Pluronic F127 soft-template (0.80 g) were dissolved in a mixture of ethanol and distilled water in the proportion of 50/50 (10 mL/10 mL). The obtained solution was stirred until it became transparent [23].

The preparation of the different composites was done by adding an appropriate quantity of non-woven waste cotton into 20 mL of resin solution in a rectangular smooth glass mold. The obtained composite was left for eight hours at room temperature in a fume hood in order to completely evaporate the ethanol/water solvent. Subsequently, the composite material received a thermo-polymerization treatment in an oven for 8 h at 80 °C, followed by 16 h at 150 °C in order to thermo-polymerize the phenolic resin. After this treatment, the sample was introduced in a horizontal oven and pyrolysed under a flow of argon by increasing the temperature from room temperature to 700 °C (heating rate of 5 °C/min, dwell time of 1 h).

As summarized in Table 1, two composites and two references samples were prepared. NC refers to the carbon with 100 wt % cotton, PhR/NC-1 composite contains 30 wt % of phenolic resin, PhR/NC-2 composite contains 70 wt % of phenolic resin and PhR was phenolic resin-derived carbon.

**Table 1.** Sample denomination and ratio content of phenolic resin and cotton composites.

| Material Name | Ratio of Phenolic Resin | Ratio of Cotton |
|:---:|:---:|:---:|
| NC | 0% | 100% |
| PhR/NC-1 | 30% | 70% |
| PhR/NC-2 | 70% | 30% |
| PhR | 100% | 0% |

### 2.2. Material Characterization

Various techniques were employed to characterize the raw materials and their composites. The thermal degradation behavior of the raw materials under an inert atmosphere was investigated using a thermo-gravimetric analyzer (TGA 851e, Mettler-Toledo, Columbus, OH, USA). A sample of

5 mg of was placed in an alumina crucible and a heating rate of 5 °C·min$^{-1}$ up to 800 °C under inert gas flow (nitrogen) was applied.

The materials' morphology was analyzed by scanning electron microscopy (FEI model Quanta 400 SEM, Philips, Andover, MA, USA).

Transmission electron microscopy (TEM) was used in order to determine the structures of the carbon materials. A small and representative sample was tested with an ARM-200F instrument working at 200 kV (Jeol, Peabody, MA, USA).

The PhR and PhR/NC composite textural properties were investigated by recording the $N_2$ adsorption isotherms at 77 K with an ASAP 2020 instrument (Micromeritics, Atlanta, GA, USA). Previous to analysis, carbon material was out-gassed overnight under vacuum at 623 K. The BET surface area ($S_{BET}$) of PhR/NC composite carbon was determined from $N_2$ adsorption isotherms in relative pressure ($P/P_0$) range of 0.05–0.30. The micropore surface ($S_{mic}$) area and the micropore volume ($V_{micro}$) were investigated by the *t*-plot method. The average pore diameter ($D_p$) was obtained according to the Density Functional Theory (DFT) method.

### 2.3. Batch Adsorption Tests

Adsorption tests were performed using ARS as a representative dye used in textile industry. The ARS was purchased from Sigma Aldrich (Lyon, France). Its chemical structure is shown in Figure 1. For all experiments, the appropriate amount of ARS dye was dissolved in distilled water to obtain solutions with distinct concentrations. Adsorption tests were performed using a shaking thermostat water bath at a constant agitation speed of 200 rpm. The effect of the experimental conditions on the removal of ARS was studied by varying the pH of solution (3–8), the dye initial concentration (5–200 mg·L$^{-1}$), the temperature of solution (18–40 °C) and the contact time (5–1440 min). The ARS concentration was measured with a double beam UV-visible spectrophotometer. All experiments were repeated three times to ensure good reproducibility. The ARS adsorption capacity was determined according to the Equation (1):

$$q\left(mg\cdot g^{-1}\right) = \frac{C_i - C_e}{M}\cdot V \tag{1}$$

where: q (mg·g$^{-1}$) is the carbon adsorption capacity, $C_i$ (mg·L$^{-1}$) represent the initial ARS concentration, $C_e$ (mg·L$^{-1}$) is the equilibrium concentration of ARS, V (L) is the volume of initial ARS solution while M (g) is the mass of carbon adsorbent.

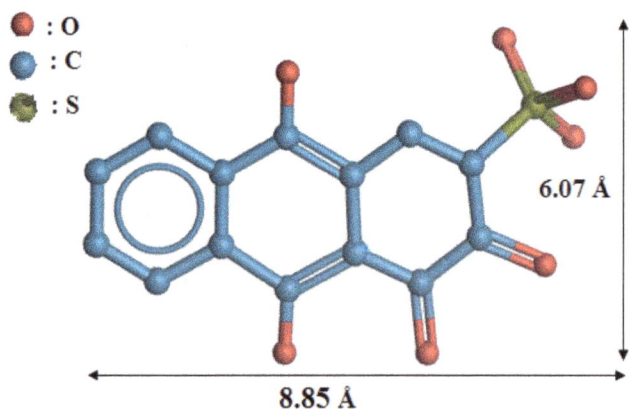

**Figure 1.** Molecular structure and sizes of Alizarin Red S (ARS) calculated with molecular modeling system Mol. Browser 3.8.

*2.4. Calorimetric Experiments*

Calorimetric adsorption is a robust method for the thermodynamic characterization of liquid-solid interfacial phenomena, as well as for the characterization of the surface of carbonaceous materials. For the calorimetric study, a Microscal 3Vi Flow Micro-Calorimeter (FMC, London, UK) coupled with UV-visible spectrometer (SPD 20 A, Shimadzu, Kyoto, Japan) developed by Microscal Ltd. (London, UK) was used. The calorimetric cell (0.17 cm$^3$) was delimited by two PTFE connectors allowing the supply and discharge of the solvent flow. It was filled in the most compact way possible with the carbon adsorbate. A permanent and controlled flow of solvent was carried out and the solution was injected into the flow via an injection loop of determined volume.

Carbon material (34 mg) was previously separated and placed in the cell for 15 min under vacuum. Then additional material was used to meet the filling and compacting conditions. A buffer solution (pH = 3) at a flow rate of 3.3 mL·h$^{-1}$ was employed as vector solvent. The probe molecule used in this experiment was a solution of ARS (10 g·L$^{-1}$) in the same buffer solution as the vector fluid. The experiments were performed at room temperature A UV-Visible downstream detector was used at a wavelength of λ = 339 nm corresponding to absorption maximum of ARS. Firstly the sample was placed under buffer solution flow (3.3 mL·h$^{-1}$) in order to obtain equilibrium. The probe molecule (ARS) is injected in the flow thanks to a determined-volume loop. When the probe molecules (ARS) come into contact with the surface of the adsorbent (carbon material: NC and PhR/NC1), there will be a release of heat and the variation of adsorption enthalpy was measured. The heat detector is based on a Wheatstone bridge system with 2 thermistors (with sensitivity of 10$^{-5}$ °C) and reference resistors. A heat change will make their electrical resistance vary, which will temporarily destabilize the Wheatstone bridge until extinction of the thermal effect. This destabilization will be observed by a change of the equilibrium electrical current (I = 0 without heat change) or electrical power in the Wheatstone bridge detector (P = U·I with a constant voltage).

FMC calibration has been realized via the resistance calibration (Pt-100). The total heating change during thermal phenomena corresponds to the integrated change in heating power. By varying the power and the emission time (E = P·Δt), we can simulate the energy release and calibrate the FMC. The uncertainly measurement of adsorption energy was estimated at ±3 mJ.

At the exit of FMC, the non-adsorbed part or/and desorbed part of ARS was measured by the associated UV-Visible spectrometer at λ = 339 nm. The calibration of the UV-Visible detector is realized by a series of loops directly connected to the apparatus and placed under the same flow conditions.

## 3. Results and Discussion

*3.1. Morphology and Structure*

After the carbonization process, the structure of obtained carbon materials can vary according to their initial composition. NC carbon, containing 100% of cotton fibers, was composed by a web of randomly oriented carbonized fibers (Figure 2a). When the precursor was the mix of cotton fibers and phenolic resin the surface structure of the obtained carbon composite changed. It can be clearly seen that the phenolic resin coats the cotton fiber and fills up the cavities in the fiber web (Figure 2b,c). The distribution of phenolic resins becomes more homogeneous as the ratio of phenolic resin increases. For PhR/NC-2 (Figure 2c) the distribution is more homogeneous and occupies more cavities in the cotton web than PhR/NC-1 (Figure 2b) composite carbon which the resin ratio is 70% and 30%, respectively. When only phenolic resin is used, no specific morphology was observed, the material is block-like and the particle size depends on the grinding conditions (data not shown).

Considering that an important component of the phenolic resin was the Pluronic F127 used as template, its increasing ratio favors the formation of ordered mesoporous structures around the carbonized cotton fibers. The structure of the carbonized phenolic resin zone was investigated in more detail by TEM. By direct observation, it can be seen clearly that the pure phenolic resin carbon (PhR) exhibit a highly ordered mesopore structure organization with hexagonal arrays (Figure 3A).

This result was in accord with previous studies [23] concerning the use of a similar precursor. However, in the presence of both carbonized cotton fibers and phenolic resin, the pore organization become less ordered as the ratio of phenolic resin decreased. Indeed, in the presence of only 30 wt % of phenolic resin (PhR/NC-1) in the carbon material, a pore array discontinuity appears and the carbon surface is composed by a mix of fiber surfaces and organized pore distribution zones caused by the template (Figure 3C). This particular structure organization was due to the fact of the fibers were somehow connected through a mesoporous carbon network. It is worth mention that the carbon fibers cannot be observed by TEM due to the limitations of this technique for thick materials.

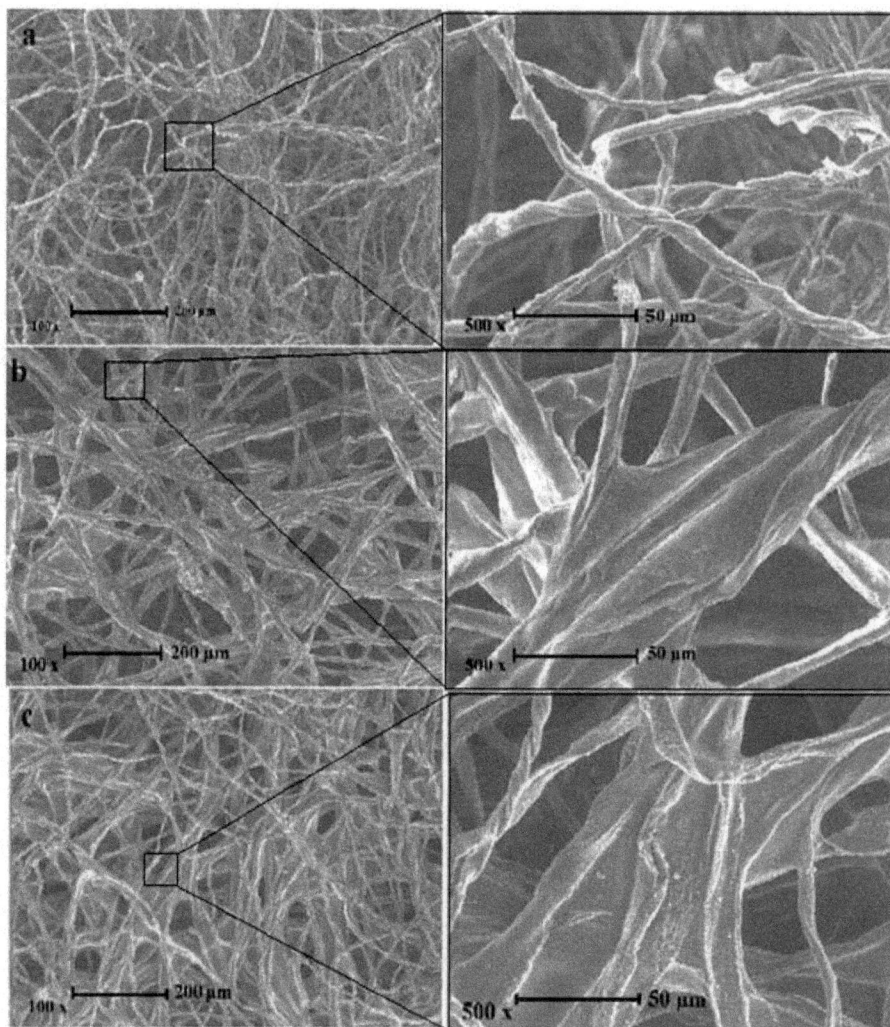

**Figure 2.** SEM images of synthesized carbon materials: (**a**) NC; (**b**) PhR/NC-1 and (**c**) PhR/NC-2.

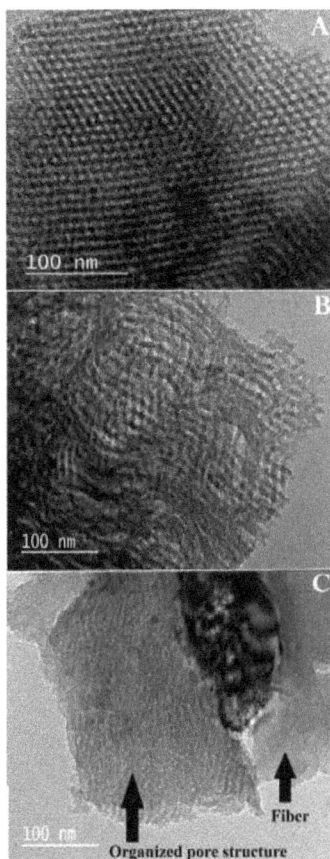

**Figure 3.** TEM pictures: (**A**) PhR; (**B**) PhR/NC-1 and (**C**) PhR/NC-2.

## 3.2. Texture/Porosity

The $N_2$ adsorption/desorption isotherms plot of carbon materials prepared with different ratios of phenolic resin are presented in Figure 4a. An increase of the quantity of nitrogen adsorbed is observed with the increase of the phenolic resin quantity in the composite material. The volume of micropores along with increasing phenolic resin proportion (Figure 4b) is also higher in the case of PhR carbon which suggests that the phenolic resin present in the composite carbon PhR/NC-1 and PhR/NC-2 favors micropores formation. At the relative pressure range of 0.98–1.00, the isotherm of PhR/NC-1 and PhR/NC-2 composite presents a high adsorbed volume compared with NC [43] carbon, highlighting an improvement in the pore volume.

It should be remarked that an increase of the phenolic resin ratio induces a change of isotherm type from the- type-I specific to microporous materials (the case of PhR/NC-1 carbon) to a mixture of type-I and type-IV, characteristic of micro and mesoporous materials (the case of both PhR/NC-2 and PhR carbon). Moreover, if phenolic resin was added to cotton fiber the hysteresis loop becomes H-1 type. This hysteresis type was probably due to the specific distributions of pore size exhibiting a narrow organization of relatively uniform (geometrical) pores. H-1 hysteresis is known as one of the intrinsic characteristics of regular pore structure materials [44,45]. In a previous study, a similar hysteresis has been shown in the presence of template in the phenolic resin and when the precursor mass exceeded the mass of template [23].

The pore size distribution (Figure 4b) shows for PhR/NC-1 materials only the presence of micropores (size < 2 nm). This is related to the low quantity of phenolic resin (30%) compared to carbon fibers (70%). This is not surprising taking into consideration that the pure NC carbon is predominantly microporous (see Table 2). For the PhR/NC-2 composite, the increase in the phenolic resin content up to 70% allows the microporosity to increase and in addition the creation of mesoporosity. The mesopore size distribution (Figure 4b) presents a maximum at ~10 nm. When only phenolic resin is used (PhR) the microporosity and mesoporosity further increase, and the pore size as well, reaching a maximum at 17 nm (Figure 4b). Therefore, mixing of phenolic resin with cotton fiber as carbon precursor generates a mesoporous structure and improves the microporosity, due mainly to the phenolic resin incorporation.

The effect of resin ratio on the textural features of the obtained carbon material has been evaluated. It can be seen that when the ratio of phenolic resin increases from 30% to 70%, the BET surface area and the total pore volume ($V_t$) of PhR/NC-1 and PhR/NC-2 composite carbon increase from 394 $m^2 \cdot g^{-1}$ to 436 $m^2 \cdot g^{-1}$ and from 0.14 $cm^3 \cdot g^{-1}$ to 0.18 $cm^3 \cdot g^{-1}$, respectively (Table 2). The increase in the specific surface area of carbon may be related to the presence of phenolic resin (which has a high specific surface area of 591 $m^2 \cdot g^{-1}$) which covers and interconnects the cotton fibers. Indeed, the specific surface area of the composites was a mixture between that of the resin and the fiber. A compromise between the BET surface area of the materials and the amount of phenolic resin used for economic and environmental reasons, the study will be limited to the use of PhR/NC-1 (which contains 30% of phenolic resin) for the adsorption tests.

**Table 2.** Textural characteristics of carbon materials determined from nitrogen sorption isotherms.

| Unit | $S_{BET}$ | $S_{mic}$ | $V_{mic}$ | $V_{me}$ | $V_t$ | $D_p$ |
|---|---|---|---|---|---|---|
| | $m^2 \cdot g^{-1}$ | $m^2 \cdot g^{-1}$ | $cm^3 \cdot g^{-1}$ | $cm^3 \cdot g^{-1}$ | $cm^3 \cdot g^{-1}$ | nm |
| NC [43] | 292 | 255 | 0.11 | 0.03 | 0.14 | <2 nm |
| PhR/NC-1 | 394 | 526 | 0.24 | 0.03 | 0.18 | <2 nm |
| PhR/NC-2 | 435 | 403 | 0.16 | 0.24 | 0.40 | <2 nm + 10 nm |
| PhR | 591 | 628 | 0.25 | 0.50 | 0.69 | >2 nm + 17 nm |

$S_{BET}$—is the surface area determined by the BET method, $V_{meso}$—is the mesopore volume, $S_{mic}$—is the micropore surface, $V_{mic}$—is the micropore volume, $V_t$—is the total pore volume and $D_p$—is the pore diameter determined according to DFT model.

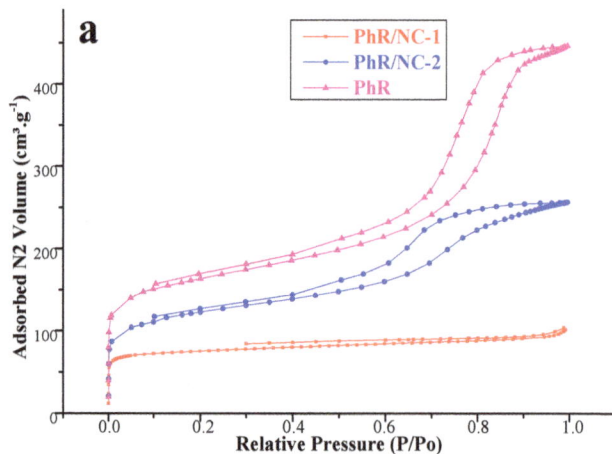

**Figure 4.** *Cont.*

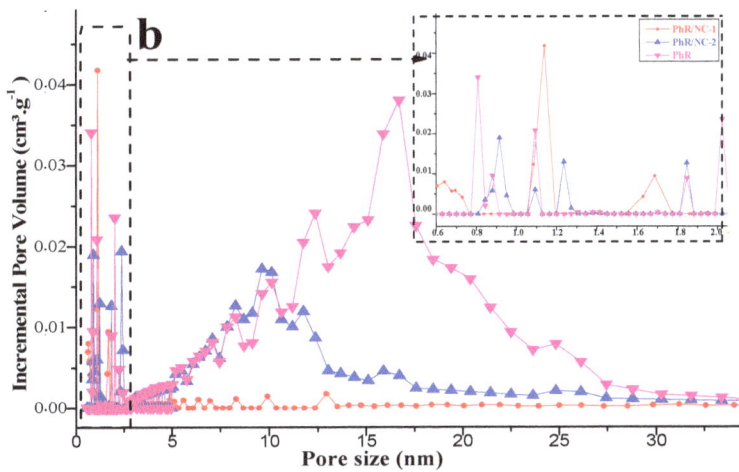

**Figure 4.** (**a**) Nitrogen adsorption/desorption isotherms of carbon materials; (**b**) Density Functional Theory (DFT) Pore size distribution.

### 3.3. TGA Results

The TGA curves and their corresponding derivatives curves of PhR/NC-1non thermally treated composite compared with PhR and NC as reference materials are shown in Figure 5. Below 100 °C a small weight loss for all materials corresponding to desorption of water, is noticed.

In the range of 100–450 °C, an intense weight loss of PhR/NC-1 is observed. This may be attributed to the pyrolysis of both cotton and phenolic resin and the remove of template which was previously demonstrated to occur around 400 °C [23]. Indeed, cotton (NC) and phenolic resin (PhR) show also the highest weight loss in this range. On the other hand, the DTG curve of PhR/NC-1 shows only one peak, however the PhR polymer exhibits two well-defined peaks at 244 °C and 375 °C. This suggests an intimate mixture between the two pristine materials. Thus, the mixing of phenolic resin with cotton fibers can decrease the weight loss of composite on this range of temperature. A similar observation was shown by Cho et al. where the mix of phenolic resin with lignocellulosic fiber (bamboo fiber) can improve its carbon yield at high temperature [46].

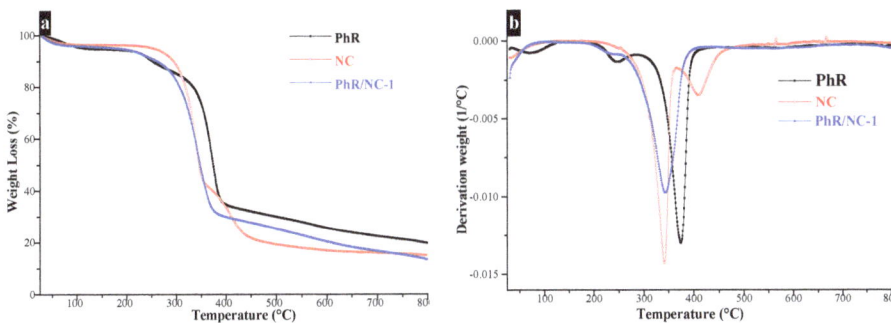

**Figure 5.** (**a**) Thermo gravimetric analysis of individual precursors (PhR and NC [43]) and the composite PhR/NC-1; (**b**) their corresponding derivative weight loss curves.

*3.4. Influence of Adsorption Parameters on the Adsorption Capcity*

Adsorption phenomena depend on several factors such as contact time, pH, temperature and initial concentration of dye. Figure 6 shows the effect of those parameters on the adsorption capacity of ARS by PhR/NC-1 composite material. It can be seen that when both initial pH and the temperature of adsorption reaction increase, the adsorption capacity decrease (Figure 6a,b). However, when the initial ARS concentration increases the adsorption capacity increases. It should be remarked that in all the cases, the adsorption capacity of PhR/NC-1 was higher than that of NC [43] carbon as reference, which was 104 mg/g and 77 mg/g, respectively. This may be related to the improved textural properties (specific surface area and pore volume) of composite material compared to NC material.

The pH of ARS solution induces an important effect on adsorption capacity of composite carbon. Indeed, when the pH increases from 3 to 8, the PhR/NC-1 adsorption capacity decreases from 84.8 to 13.4 mg·g$^{-1}$. This result suggests that the adsorption capacity of composite carbon was closely depend on the electrostatic charge and the degree of ionization.

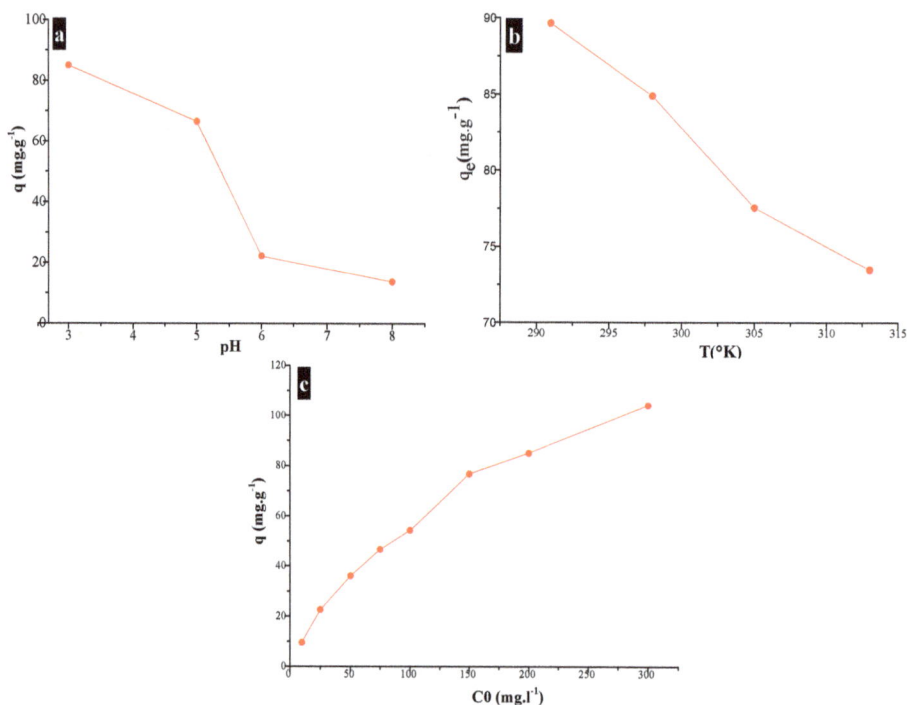

**Figure 6.** Influence of adsorption parameters on the adsorption capacity of PhR/NC-1: (**a**) effect of initial pH; (**b**) effect of temperature and (**c**) effect of initial concentration of ARS.

As regards the effect of temperature, a major difference between the adsorption capacities of PhR/NC-1 carbon is noticed. With an increase of temperature, the adsorption capacity of the composite carbon becomes less favorable.

With the increasing value of initial ARS concentration, there is a generation of more and more force that favors the mass transfer of ARS molecule into PhR/NC-1 carbon (Figure 6c).

In summary, the interaction between ARS molecules and NC or PhR/NC-1 composite shows an important dependence on the electrostatic energy and the degree of ionization that was affected by the pH of the dye solution. The initial dye concentration and the temperature of reaction have a direct impact on the adsorption capacity of the composite carbons. The results were in agreement with previously reported data on ARS adsorption by Fayazi et al. [36]. Thaking into account the obtained results, ARS adsorption at 25 °C, pH 3 and with initial concentration of 200 mg·L$^{-1}$ was selected as optimum adsorption conditions for further investigation.

### 3.4.1. Kinetic Model

Kinetics models have been widely investigated to describe the adsorption of textile dyes by activated carbons. The Lagergren pseudo-first-order [47], pseudo-second-order [48] and intra-particle models [49] were used in this study to evaluate the adsorption of ARS on PhR/NC-1 as composite carbon. The respective equations are reported in Table 3, where $q_e$ and $q_t$ were the adsorption capacity of PhR/NC-1 at equilibrium and during a contact time t. $K_1$ and $K_2$ are the pseudo-first-order and pseudo-second-order kinetics constants, respectively. $K_{id}$ was the intra-particle diffusion constant. All these constants are summarized in Table 3. The correlation coefficient ($R^2$) was calculated in order to evaluate the correlation between predicted and experimental data.

The obtained correlation coefficients of the pseudo-first-order, pseudo-second-order and intra-particle-diffusion are 0.95, 0.99 and 0.96, respectively (Table 3). This results suggest that both pseudo-first-order and intra-particle-diffusion models have a poor fit to the experimental data. However, the corresponding linear correlation coefficient $R^2$ of pseudo-second-order model is close to the unit ($R^2 = 0.99\sim1$), which suggest that this kinetic model is the best fit of the experimental data. This result indicates that there is an electrostatic attraction between ARS as an acidic dye and PhR/NC-1 carbon through chemisorption phenomena. In our previous study [43], similar results were shown by using activated carbon as an adsorbent for ARS. As reported previously [50] a pseudo-second-order model was the main kinetic model that fitted the adsorption of ARS dye by activated carbon.

**Table 3.** Kinetic parameters of ARS adsorption.

| Kinetic Model | Pseudo-First-Order | | Pseudo-Second-Order | | Intra-Particle-Diffusion | |
|---|---|---|---|---|---|---|
| Equation | $\log(q_e - q_t) = \log(q_e) - \frac{k_1}{2.303} t$ [31] | | $\frac{t}{q_t} = \frac{1}{k_2 q_{eq}} + \frac{1}{q_{eq}} t$ [32] | | $q_t = K_{id} \cdot t^{\frac{1}{2}} + C$ [33] | |
| Parameters | $q_e$ (mg·g$^{-1}$) 69.57 | $K_1$ (min$^{-1}$) 12 × 10$^{-3}$ | $q_e$ (mg·g$^{-1}$) 100.71 | $K_2$ (L·mg$^{-1}$·min$^{-1}$) 5.40 × 10$^{-5}$ | $K_{id}$ (mg·g$^{-1}$·min$^{-0.5}$) 0.025 | C 6.23 |
| $R^2$ | 0.95 | | 0.99 | | 0.96 | |

### 3.4.2. Adsorption Isotherms of Synthesized Carbon

Experiments were performed at 25 °C, pH 3 and with initial dye concentrations of ARS between 5 and 300 mg·L$^{-1}$. According to the linear regression factor $R^2$ (Table 4), both Langmuir and Freundlich models show good agreement with the experimental results. However, the Langmuir model was the most suitable to fit experimental data ($R^2 = 0.987$). Such a model exhibited a maximum adsorption capacity of $q_m = 132.63$ mg·g$^{-1}$. The best fit of the Langmuir model as the isotherm model in the present process shows that ARS molecule was attached to binding sites of PhR/NC-1 by homogeneous levels of energy which means a monolayer adsorption phenomenon. In our previous study about the adsorption of ARS by AC derived from cotton waste [43], the Langmuir isotherm also showed the best agreement with the experimental data. As compared with the previous study, the Langmuir model was the most suitable mathematical isotherm model that describes both the adsorption of ARS by activated carbon [43] and the adsorption of organic dye by mesoporous ordered carbon [51]. The value of $K_L$ as Langmuir constant is equal to 0.012 L·mg$^{-1}$, which suggest that the adsorption of ARS by PhR/NC-1 was favorable and probably irreversible in nature. The 'n' value of the Freundlich model is 1.64, therefore laying between 1 and 2, meaning that the adsorption of ARS by PhR/NC-1 composite carbon was favorable.

**Table 4.** Isotherm parameters of ARS adsorption.

| Model | Equation | Parameters | | R$^2$ |
|---|---|---|---|---|
| Langmuir | $\frac{C_e}{q_e} = \frac{1}{q_m} \cdot C_e + \frac{1}{K_L \cdot q_m}$ [35] | $q_m$ (mg·g$^{-1}$) 132.630 | $K_L$ (L·mg$^{-1}$) 0.012 | 0.987 |
| Freundlich | $\log(q_e) = \log(K_F) + \frac{1}{n} \cdot \log(C_e)$ [36] | $K_f$ (mg$^{(1-n)}$ L$^n$·g$^{-1}$) 4.030 | $n$ 1.644 | 0.985 |

### 3.5. Calorimetric Results

The release of heat of ARS adsorption is reported in Figure 7 for NC and PhR/NC-1. It can be seen that the calorimetric signal of composite carbon (PhR/NC-1) presented an intense exothermic peak relatively finely resolved over time (between 17,640 and 18,000 s) followed by a much longer endothermic peak (18,000 to 18,900 s) with lower amplitude. The exothermic peak is attributed to the adsorption stage while the endothermic peak is attributed to the desorption of ARS from the surface of the carbonaceous material. The existence of an exothermic peak and an endothermic one, with different intensity, reflects that the adsorption of ARS is partially reversible for both carbons.

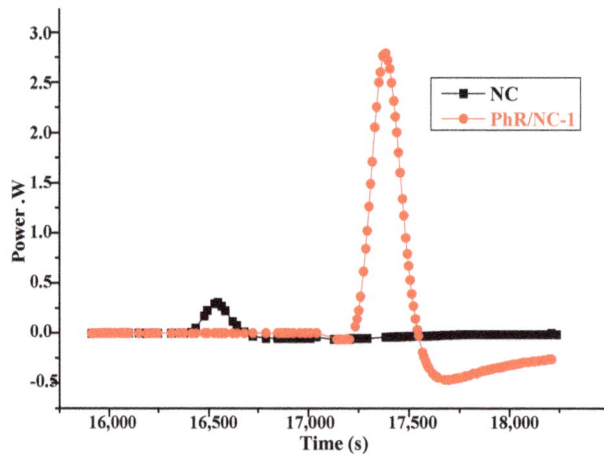

**Figure 7.** Heating power of the calorimeter during release of heat of ARS adsorption on NC carbon and PhR/NC-1 composite.

For equivalent times of the phenomena (FWHM = 149 and 182 s for composite and NC respectively), we can observe that the energy is higher with the composite carbon PhR/NC-1. This seems to indicate a higher interaction with the composite carbon than NC and higher kinetics with the first carbon fibers.

Several successive injections of ARS were realized and the cumulative behavior was followed. The FMC energy appears to depend also on the injected quantity of ARS for both carbons. This was obviously visible in Figure 8 which illustrates the evolution of the cumulative total energy as a function of cumulative amount of injected ARS. It can be seen that the cumulative energy increases with the increasing quantity of injected ARS. Moreover, this result confirms that the composite carbon exhibits more energy than NC. The difference between the total energy could be due to the presence of phenolic resin in the composite carbon which increases the number of contact points on the carbon composite surface, increases its polarity and promotes ARS-carbon interactions. The respective phases of the adsorption process can be described using the Hill model [52] as follows:

$$\log(\theta/(1-\theta)) = n \cdot \log[L] - n \cdot \log(K_d), \qquad (2)$$

where, $\theta$ is the degree of cooperatively, n is the Hill coefficient, L is the ligand concentration and $K_d$ is the microscopic association coefficient.

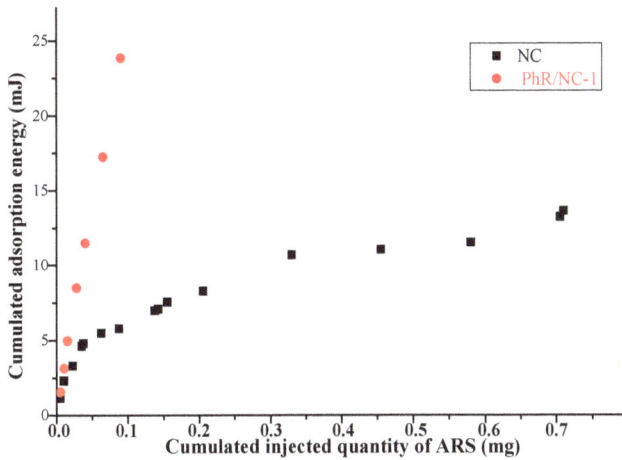

**Figure 8.** Effect of contact time on the adsorption capacity of ARS on NC and PhR/NC-1 carbons (initial concentration: 200 mg·L$^{-1}$, 25 °C, pH: 3).

At the adsorption equilibrium, the covering surface of adsorbent can be considered as directly proportional to the cumulative energy using an analogy between the concentration (L) and cumulative injected volume of ARS. Under this hypothesis, the Hill equation was modified as follows:

$$\log\left(\frac{\varepsilon}{1-\varepsilon}\right) = n \cdot \log(V_{cum}) - n \cdot \log(K_d), \qquad (3)$$

$$\varepsilon = \frac{E_{cum}}{E_{cum_{max}}}, \qquad (4)$$

where, $V_{cum}$ is the cumulative injected volume, $E_{cum}$ is the cumulative energy and $E_{cum_{max}}$ is the maximum value of the cumulative energy.

It can be seen that the Hill plots have a linear form (Figure 9). Given the fact the Hill coefficient (slope of the Hill plot) of the adsorption phase was greater than 1 (Table 5), the binding of one molecule of ARS promotes the binding of additional ones. This result suggests a perfectly free and independent adsorption of ARS on the surface of the composite carbon.

The desorption Hill coefficient was practically the same as the adsorption coefficient, meaning that the desorption phase involves increased affinity with ARS molecules. However, the desorption microscopic association coefficient ($K_d$ = 130) was less than that of adsorption ($K_d$ = 142), likely reflecting domination of ARS adsorption phase.

Total phase behavior with Hill coefficient was greater than 1 (*n* = 1.37, Table 5), but for the reaction between ARS and NC, the Hill coefficient was less than 1 (*n* = 0.79 < 1; Table 5). That is the reason why the cooperative adsorption was considered negative. The binding of one molecule of ARS inhibits supplementary binding of another one, suggesting that the reaction between ARS and NC was probably very limited and reversible (the system limits the adsorption and makes a large part of adsorbed ARS (or almost all the adsorbed molecules) to desorb.

From a quantitative perspective, ARS molecules show good affinity for composite carbon and at the same time higher than that for NC carbon. This result may be due to the contribution given by the presence of phenolic resin having higher porosity with interconnected pores in composite carbon.

The quantitative results were completed by an energy analysis, in order to investigate the energy potential between ARS molecules and carbon material.

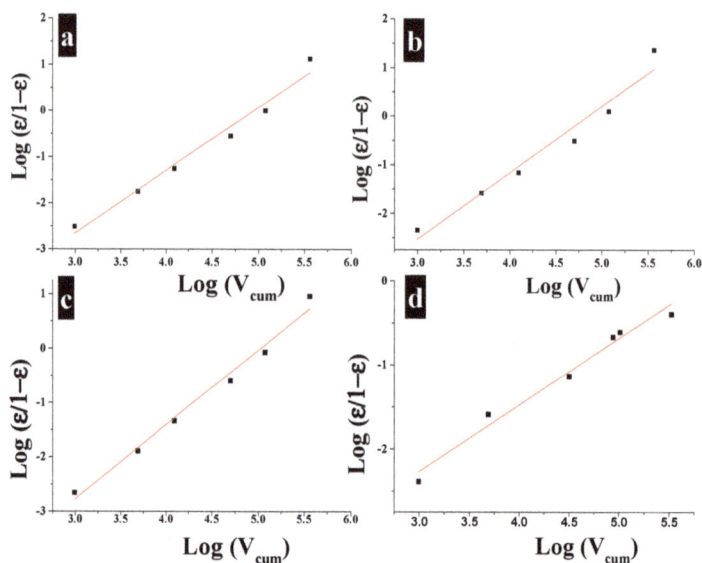

**Figure 9.** Hill plot: (**a**) adsorption of ARS by composite carbon; (**b**) desorption of ARS by composite carbon; (**c**) total phase between ARS and composite carbon; and (**d**) total phase between ARS and NC.

**Table 5.** Hill parameters.

| Material | Phase | $n$ | $R_d$ | $R^2$ |
|---|---|---|---|---|
| | Adsorption | 1.36 | 142.01 | 0.98 |
| PhR/NC-1 | Desorption | 1.36 | 130.01 | 0.97 |
| | Total | 1.37 | 152.81 | 0.99 |
| NC | Total | 0.79 | 359.99 | 0.98 |

## 4. Conclusions

An original eco-friendly composite was successfully synthesized in this study and employed as an absorbent of ARS dye from aqueous solution. A mix of green phenolic resin and needle-punched non-woven cotton waste was investigated and resulted in carbon fibers embedded in a mesoporous carbon network composite. The increasing ratio of phenolic resin in the composite precursor from 0% (NC as reference) to 70% causes a significant change on the morphology and mesoporosity arrangement, which become more homogenous and more organized. The surface area and the pore volumes are improved as well, resulting in better ARS adsorption capacities of composite materials compared to NC reference material. The increase of pH and temperature reduces the adsorption capacity of ARS but the increases in the initial concentration of ARS improve its adsorption capacity by the composite carbon. FMC analysis suggested an independent and free adsorption of ARS on the surface of the composite carbon. A pseudo-second-ordered model was the best fitted to the experimental kinetics while the Freundlich model was the most suitable model that describes the ARS adsorption isotherm.

*Energies* **2017**, *10*, 1321

In summary, a green composite carbon prepared by a simple approach show good ability to remove an anionic dye from textile effluent. This study may be completed by a study of the competitive adsorption of dyes.

**Acknowledgments:** We would like to thank Gautier Schrodj, Ludovic Josien and Loïc Vidal for the help provided with the TGA, TEM and SEM analyses.

**Author Contributions:** This study is the result of the cooperative research by all the authors. Béchir Wanassi and Ichrak Ben Hariz wrote the paper and performed together with Cyril Vaulot the experimental assays. Camélia Matei Ghimbeu and Mejdi Jeguirim supervised the work and the manuscript preparation. All authors have discussed and agreed to submit the manuscript.

**Conflicts of Interest:** The authors declare no conflict of interest.

## References

1. Zaharia, M.; Pătrașcu, A.; Gogonea, M.R.; Tănăsescu, A.; Popescu, C. A Cluster Design on the Influence of Energy Taxation in Shaping the New EU-28 Economic Paradigm. *Energies* **2017**, *10*, 257. [CrossRef]
2. Feng, Z.-K.; Niu, W.-J.; Zhou, J.-Z.; Cheng, C.-T.; Qin, H.; Jiang, Z.-Q. Parallel Multi-Objective Genetic Algorithm for Short-Term Economic Environmental Hydrothermal Scheduling. *Energies* **2017**, *10*, 163. [CrossRef]
3. Botkin, D.B.; Keller, E.A. *Environmental Science: Earth as a Living Planet*; Willey: Hoboken, NJ, USA, 2007.
4. Hamawand, I.; Ghadouani, A.; Bundschuh, J.; Hamawand, S.; Al Juboori, R.A.; Chakrabarty, S.; Yusaf, T. A Critical Review on Processes and Energy Profile of the Australian Meat Processing Industry. *Energies* **2017**, *10*, 731. [CrossRef]
5. Marcos, A.C.; Al-Kassir, A.; Cuadros, F.; Yusaf, T. Treatment of Slaughterhouse Waste Water Mixed with Serum from Lacteal Industry of Extremadura in Spain to Produce Clean Energy. *Energies* **2017**, *10*, 765. [CrossRef]
6. Meksi, N.; Moussa, A. A review of progress in the ecological application of ionic liquids in textile processes. *J. Clean. Prod.* **2009**. [CrossRef]
7. Rosa, J.M.; Fileti, A.M.F.; Tambourgi, E.B.; Santana, J.C.C. Dyeing of cotton with reactive dyestuffs: The continuous reuse of textile wastewater effluent treated by Ultraviolet/Hydrogen peroxide homogeneous photocatalysis. *J. Clean. Prod.* **2015**, *90*, 60–65. [CrossRef]
8. Allègre, C.; Moulin, P.; Maisseu, M.; Charbit, F. Treatment and reuse of reactive dyeing effluents. *J. Membr. Sci.* **2006**, *269*, 15–34. [CrossRef]
9. El-Tayeb, A.; El-Shazly, A.H.; Elkady, M.F. Investigation the Influence of Different Salts on the Degradation of Organic Dyes Using Non-Thermal Plasma. *Energies* **2016**, *9*, 874. [CrossRef]
10. Soares, P.A.; Souza, R.; Soler, J.; Silva, T.F.C.V.; Souza, S.M.A.G.U.; Boaventura, R.A.R.; Vilar, V.J.P. Remediation of a synthetic textile wastewater from polyester-cotton dyeing combining biological and photochemical oxidation processes. *Sep. Purif. Technol.* **2017**, *172*, 450–462. [CrossRef]
11. Holkar, C.R.; Jadhav, A.J.; Pinjari, D.V.; Mahamuni, N.M.; Pandit, A.B. A critical review on textile wastewater treatments: Possible approaches. *J. Environ. Manag.* **2016**, *182*, 351–366. [CrossRef] [PubMed]
12. Güyer, G.T.; Nadeem, K.; Dizge, N. Recycling of pad-batch washing textile wastewater through advanced oxidation processes and its reusability assessment for Turkish textile industry. *J. Clean. Prod.* **2016**, *139*, 488–494. [CrossRef]
13. Polat, D.; Balcı, İ.; Özbelge, T.A. Catalytic ozonation of an industrial textile wastewater in a heterogeneous continuous reactor. *J. Environ. Chem. Eng.* **2015**, *3*, 1860–1871. [CrossRef]
14. Peláez-Cid, A.-A.; Herrera-González, A.-M.; Salazar-Villanueva, M.; Bautista-Hernández, A. Elimination of textile dyes using activated carbons prepared from vegetable residues and their characterization. *J. Environ. Manag.* **2016**, *181*, 269–278. [CrossRef] [PubMed]
15. Belaid, K.D.; Kacha, S.; Kameche, M.; Derriche, Z. Adsorption kinetics of some textile dyes onto granular activated carbon. *J. Environ. Chem. Eng.* **2013**, *1*, 496–503. [CrossRef]
16. Duan, X.; Srinivasakannan, C.; Wang, X.; Wang, F.; Liu, X. Synthesis of activated carbon fibers from cotton by microwave induced $H_3PO_4$ activation. *J. Taiwan Inst. Chem. Eng.* **2017**, *70*, 374–381. [CrossRef]
17. Zheng, J.; Zhao, Q.; Ye, Z. Preparation and characterization of activated carbon fiber (ACF) from cotton woven waste. *Appl. Surf. Sci.* **2014**, *299*, 86–91. [CrossRef]

18. Marco-Lozar, J.P.; Juan-Juan, J.; Suárez-García, F.; Cazorla-Amorós, D.; Linares-Solano, A. MOF-5 and activated carbons as adsorbents for gas storage. *Int. J. Hydrogen Energy* **2012**, *37*, 2370–2381. [CrossRef]
19. Kyotani, T.; Nagai, T.; Inoue, S.; Tomita, A. Formation of New Type of Porous Carbon by Carbonization in Zeolite Nanochannels. *Chem. Mater.* **1997**, *9*, 609–615. [CrossRef]
20. Su, F.; Zhao, X.S.; Lv, L.; Zhou, Z. Synthesis and characterization of microporous carbons templated by ammonium-form zeolite Y. *Carbon* **2004**, *42*, 2821–2831. [CrossRef]
21. Kyotani, T.; Ma, Z.; Tomita, A. Template synthesis of novel porous carbons using various types of zeolites. *Carbon* **2003**, *41*, 1451–1459. [CrossRef]
22. Kyotani, T. Synthesis of Various Types of Nano Carbons Using the Template Technique. *Bull. Chem. Soc. Jpn.* **2006**, *79*, 1322–1337. [CrossRef]
23. Ghimbeu, C.M.; Vidal, L.; Delmotte, L.; Meins, J.-M.L.; Vix-Guterl, C. Catalyst-free soft-template synthesis of ordered mesoporous carbon tailored using phloroglucinol/glyoxylic acid environmentally friendly precursors. *Green Chem.* **2014**, *16*, 3079–3088. [CrossRef]
24. He, X.; Yu, H.; Fan, L.; Yu, M.; Zheng, M. Honeycomb-like porous carbons synthesized by a soft template strategy for supercapacitors. *Mater. Lett.* **2017**, *195*, 31–33. [CrossRef]
25. Shu, D.; Cheng, H.; Lv, C.; Asi, M.A.; Long, L.; He, C.; Zou, X.; Kang, Z. Soft-template synthesis of vanadium oxynitride-carbon nanomaterials for supercapacitors. *Int. J. Hydrogen Energy* **2014**, *39*, 16139–16150. [CrossRef]
26. Ma, C.; Kang, L.; Shi, M.; Lang, X.; Jiang, Y. Preparation of Pt-mesoporous tungsten carbide/carbon composites via a soft-template method for electrochemical methanol oxidation. *J. Alloys Compd.* **2014**, *588*, 481–487. [CrossRef]
27. Balach, J.; Tamborini, L.; Sapag, K.; Acevedo, D.F.; Barbero, C.A. Facile preparation of hierarchical porous carbons with tailored pore size obtained using a cationic polyelectrolyte as a soft template. *Colloids Surf. A Physicochem. Eng. Asp.* **2012**, *415*, 343–348. [CrossRef]
28. Wang, T.; Kailasam, K.; Xiao, P.; Chen, G.; Chen, L.; Wang, L.; Li, J.; Zhu, J. Adsorption removal of organic dyes on covalent triazine framework (CTF). *Microporous Mesoporous Mater.* **2014**, *187*, 63–70. [CrossRef]
29. Konicki, W.; Cendrowski, K.; Chen, X.; Mijowska, E. Application of hollow mesoporous carbon nanospheres as an high effective adsorbent for the fast removal of acid dyes from aqueous solutions. *Chem. Eng. J.* **2013**, *228*, 824–833. [CrossRef]
30. Shi, T.; Wen, Y.; Ma, C.; Jia, S.; Wang, Z.; Zou, S. Adsorption Characteristics of Phenol and Reactive Dyes from Aqueous Solution onto Ordered Mesoporous Carbons Prepared via a Template Synthesis Route. *Adsorpt. Sci. Technol.* **2009**, *27*, 643–659. [CrossRef]
31. Regti, A.; Laamari, M.R.; Stiriba, S.-E.; El Haddad, M. Potential use of activated carbon derived from Persea species under alkaline conditions for removing cationic dye from wastewaters. *J. Assoc. Arab Univ. Basic Appl. Sci.* **2017**. [CrossRef]
32. De Souza Macedo, J.; Da Costa Júnior, N.B.; Almeida, L.E.; da Silva Vieira, E.F.; Cestari, A.R.; de Fátima Gimenez, I.; Villarreal Carreño, N.L.; Barreto, L.S. Kinetic and calorimetric study of the adsorption of dyes on mesoporous activated carbon prepared from coconut coir dust. *J. Colloid Interface Sci.* **2006**, *298*, 515–522. [CrossRef]
33. Noorimotlagh, Z.; Darvishi Cheshmeh Soltani, R.; Khataee, A.R.; Shahriyar, S.; Nourmoradi, H. Adsorption of a textile dye in aqueous phase using mesoporous activated carbon prepared from Iranian milk vetch. *J. Taiwan Inst. Chem. Eng.* **2014**, *45*, 1783–1791. [CrossRef]
34. Georgin, J.; Dotto, G.L.; Mazutti, M.A.; Foletto, E.L. Preparation of activated carbon from peanut shell by conventional pyrolysis and microwave irradiation-pyrolysis to remove organic dyes from aqueous solutions. *J. Environ. Chem. Eng.* **2016**, *4*, 266–275. [CrossRef]
35. Singh, P.; Vishnu, M.C.; Sharma, K.K.; Borthakur, A.; Srivastava, P.; Pal, D.B.; Tiwary, D.; Mishra, P.K. Photocatalytic degradation of Acid Red dye stuff in the presence of activated carbon-$TiO_2$ composite and its kinetic enumeration. *J. Water Process Eng.* **2016**, *12*, 20–31. [CrossRef]
36. Fayazi, M.; Ghanei-Motlagh, M.; Taher, M.A. The adsorption of basic dye (Alizarin red S) from aqueous solution onto activated carbon/$\gamma$-$Fe_2O_3$ nano-composite: Kinetic and equilibrium studies. *Mater. Sci. Semicond. Process.* **2015**, *40*, 35–43. [CrossRef]

37. Sandeman, S.R.; Gun'ko, V.M.; Bakalinska, O.M.; Howell, C.A.; Zheng, Y.; Kartel, M.T.; Phillips, G.J.; Mikhalovsky, S.V. Adsorption of anionic and cationic dyes by activated carbons, PVA hydrogels, and PVA/AC composite. *J. Colloid Interface Sci.* **2011**, *358*, 582–592. [CrossRef] [PubMed]

38. Natalio, F.; Tahir, M.N.; Friedrich, N.; Köck, M.; Fritz-Popovski, G.; Paris, O.; Paschke, R. Structural analysis of Gossypium hirsutum fibers grown under greenhouse and hydroponic conditions. *J. Struct. Biol.* **2016**, *194*, 292–302. [CrossRef] [PubMed]

39. Pradhan, N.; Rene, E.R.; Lens, P.N.L.; Dipasquale, L.; D'Ippolito, G.; Fontana, A.; Panico, A.; Esposito, G. Adsorption Behaviour of Lactic Acid on Granular Activated Carbon and Anionic Resins: Thermodynamics, Isotherms and Kinetic Studies. *Energies* **2017**, *10*, 665. [CrossRef]

40. Garcia-Cuello, V.S.; Giraldo, L.; Moreno-Pirajan, J.C. Textural Characterization and Energetics of Porous Solids by Adsorption Calorimetry. *Energies* **2011**, *4*, 928–947. [CrossRef]

41. Román, S.; Ledesma, B.; Álvarez-Murillo, A.; Al-Kassir, A.; Yusaf, T. Dependence of the Microporosity of Activated Carbons on the Lignocellulosic Composition of the Precursors. *Energies* **2017**, *10*, 542. [CrossRef]

42. Qian, K.; Kumar, A.; Patil, K.; Bellmer, D.; Wang, D.; Yuan, W.; Huhnke, R.L. Effects of Biomass Feedstocks and Gasification Conditions on the Physiochemical Properties of Char. *Energies* **2013**, *6*, 3972–3986. [CrossRef]

43. Wanassi, B.; Hariz, I.B.; Ghimbeu, C.M.; Vaulot, C.; Hassen, M.B.; Jeguirim, M. Carbonaceous adsorbents derived from textile cotton waste for the removal of Alizarin S dye from aqueous effluent: Kinetic and equilibrium studies. *Environ. Sci. Pollut. Res.* **2017**, 1–15. [CrossRef] [PubMed]

44. Thommes, M.; Kaneko, K.; Neimark, A.V.; Olivier, J.P.; Rodriguez-Reinoso, F.; Rouquerol, J.; Sing, K.S.W. Physisorption of gases, with special reference to the evaluation of surface area and pore size distribution (IUPAC Technical Report). *Pure Appl. Chem.* **2015**, *87*, 1051–1069. [CrossRef]

45. Giesche, H.; Unger, K.K.; Müller, U.; Esser, U. Hysteresis in nitrogen sorption and mercury porosimetry on mesoporous model adsorbents made of aggregated monodisperse silica spheres. *Colloids Surf.* **1989**, *37*, 93–113. [CrossRef]

46. Cho, D.; Myung Kim, J.; Kim, D. Phenolic resin infiltration and carbonization of cellulose-based bamboo fibers. *Mater. Lett.* **2013**, *104*, 24–27. [CrossRef]

47. Lagergren, S. About the theory of so-called adsorption of soluble substances. *K. Sven. Vetenskapsakademiens Handl.* **1898**, *24*, 1–39.

48. Ho, Y.S.; McKay, G. Pseudo-second order model for sorption processes. *Process Biochem.* **1999**, *34*, 451–465. [CrossRef]

49. Belaid, K.; Kacha, S. Étude cinétique et thermodynamique de l'adsorption d'un colorant basique sur la sciure de bois. *Revue Des Sci. De L'eau* **2011**, *24*, 131–144. [CrossRef]

50. Gautam, R.K.; Mudhoo, A.; Chattopadhyaya, M.C. Kinetic, equilibrium, thermodynamic studies and spectroscopic analysis of Alizarin Red S removal by mustard husk. *J. Environ. Chem. Eng.* **2013**, *1*, 1283–1291. [CrossRef]

51. Qiang, Z.; Gurkan, B.; Ma, J.; Liu, X.; Guo, Y.; Cakmak, M.; Cavicchi, K.A.; Vogt, B.D. Roll-to-roll fabrication of high surface area mesoporous carbon with process-tunable pore texture for optimization of adsorption capacity of bulky organic dyes. *Microporous Mesoporous Mater.* **2016**, *227*, 57–64. [CrossRef]

52. Hill, T.L. Statistical Mechanics of Multimolecular Adsorption II. Localized and Mobile Adsorption and Absorption. *J. Chem. Phys.* **1946**, *14*, 441–453. [CrossRef]

![energies logo] *energies*

MDPI

*Article*

# Analysis of Micronized Charcoal for Use in a Liquid Fuel Slurry

**John M. Long [1],\* and Michael D. Boyette [2]**

[1]  Department of Biosystems and Agricultural Engineering, Oklahoma State University, 111 Agricultural Hall, Stillwater, OK 74078, USA
[2]  Department of Biological and Agricultural Engineering, North Carolina State University, Campus Box 7625, Raleigh, NC 27695, USA; boyette@ncsu.edu
\*  Correspondence: john.m.long@okstate.edu; Tel.: +1-405-744-7893

Academic Editor: Mejdi Jeguirim
Received: 21 September 2016; Accepted: 19 December 2016; Published: 27 December 2016

**Abstract:** Yellow poplar (*Liriodendron tulipifera*) was chosen as the woody biomass for the production of charcoal for use in a liquid fuel slurry. Charcoal produced from this biomass resulted in a highly porous structure similar to the parent material. Micronized particles were produced from this charcoal using a multi-step milling process and verified using a scanning electron microscope and laser diffraction system. Charcoal particles greater than 50 µm exhibited long needle shapes much like the parent biomass while particles less than 50 µm were produced with aspect ratios closer to unity. Laser diffraction measurements indicated D10, D50, and D90 values of 4.446 µm, 15.83 µm, and 39.69 µm, respectively. Moisture content, ash content, absolute density, and energy content values were also measured for the charcoal particles produced. Calculated volumetric energy density values for the charcoal particles exceeded the No. 2 diesel fuel that would be displaced in a liquid fuel slurry.

**Keywords:** charcoal; slurry; diesel; biomass; biofuel; biochar

---

## 1. Introduction

Rudolph Diesel claimed in his original German patent the capability to operate his engine design on a variety of liquid, gaseous, and solid fuels [1]. Diesel conducted several experiments using dry coal dust as a solid fuel to support these claims [2]. Researchers further explored the use of solid fuels, primarily coal dust, throughout the early 20th century as an alternative to liquid fuels of the time [3]. German researchers were particularly interested in coal dust operating diesel engines as an alternative during the Second World War fuel shortages [2]. In the United States during the 1960s, researchers focused on carburation and injection issues plaguing the usage of coal dust as a safe and reliable fuel for diesel engines [4]. Many of these issues were resolved by using water as a carrier for the coal dust instead of air. The US Department of Energy funded several research grants aimed at developing usable coal-water slurries for diesel engines tied mainly to locomotives [5]. The culmination of this research suggested that the optimum slurry composition (mass basis) was 48% percent coal particles with an average size of 3 µm, 2% additives, and the balance water with less than one percent ash [5]. Using larger coal particles decreased time to sedimentation and required agitation before use [6]. Coal-water slurries required engine designs to achieve in-cylinder gas temperatures above 890 K for proper ignition and combustion [7]. Although coal-water slurries were successfully used as a fuel, the material properties of the coal particles created major wear issues. Engines running coal-water slurries exhibited component wear that was 2 to 20 times faster than identical engines running on No. 2 diesel fuel [5]. More recent studies have continued this research using other solid carbon sources such as carbon black from automotive tire waste and other petroleum based wastes [8–12].

Studies on alternative fuel sources have placed focus on renewable and bio-based fuel sources in recent years. Biomass derived charcoal or biochar is one renewable solid fuel that is similar to coal [13]. Some charcoal slurry properties have been investigated for use in diesel engines. Efforts to incorporate dry biomass and biochar dust have been attempted in recent years, but most of the work focused on incorporating these solid particles into a liquid [14]. N'kpomin et al. [15] evaluated ternary mixtures of charcoal, oil, and water through viscosity measurements. For a particular mass percent of charcoal, particle sizes below 4 μm significantly increased viscosity due to the increased effect of surface forces. Recent studies examined ternary mixtures of biochar-glycerol-water and determined that the slurry was most stable when the liquid phase of the slurry consisted of 90% by mass glycerol with the remainder as water [16]. Biochar sedimentation was much lower for slurries with less water in the liquid phase.

Biomass derived charcoal ash studied by Ellem and Mulligan [17] were found to contain mostly silicon ($SiO_2$), aluminum ($Al_2O_3$), iron ($Fe_2O_3$), calcium ($CaO$), magnesium ($MgO$), sodium ($NaO_2$), potassium ($K_2O$), and phosphorus ($P_2O_5$). It was noted that mechanical separation and leaching would be effective in removing the majority of ashes within biomass charcoals. Compared to coal, ash contained within biomass derived charcoal slurries produced much lower levels of abrasive wear in many studies [18]. The easy removal of ash and lower abrasive wear of charcoal slurries open the possibility of producing an energy-dense, low-ash fuel slurry that could be used in multiple applications. Reciprocating piston diesel engines such as medium and low speed diesels used for locomotion would be a likely candidate that would benefit from such a fuel, but even gas turbines could operate on a biomass derived charcoal slurry if ash removal could achieve acceptable levels to minimize slagging and other related issues [19].

The goal of this research was to develop and analyze charcoal particles that could be used to displace current medium speed diesel engine liquid fuels, such as No. 2 diesel fuel, as part of a liquid fuel slurry. Experiments were conducted to answer the following questions:

- Can charcoal be effectively milled to produce micronized particles (<1 mm diameter)?
- How does the size of the charcoal particle affect the shape and structure?
- What are typical values for the physical and chemical properties of biomass derived charcoal particles?
- How does the displacement of liquid fuel with charcoal particles affect the liquid fuel's energy density?

## 2. Materials and Methods

Woody biomass was chosen as the feedstock for the charcoal produced during this research. Yellow poplar (*Liriodendron tulipifera*) was chosen as the species for carbonization. The yellow poplar tree has a characteristically long limbless trunk that has the potential to produce a much greater volume of dry wood than other denser species such as oak [20]. Three logs were cut from Fishel Family Farms (Clemmons, NC, USA) and were allowed to lay in direct sunlight outdoors for three hours with an average ambient temperature of 32 °C. This short warming period heated the sap and tissue in the cambium layer of the wood which aided in the removal of the bark from the wood. Bark separation is a mechanical process used to remove any windblown silicates and other ashes present in the bark from the material used for charcoal. A 3 cm wide strip of bark was removed along the length of each log using a drawknife. A small flat edge such as a putty knife was then inserted along the cambium later between the inner wood and the bark. The bark was then peeled from the log by hand in one solid sheet as shown in Figure 1 and discarded.

After bark separation, the logs were cut to lengths of roughly 430 mm and were either halved or quartered depending on initial diameter to produce pieces that were similar in overall volume. These pieces were then placed in a forced-air dryer at 108 °C for 16 h. The pieces were then allowed to equilibrate to room temperature before handling and recording their mass. The cut and dried wood pieces were then placed in a custom stainless steel chamber designed for lab-scale charcoal production (Figure 2).

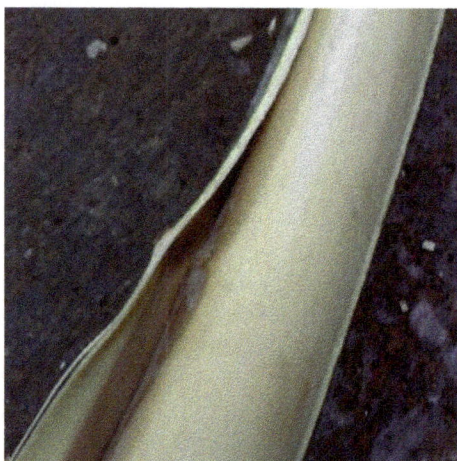

**Figure 1.** Mechanical bark separation.

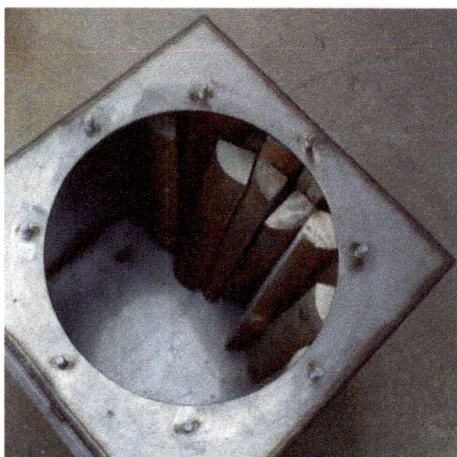

**Figure 2.** Loading the carbonization chamber with dried wood pieces.

The chamber measured 407 mm × 407 mm × 458 mm ($W \times D \times H$) and contained a 330 mm diameter round opening on the top. Two 3.175 mm nominal National Pipe Taper (0.125-27 NPT) couplers were welded into one side of the chamber to allow for insertion of pipe nipples for exhausting combustible gases during carbonization. The chamber was sealed and placed into a high temperature furnace (Iguana, Paragon Industries, L.P., Mesquite, TX, USA). The heating schedule programmed into the furnace controller consisted of a full rate heating until a temperature of 250 °C was achieved and then a rate of 100 °C per hour until 700 °C was reached. The furnace was then programmed to turn off at 700 °C. The furnace doors were then opened and the chamber was allowed to cool overnight. Since pyrolysis of the parent biomass should near completion at 700 °C, this high temperature process was selected to produce an energy dense, low-oxygen charcoal product [21]. A higher temperature would favor carbon reduction reactions that decrease charcoal yield with little gain in charcoal energy density, and a lower temperature would not allow complete pyrolysis and would produce a higher charcoal yield at the cost of charcoal energy density [22].

The charcoal was then processed using a multi-step milling process (Figure 3).

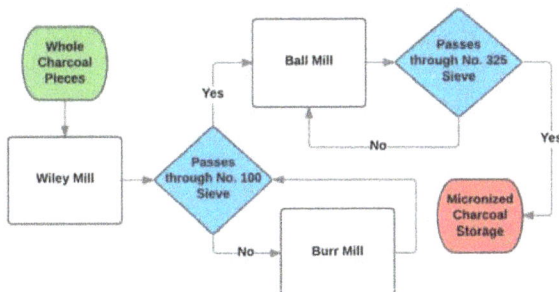

**Figure 3.** Multi-step milling process flow chart.

The process began with whole charcoal as produced by the furnace. The whole charcoal was fed into a Wiley mill (Model 4, Thomas Scientific, Sweedsboro, NJ, USA) that used a 1 mm screen. Charcoal that exited the Wiley mill was placed into a No. 100 (149 μm) sieve. Charcoal particles that could not pass through the sieve were processed further through a burr mill until the particles were able to pass through the No. 100 sieve. Charcoal particles that passed through the No. 100 sieve were loaded into a custom lab-scale ball mill. The lab-scale ball mill consisted of a three-liter polypropylene bottle as the tumbling cylinder placed between two electric conveyor rollers mounted to a frame. The three-liter bottle was loaded with twelve 12.7 mm steel spheres and filled to half capacity with ground charcoal. The three-liter bottle was rotated at a constant speed of 1 Hz. After a minimum residence time of 24 h, the charcoal particles were removed and placed into a No. 325 (44 μm) sieve. Charcoal particles that passed through the No. 325 sieve were placed into storage for further analysis and the remainder was placed back into the ball mill.

Initial evaluation of the milling process was conducted by hand sieving a random sample of the material leaving the first process (Wiley mill). Since the Wiley mill incorporated a 1000 μm screen, a No. 35 (500 μm) and No. 100 sieve were chosen for quick evaluation of the sample. The No. 35 sieve was placed upon the No. 100 sieve and installed over a receiver pan. A sample was placed into the No. 35 sieve and agitated by hand for 60 s. The mass of the material remaining on each sieve and within the receiver was recorded.

## 2.1. Particle Analysis

Samples were randomly selected from the charcoal particles produced by the multi-step milling process. Initial particle analysis involved sending samples to be analyzed by a scanning electron microscope (SEM) with an incorporated energy-dispersive X-ray detector. The high resolution images produced by the SEM were used to describe the relative shape and size of the particles produced by the multi-step milling process. The dispersive X-ray detector allowed for elemental analysis of portions of the viewable image. Areas of the sample containing ash were selected for this elemental analysis.

Particle size and distribution were also quantified using a laser diffraction system (Spraytec, Malvern Instruments Ltd., Worcestershire, UK). The system was calibrated to measure particle sizes from 0.1 μm to 1000 μm. Sample charcoal particles were entrained in an air stream that dispersed the particles as they passed between the laser emitter and receiver. Five test replications were performed.

*2.2. Physical and Chemical Property Analysis*

One of the goals of this research was to quantify a few of the physical and chemical properties of biomass derived charcoal particles. Moisture content, ash content, absolute density, and energy content were the four properties quantified. Moisture and ash content were selected to quantify the amount of ash and water that the charcoal would contribute to a fuel slurry. Absolute density was measured to determine the volume displaced by the charcoal particles when added to a liquid fuel slurry.

The moisture content of the charcoal particles was measured using an analytical balance and muffle furnace (Isotemp Muffle Furnace 10-550-126, Thermo Fisher Scientific, Waltham, MA, USA). Empty crucibles were placed in the muffle furnace for one hour at 750 °C to remove any residual surface deposits. After cooling in a desiccator for an hour, the crucible mass was recorded to the nearest 0.0001 g. Each crucible was filled with a one-gram sample of charcoal particles and a new crucible mass was recorded. The crucibles were placed in the muffle furnace at 105 °C for 24 h. The crucibles were then removed and placed in the desiccator for one hour to cool. A new oven-dry mass was then recorded for each crucible. The wet-basis (w.b.) moisture content was calculated using the standard method [23]. The moisture content analysis was replicated four times for each charcoal particle sample.

The ash content of the charcoal particles was measured using the analytical balance and muffle furnace as a continuation of the moisture content analysis procedure. The oven-dry charcoal particles were removed from the desiccator and placed in the muffle furnace at 750 °C for 20 h. The crucibles were then removed and placed in a desiccator to cool for one hour. A new mass value was recorded for the cooled crucible and remaining ash. The ash content was calculated as a percentage of the oven-dry charcoal particle sample mass. Used crucibles were wiped clean and placed in the muffle furnace at 750 °C for one hour and then allowed to cool in a desiccator for reuse. The ash content analysis was replicated four times for each charcoal particle sample.

Absolute density of the charcoal particles was measured using a helium pycnometer (Accupyc 1330, Micrometrics, Norcross, GA, USA). The pycnometer was set to purge the sample and reference chambers before each volumetric measurement. The sample chamber insert was filled with charcoal particles and the mass of the sample was recorded to the nearest 0.0001 g. For each sample run, the pycnometer was set to perform 10 volumetric measurements and record the average. Absolute density was calculated by the pycnometer using the measured mass and absolute volume for each sample run. All seals and surfaces were wiped clean and were properly greased between each sample run. The absolute density analysis was replicated three times for each charcoal particle sample selected.

Energy content or higher heating value (HHV) of the charcoal particles was measured using a bomb calorimeter (CAL-2KECO, Digital Data Systems Pty. Ltd., Johannesburg, South Africa). A new steel crucible was fired in the muffle furnace at 800 °C for 10 min to remove any residual oils or other surface deposits. After cooling, the crucible mass was recorded to the nearest 0.0001 g. A 0.5 g sample of charcoal particles was selected for each run and placed into the crucible. A new mass was recorded and the actual sample mass was calculated. Firing wires and seals in the bomb calorimeter vessel were examined before each run and replaced if necessary. A cotton string was tied to the firing wire and placed against the charcoal particles in the crucible to act as an ignition source. The completed crucible assembly was placed into the bomb calorimeter vessel and charged with pure oxygen to a pressure of 2068 kPa. The HHV was measured by the bomb calorimeter for each sample in units of "BTU/pound". SI units were calculated using ASAE Standard EP285.7 [24]. The energy content analysis was replicated three times for each charcoal particle sample.

### 3. Results and Discussions

Yellow poplar biomass was successfully prepared for carbonization and 14.8 kg was loaded into the furnace chamber. The carbonization process yielded 3.5 kg of yellow poplar charcoal or 24% of the weight of the parent biomass. This type of yield is consistent with low-oxygen biochars where overall yield is sacrificed to increase the energy density of the char.

The charcoal produced retained much of the original structure of the biomass as shown in Figure 4.

**Figure 4.** Comparison of oven-dried yellow poplar biomass (**top**) to the charcoal produced (**bottom**).

The lower mass value of the charcoal combined with little visual reduction in overall volume, suggested a low bulk density material. The low bulk density was due to the increase in the size of internal void spaces within the charcoal structure. This reduced the thickness of the structural walls resulting in a very porous structure requiring lower shearing forces than the parent biomass.

A multi-step milling process was developed to generate micronized charcoal. Initial evaluation of the milling process was conducted to determine the effectiveness of the initial step (Wiley mill) using a simple hand sieving method. A sample was collected after processing the charcoal through the Wiley mill. The recorded mass values for this process are presented in Table 1.

**Table 1.** Weight distribution of a selected charcoal particle sample after processing through the Wiley mill.

| Sieve | Total Mass | Empty Mass | ΔMass | Mass |
|-------|------------|------------|-------|------|
|       | g          | g          | g     | %    |
| No. 35 | 397.9 | 397.5 | 0.4 | 0.9 |
| No. 100 | 439.9 | 431.1 | 8.8 | 20.6 |
| Receiver | 395.8 | 362.2 | 33.6 | 78.5 |

Almost the entire charcoal particle sample was able to pass through a sieve that is half the size of the screen within the mill and almost 80% of the charcoal particles passed through a No. 100 screen. Although this data was measured from a single sample selected from the charcoal exiting the first milling step, the analysis suggests that the charcoal material produced required less milling effort than similar materials, such as coal found in the literature.

*3.1. Particle Analysis*

Initial sizing of the charcoal particles was obtained using images produced by a SEM. Figure 5 shows one of the images produced by the SEM using a magnification power of 400×.

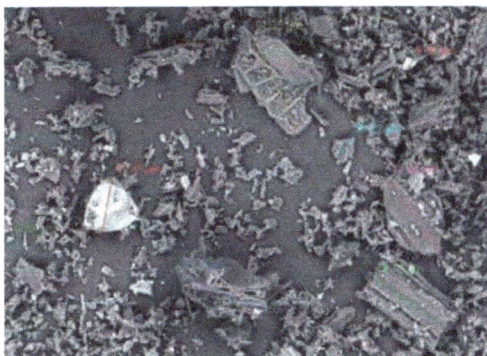

**Figure 5.** SEM image of the charcoal particles with ash particles present.

Many of the ash particles contrast the charcoal particles in this image with a lighter color. The largest dimension of the particles was measured within the image as a qualitative measure of particle size. The four largest charcoal particles pictured in Figure 5 each measured approximately 90 μm while the largest ash particle measured approximately 50 μm. The smaller charcoal particles averaged 5 μm and the smaller ash particles averaged 7 μm. The larger charcoal particles tended to retain the cylindrical and porous structural arrangement of the biomass feedstock. Aspect ratios of these particles were much smaller, with needle-like shapes at the smallest values. The smaller particles (<50 μm) tended to be described as irregularly shaped with aspect ratios near unity. Similar observations have been made concerning aspect ratio measurements for ground wood particles [25] and biochar [26].

A portion of the SEM imagery was selected that contained several identifiable ash particles as a means to identify the elemental composition of the ash present in the sample. An energy-dispersive X-ray spectrometer attached to the SEM was used to analyze a selected portion of the sample for elemental composition. A sample area was selected that contained a large number of visible ash particles. The mass fractions determined by this analysis were highly dependent on the selected sample area and was not representative of the actual elemental mass fraction composition in the overall charcoal sample. This analysis did however identify that the ash present involved molecules containing calcium, potassium, and silicon, with some type of oxide being the most probable form. These are all common types of elemental ash found in biomass [17].

The size of the charcoal particles was quantified using the laser diffraction system. The system analyzed the particles measuring the volume frequency of particles falling within discrete size steps up to 1000 μm. The percentage of the overall sample volume occupied by particles within a particular size step was recorded for each sample run. The average percentage value and standard error for each size step were calculated for all of the sample runs. A histogram of this data is shown in Figure 6.

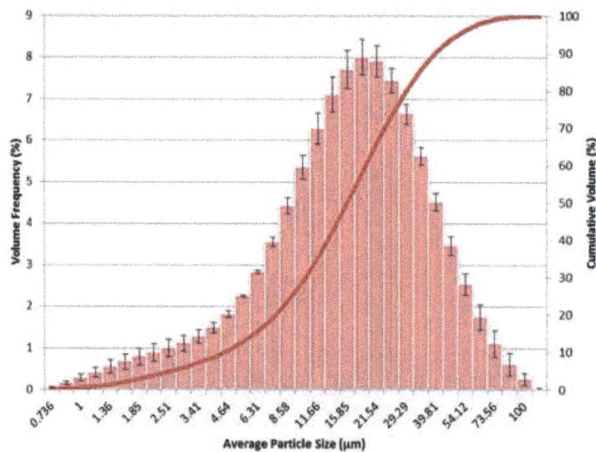

**Figure 6.** Average particle size histogram of percent sample volume. Standard Error bars are shown for each bin. The solid line represents the cumulative sample volume for a given step size and smaller.

Measured sample particle sizes ranged from 0.736 μm to 116.59 μm with a median value or D50 of 15.83 μm. Two different types of mean diameters were calculated for the data. The Sauter mean diameter (D[3,2]) was measured as 9.166 μm and represents the mean particle size when taking into account the surface area of the particles. The De Brouckere mean diameter (D[4,3]) was measured as 19.59 μm and represents the mean particle size when taking into account the volume of the particles. The De Brouckere mean diameter is the more conservative mean value since it is very sensitive to the presence of larger particles in the sample. Since this research is concerned with the displacement of liquid fuel, the De Brouckere mean diameter was the one chosen to represent the mean particle size for the sample. The D10 and D90 values were determined as 4.446 μm and 39.69 μm, respectively. These values represent the points at which 10% and 90% of sample volume consists of particles that are of that particular size and smaller. 0% of the sample volume can be described as a particle sized within the range of these two values.

### 3.2. Physical and Chemical Property Analysis

Charcoal particle samples were randomly selected from the charcoal produced for this research after completing the multi-step milling process. Moisture content, ash content, absolute density, and energy content were measured for each of the samples. Moisture content was measured using a dry-basis (d.b.) calculation, but is presented as a w.b. value (Figure 7a).

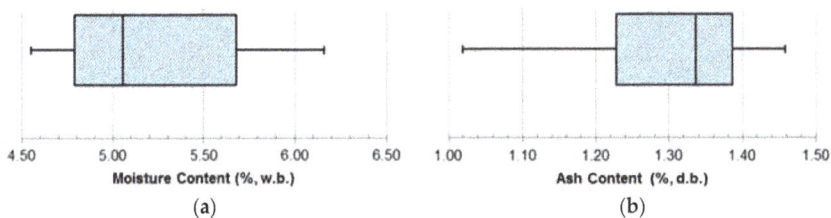

**Figure 7.** The distribution of measured values for (**a**) wet-basis moisture content; and (**b**) dry-basis ash content of the micronized charcoal particles. The box-and-whisker plot represents the minimum, first quartile, median, third quartile, and maximum values from left to right, respectively.

The w.b. moisture content ranged from 4.55% to 6.16% with a median value of 5.06%. The overall distribution was skewed towards the lower values. Ash content (Figure 7b) held a much tighter distribution of measured values. The d.b. ash content ranged from 1.02% to 1.46% with a median value of 1.34%. Beyond a few low ash samples, the majority of the measured values were towards the higher end of the data range. These values agreed with the values found in the literature [27].

Absolute density was measured to describe the density of the individual charcoal particles. When added to a liquid fuel slurry, the volume of the particles would displace an equal volume of the liquid fuel. The absolute density was calculated through the measurement of the volume of the particles in a sample and the measured mass of the sample. The calculated absolute density for the charcoal samples is presented in Figure 8a. The absolute density ranged from 1525 kg/m$^3$ to 1667 kg/m$^3$ with a median value of 1596 kg/m$^3$.

The energy content of the charcoal particles was measured with the bomb calorimeter to determine the energy that would evolve within an engine combustion chamber (Figure 8b). The energy content ranged from 34,107.8 kJ/kg to 36,329.6 kJ/kg with a median value of 35,261.3 kJ/kg. The energy density measurement data were symmetrically distributed about the median value. This energy content was determined as a mass basis but can be converted to a volumetric basis using the measured density of the charcoal particle. Energy per unit volume is much more descriptive of the energy density of a liquid fuel. Taking into consideration the full range of values for the absolute density, the volumetric energy content for charcoal was determined to be within the range of 52,014.4 MJ/m$^3$ to 60,560.4 MJ/m$^3$ for the measured samples. Considering that No. 2 diesel fuel has an average density of 851 kg/m$^3$ and a HHV of 38,362.5 MJ/m$^3$ on a volumetric basis, the charcoal was found to have a much higher volumetric energy density [28].

This higher energy density is a direct result of the high carbon charcoal produced through the high temperature carbonization process. According to the literature, the energy density (HHV) of the parent yellow poplar biomass is 18.12 MJ/kg [29]. Using the lower range of the measured energy density

(HHV) for the yellow poplar charcoal particles and the process yield, only 45% of the energy contained in the parent biomass remains in the charcoal product. A lower temperature carbonization process would yield more charcoal material with a higher overall amount of energy from the parent biomass, but the energy density of the material would be much lower. For high energy density charcoal production to be a viable option, the parent biomass energy fraction contained in the syngas and pyrolysis oil must have value in the production process or as a product elsewhere.

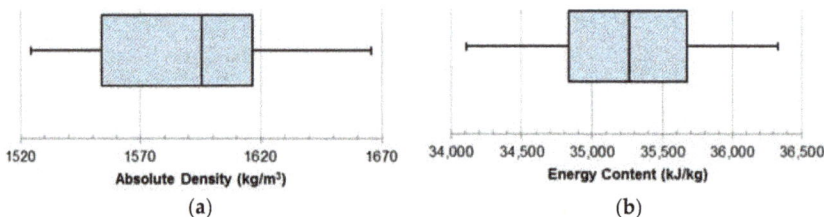

(a)          (b)

**Figure 8.** The distribution of measure values for (**a**) absolute density; and (**b**) energy content of the micronized charcoal particles. The box-and-whisker plot represents the minimum, first quartile, median, third quartile, and maximum values from left to right, respectively.

## 4. Conclusions

Biomass derived charcoal was produced, milled, and analyzed for its potential as a micronized particle to be used in a liquid fuel slurry. Particle shape and size was determined using both qualitative and quantitative methods. Conclusions resulting from this study were:

- A multi-step milling process was developed that successfully produced micronized particles with 90% of the sample volume consisting of particles smaller than 50 μm. Charcoal particles less than 1000 μm could be effectively generated in a single milling step, but charcoal particles less than 50 μm required further processing steps.
- Particles produced that were greater than 50 μm exhibited longer needle shapes with high aspect ratios when viewed at high magnification levels. Particles smaller than 50 μm exhibited aspect ratios that were much closer to unity. Since liquid fuels form spherical droplets with aspect ratios close to unity, a successful milling process should produce charcoal particle with sizes less than 50 μm.
- Physical and chemical property analyses of the charcoal particles determined that the charcoal particle density was much higher than No. 2 diesel fuel. This increase in energy density is achieved through a high-temperature, lower-yielding carbonization process. Therefore, displacing a liquid fuel with a charcoal particle would increase the theoretical volumetric energy density of the fuel mixture. The resulting slurry would be appropriate for use in medium speed and low speed diesel engines whenever volumetric energy density is desirable. Further processing would be required to produce a very low-ash, volumetrically energy dense slurry capable of meeting the necessary requirements for gas turbine systems.

**Acknowledgments:** This material was based upon work made possible through support by the North Carolina Agricultural Foundation, Inc. and Carolina Greenhouses, Inc.

**Author Contributions:** John M. Long and Michael D. Boyette conceived and designed the experiments; John M. Long performed the experiments and analyzed the data; Michael D. Boyette contributed reagents/materials/analysis tools; John M. Long and Michael D. Boyette wrote the paper.

**Conflicts of Interest:** The authors declare no conflict of interest. The founding sponsors had no role in the design of the study; in the collection, analyses, or interpretation of data; in the writing of the manuscript, and in the decision to publish the results.

## References

1.  Diesel, R. Arbeitsverfahren und Ausführungsart für Verbrennungskraftmaschinen. DE67207A, 28 February 1982. (In German)
2.  Soehngen, E. *Development of Coal-Burning Diesel Engines in Germany*; National Technical Information Service: Alexandria, VA, USA, 1976.
3.  McMillian, M.; Webb, H. Coal-fueled diesels: Systems development. *J. Eng. Gas Turbines Power* **1989**, *111*, 485–490. [CrossRef]
4.  Ryan, T. Coal-fueled diesel development: A technical review. *J. Eng. Gas Turbines Power* **1994**, *116*, 740–748. [CrossRef]
5.  Caton, J.A.; Hsu, B.D. General electric coal-fueled diesel engine program (1982–1993): A technical review. *J. Eng. Gas Turbines Power* **1994**, *116*, 749–757. [CrossRef]
6.  Chen, X.; Zhao, L.; Zhang, X.; Qian, C. An investigation on characteristics of coal–water slurry prepared from the solid residue of plasma pyrolysis of coal. *Energy Convers. Manag.* **2012**, *62*, 70–75. [CrossRef]
7.  Hsu, B.; Flynn, P.L. Preliminary study of using coal-water slurry fuel in GE-7FDL medium speed diesel engine. In Proceedings of the 18th International Congress on Combustion Engines, Tianjin, China, 5–8 June 1989.
8.  Wamankar, A.K.; Murugan, S. Experimental investigation of carbon black–water–diesel emulsion in a stationary DI diesel engine. *Fuel Process. Technol.* **2014**, *125*, 258–266. [CrossRef]
9.  Wamankar, A.K.; Murugan, S. Combustion, performance and emission of a diesel engine fuelled with diesel doped with carbon black. *Energy* **2015**, *86*, 467–475. [CrossRef]
10. Wamankar, A.K.; Murugan, S. Combustion, performance and emission characteristics of a diesel engine with internal jet piston using carbon black-water-diesel emulsion. *Energy* **2015**, *91*, 1030–1037. [CrossRef]
11. Wamankar, A.K.; Satapathy, A.K.; Murugan, S. Experimental investigation of the effect of compression ratio, injection timing & pressure in a DI (direct injection) diesel engine running on carbon black-water-diesel emulsion. *Energy* **2015**, *93*, 511–520.
12. Molino, A.; Erto, A.; Di Natale, F.; Donatelli, A.; Iovane, P.; Musmarra, D. Gasification of granulated scrap tires for the production of syngas and a low-cost adsorbent for Cd(II) removal from wastewaters. *Ind. Eng. Chem. Res.* **2013**, *52*, 12154–12160. [CrossRef]
13. Liu, Z.; Quek, A.; Hoekman, S.K.; Balasubramanian, R. Production of solid biochar fuel from waste biomass by hydrothermal carbonization. *Fuel* **2013**, *103*, 943–949. [CrossRef]
14. Piriou, B.; Vaitilingom, G.; Veyssière, B.; Cuq, B.; Rouau, X. Potential direct use of solid biomass in internal combustion engines. *Prog. Energy Combust. Sci.* **2013**, *39*, 169–188. [CrossRef]
15. N'kpomin, A.; Boni, A.; Antonini, G.; François, O. The deashed charcoal-oil-water mixture: A liquid fuel for biomass energetical valorization. *Chem. Eng. J. Biochem. Eng. J.* **1995**, *60*, 49–54. [CrossRef]
16. Liu, P.; Zhu, M.; Zhang, Z.; Leong, Y.K.; Zhang, Y.; Zhang, D. An experimental study of rheological properties and stability characteristics of biochar-glycerol-water slurry fuels. *Fuel Process. Technol.* **2016**, *153*, 37–42. [CrossRef]
17. Ellem, G.K.; Mulligan, C.J. Biomass char as a fuel for internal combustion engines. *Asia-Pac. J. Chem. Eng.* **2012**, *7*, 769–776. [CrossRef]
18. Wamankar, A.K.; Murugan, S. Review on production, characterisation and utilisation of solid fuels in diesel engines. *Renew. Sustain. Energy Rev.* **2015**, *51*, 249–262. [CrossRef]
19. Mehr-Homji, C.; Zachary, J.; Bromley, A. Gas turbine fuels-system design, combustion and operability. In Proceedings of the 39th Turbomachinery Symposium, Houston, TX, USA, 4–7 October 2010; pp. 4–7.
20. Beck, D.E. *Liriodendron tulipifera* L. In *Silvics of North America: 2. Hardwoods*; Forest Service: Washington, DC, USA, 1990; pp. 406–416.
21. Chen, T.; Wu, J.; Zhang, Z.; Zhu, M.; Sun, L.; Wu, J.; Zhang, D. Key thermal events during pyrolysis and $CO_2$-gasification of selected combustible solid wastes in a thermogravimetric analyser. *Fuel* **2014**, *137*, 77–84. [CrossRef]
22. Molino, A.; Chianese, S.; Musmarra, D. Biomass gasification technology: The state of the art overview. *J. Energy Chem.* **2016**, *25*, 10–25. [CrossRef]
23. Wilhelm, L.R.; Suter, D.A.; Brusewitz, G.H. Drying and Dehydration. In *Food & Process Engineering Technology*; American Society of Agricultural Engineers: St. Joseph, MI, USA, 2004.

24. American Society of Agricultural and Biological Engineers (ASABE). *ASAE EP285.8 Use of SI (Metric) Units*; American Society of Agricultural and Biological Engineers (ASABE): St. Joseph, MI, USA, 2014.
25. Tannous, K.; Lam, P.; Sokhansanj, S.; Grace, J. Physical properties for flow characterization of ground biomass from Douglas Fir Wood. *Part. Sci. Technol.* **2013**, *31*, 291–300. [CrossRef]
26. Shivaram, P.; Leong, Y.; Yang, H.; Zhang, D. Flow and yield stress behaviour of ultrafine Mallee biochar slurry fuels: The effect of particle size distribution and additives. *Fuel* **2013**, *104*, 326–332. [CrossRef]
27. Molino, A.; Nanna, F.; Villone, A. Characterization of biomasses in the southern Italy regions for their use in thermal processes. *Appl. Energy* **2014**, *131*, 180–188. [CrossRef]
28. Tyson, K.S.; McCormick, R. *Biodiesel Handling and Use Guide*; National Renewable Energy Laboratory (NREL): Golden, CO, USA, 2006.
29. Na, B.; Ahn, B.; Lee, J. Changes in chemical and physical properties of yellow poplar (*Liriodendron tulipifera*) during torrefaction. *Wood Sci. Technol.* **2015**, *49*, 257–272. [CrossRef]

MDPI AG

St. Alban-Anlage 66

4052 Basel, Switzerland

Tel. +41 61 683 77 34

Fax +41 61 302 89 18

http://www.mdpi.com

*Energies* Editorial Office

E-mail: energies@mdpi.com

http://www.mdpi.com/journal/energies

www.ingramcontent.com/pod-product-compliance
Lightning Source LLC
Chambersburg PA
CBHW051842210326
41597CB00033B/5746

*9783038426905*